Materials Science

Materials Science

Sachin Vijay

Pratap Book Services
Ghaziabad

Edition : 2015

ISBN : 978–81–930488–2–5

Published by :
Pratap Book Services
495, Adarsh Colony
Ghaziabad (UP)

Printed in India.

Preface

Materials science, also commonly known as materials science and engineering, is an interdisciplinary field which deals with the discovery and design of new materials. This relatively new scientific field involves studying materials through the materials paradigm (synthesis, structure, properties and performance). It incorporates elements of physics and chemistry, and is at the forefront of nanoscience and nanotechnology research. In recent years, materials science has become more widely known as a specific field of science and engineering. It is an important part of forensic engineering (the investigation of materials, products, structures or components that fail or do not operate or function as intended, causing personal injury or damage to property) and failure analysis, the latter being the key to understanding, for example, the cause of various aviation accidents. Many of the most pressing scientific problems that are faced today are due to the limitations of the materials that are available and, as a result, breakthroughs in this field are likely to have a significant impact on the future of technology.

Materials science evolved—starting from the 1960s—because it was recognized that to create, discover and design new materials, one had to approach it from a unified manner. Thus, materials science and engineering emerged at the intersection of various fields such as metallurgy, solid state physics, chemistry, chemical engineering, mechanical engineering and electrical engineering. The field is inherently interdisciplinary, and the materials scientists/engineers must be aware and make use of the methods of the physicists, chemists and engineers. The field thus, maintains close relationships with these fields. Also, many physicists, chemists and engineers also find themselves working in materials science. The overlap between physics and materials science has led to the offshoot field of materials physics, which is concerned with the physical properties of materials. The approach is generally more macroscopic and applied than in condensed matter physics. Important publications in materials physics for more details on this field of study can be perused. The field of materials science and engineering is important both from a scientific perspective, as well as from an engineering one. When discovering new materials, one encounters new phenomena that may not have been observe before. Hence, there is lot of science to be discovered when working with materials. Materials science also provides test for theories in condensed matter physics. Radical materials advances can drive the creation of new products or even new industries, but stable industries also employ materials scientists to make incremental improvements and troubleshoot issues with currently used materials. Industrial applications of materials science include materials design, cost-benefit tradeoffs in industrial production of materials, processing techniques (casting, rolling, welding, ion implantation, crystal growth, thin-film deposition, sintering, glassblowing, etc.), and analytical techniques (characterization techniques such as electron microscopy, x-ray diffraction, calorimetry, nuclear microscopy (HEFIB), Rutherford backscattering, neutron diffraction, small-angle X-ray scattering (SAXS), etc.). Materials science has received much attention from researchers. In most universities, many departments ranging from physics to chemistry to chemical engineering—in addition to materials science departments—are involved in materials research. Research in materials science is vibrant and consists of many avenues.

The present textbook provides students with a solid understanding of the relationship between the structure, processing, and properties of materials.

Book writing-whether original or compilation is a laborious complicated job requiring academic labour.

The present work is purely a research compilation based on authoritative publications and researches authored by eminent scientists and subject experts.

Assistance has also been sought from Websites through Internet.

The author, humbly acknowledges the contributions of all those eminent scientists and subject experts alongwith their respective publishers and Websites in preparation of this book.

—Sachin Vijay

Contents

Chapter 1

Crystal Structure

Introduction

Historical Perspective

Materials are so important in the development of civilization that we associate Ages with them. In the origin of human life on Earth, the Stone Age, people used only natural materials, like stone, clay, skins, and wood. When people found copper and how to make it harder by alloying, the Bronze Age started about 3000 BC. The use of iron and steel, a stronger material that gave advantage in wars started at about 1200 BC. The next big step was the discovery of a cheap process to make steel around 1850, which enabled the railroads and the building of the modern infrastructure of the industrial world.

Materials Science and Engineering

Understanding of how materials behave like they do, and why they differ in properties was only possible with the atomistic understanding allowed by quantum mechanics, that first explained atoms and then solids starting in the 1930s. The combination of physics, chemistry, and the focus on the relationship between the properties of a material and its microstructure is the domain of Materials Science. The development of this science allowed designing materials and provided a knowledge base for the engineering applications (Materials Engineering).

Structure:

- At the atomic level: arrangement of atoms in different ways. (Gives different properties for graphite than diamond both forms of carbon.)
- At the microscopic level: arrangement of small grains of material that can be identified by microscopy. (Gives different optical properties to transparent vs. frosted glass.)

Properties are the way the material responds to the environment. For instance, the *mechanical, electrical and magnetic* properties are the responses to mechanical, electrical and magnetic forces, respectively. Other important properties are *thermal* (transmission of heat, heat capacity), *optical* (absorption, transmission and scattering of light), and the *chemical stability* in contact with the environment (like corrosion resistance). Processing of materials is the application of heat (heat treatment), mechanical forces, etc. to affect their microstructure and, therefore, their properties.

Why Study Materials Science and Engineering?

- To be able to select a material for a given use based on considerations of cost and performance.
- To understand the limits of materials and the change of their properties with use.
- To be able to create a new material that will have some desirable properties.

All engineering disciplines need to know about materials. Even the most "immaterial", like software or system engineering depend on the development of new materials, which in turn alter the economics, like software-hardware trade-offs. Increasing applications of system engineering are in materials manufacturing (industrial engineering) and complex environmental systems.

Classification of Materials

Like many other things, materials are classified in groups, so that our brain can handle the complexity. One could classify them according to structure, or properties, or use. The one that we will use is according to the way the atoms are bound together:

Metals: valence electrons are detached from atoms, and spread in an 'electron sea' that "glues" the ions together. Metals are usually strong, conduct electricity and heat well and are opaque to light (shiny if polished). Examples: aluminum, steel, brass, gold.

Semiconductors: the bonding is covalent (electrons are shared between atoms). Their electrical properties depend extremely strongly on minute proportions of contaminants. They are opaque to visible light but transparent to the infrared. Examples: Si, Ge, GaAs.

Ceramics: atoms behave mostly like either positive or negative ions, and are bound by Coulomb forces between them. They are usually combinations of metals or semiconductors with oxygen, nitrogen or carbon (oxides, nitrides, and carbides). Examples: glass, porcelain, many minerals.

Polymers: are bound by covalent forces and also by weak van der Waals forces, and usually based on H, C and other non-metallic elements. They decompose at moderate temperatures (100 – 400 C), and are lightweight. Other properties vary greatly. Examples: plastics (nylon, Teflon, polyester) and rubber.

Other categories are not based on bonding. A particular microstructure identifies composites, made of different materials in intimate contact (example: fibreglass, concrete, wood) to achieve specific properties. Biomaterials can be any type of material that is biocompatible and used, for instance, to replace human body parts.

Advanced Materials

Materials used in "High-Tec" applications, usually designed for maximum performance, and normally expensive. Examples are titanium alloys for supersonic airplanes, magnetic alloys for computer disks, special ceramics for the heat shield of the space shuttle, etc.

Modern Material's Needs

- Engine efficiency increases at high temperatures: requires high temperature structural materials
- Use of nuclear energy requires solving problem with residues, or advances in nuclear waste processing.
- Hypersonic flight requires materials that are light, strong and resist high temperatures.
- Optical communications require optical fibres that absorb light negligibly.
- Civil construction – materials for unbreakable windows.
- Structures: materials that are strong like metals and resist corrosion like plastics.

Atomic Structure and Bonding

Fundamental Concepts

Atoms are composed of electrons, protons, and neutrons. Electron and protons are negative and positive charges of the same magnitude, 1.6×10^{-19} Coulombs.

The mass of the electron is negligible with respect to those of the proton and the neutron, which form the *nucleus* of the atom. The unit of mass is an atomic mass unit (amu) = 1.66×10^{-27} kg, and equals 1/12 the mass of a carbon atom. The Carbon nucleus has Z=6, and A=6, where Z is the number of protons, and A the number of neutrons. Neutrons and protons have very similar masses, roughly equal to 1 amu. A neutral atom has the same number of electrons and protons, Z.

A mole is the amount of matter that has a mass in grams equal to the atomic mass in amu of the atoms. Thus, a mole of carbon has a mass of 12 grams. The number of atoms in a mole is called the Avogadro number, $N_{av} = 6.023 \times 10^{23}$. Note that N_{av} = 1 gram/1 amu.

Calculating n, the number of atoms per cm^3 in a piece of material of density d (g/cm^3).

$$n = N_{av} \times d / M$$

where M is the atomic mass in amu (grams per mol). Thus, for graphite (carbon) with a density d = 1.8 g/cm^3, M =12, we get 6×10^{23} atoms/mol $\times$ 1.8 g/cm^3 / 12 g/mol) = 9×10^{22} C/cm^3.

For a molecular solid like ice, one uses the molecular mass, $M(H_2O)$ = 18. With a density of 1 g/cm^3, one obtains n = 3.3×10^{22} H_2O/cm^3. Note that since the water molecule contains 3 atoms, this is equivalent to 9.9×10^{22} $atoms/cm^3$.

Most solids have atomic densities around 6×10^{22} $atoms/cm^3$. The cube root of that number gives the number of atoms per centimeter, about 39 million. The mean distance between atoms is the inverse of that, or 0.25 nm. This is an important number that gives the scale of atomic structures in solids.

Electrons in Atoms

The forces in the atom are repulsions between electrons and attraction between electrons and protons. The neutrons play no significant role. Thus, Z is what characterizes the atom. The electrons form a cloud around the neutron, of radius of 0.05 – 2 nanometers. Electrons do not move in circular orbits, as in popular drawings, but in 'fuzzy' orbits. We cannot tell how it moves, but only say what is the probability of finding it at some distance from the nucleus. According to quantum mechanics, only certain orbits are allowed (thus, the idea of a mini planetary system is not correct). The orbits are identified by a principal quantum number *n*, which can be related to the size, n = 0 is the smallest; n = 1, 2 .. are larger. (They are "quantized" or discrete, being specified by integers). The angular momentum *l* is quantized, and so is the projection in a specific direction *m*. The structure of the atom is determined by the *Pauli exclusion principle*, only two electrons can be placed in an orbit with a given *n, l, m* – one for each spin.

The Periodic Table

Elements are categorized by placing them in the periodic table. Elements in a column share similar properties. The noble gases have *closed shells*, and so they do not gain or lose electrons near another atom. Alkalis can easily lose an electron and become a closed shell; halogens can easily gain one to form a negative ion, again with a closed shell. The propensity to form closed shells occurs in molecules, when they share electrons to close a molecular shell. Examples are H_2, N_2, and NaCl.

The ability to gain or lose electrons is termed electronegativity or electropositivity, an important factor in ionic bonds.

Bonding Forces and Energies

The Coulomb forces are simple: attractive between electrons and nuclei, repulsive between electrons and between nuclei. The force between atoms is given by a sum of all the individual forces, and the fact that the electrons are located outside the atom and the nucleus in the centre.

When two atoms come very close, the force between them is always repulsive, because the electrons stay outside and the nuclei repel each other. Unless both atoms are ions of the same charge (e.g., both negative) the forces between atoms is always attractive at large internuclear distances r. Since the force is repulsive at small r, and attractive at small r, there is a distance at which the force is zero. This is the equilibrium distance at which the atoms prefer to stay.

The interaction energy is the *potential* energy between the atoms. It is negative if the atoms are bound and positive if they can move away from each other. The interaction energy is the integral of the force over the separation distance, so these two quantities are directly related. The interaction energy is a *minimum* at the equilibrium position. This value of the energy is called the *bond energy*, and is the energy needed to separate completely to infinity (the work that needs to be done to overcome the attractive force.) The strongest the bond energy, the hardest is to move the atoms, for instance the hardest it is to melt the solid, or to evaporate its atoms.

Primary Interatomic Bonds

Ionic Bonding: This is the bond when one of the atoms is negative (has an extra electron) and another is positive (has lost an electron). Then there is a strong, direct Coulomb attraction. An example is NaCl.

In the molecule, there are more electrons around Cl, forming Cl^- and less around Na, forming Na^+. Ionic bonds are the strongest bonds. In real solids, ionic bonding is usually combined with covalent bonding. In this case, the fractional ionic bonding is defined as %ionic = 100 $\times$ [1 – exp(-0.25 $(X_A - X_B)^2$], where X_A and X_B are the electronegativities of the two atoms, A and B, forming the molecule.

Covalent Bonding

In covalent bonding, electrons are shared between the molecules, to saturate the valency. The simplest example is the H_2 molecule, where the electrons spend more time in between the nuclei than outside, thus producing bonding.

Metallic Bonding

In metals, the atoms are ionized, loosing some electrons from the valence band. Those electrons form a *electron sea*, which binds the charged nuclei in place, in a similar way that the electrons in between the H atoms in the H_2 molecule bind the protons.

Secondary Bonding (Van der Waals)

Fluctuating Induced Dipole Bonds: Since the electrons may be on one side of the atom or the other, a dipole is formed: the + nucleus at the centre, and the electron outside. Since the electron moves, the dipole fluctuates. This fluctuation in atom A produces a fluctuating electric field that is felt by the electrons of an adjacent atom, B. Atom B then polarizes so that its outer electrons are on the side of the atom closest to the + side (or opposite to the – side) of the dipole in A. This bond is called van der Waals bonding.

Polar Molecule-Induced Dipole Bonds

A polar molecule like H_2O (Hs are partially +, O is partially –), will induce a dipole in a nearby atom, leading to bonding.

Permanent Dipole Bonds

This is the case of the hydrogen bond in ice. The H end of the molecule is positively charged and can bond to the negative side of another dipolar molecule, like the O side of the H_2O dipole.

Molecules

If molecules formed a closed shell due to covalent bonding (like H_2, N_2) then the interaction between molecules is weak, of the van der Waals type. Thus, molecular solids usually have very low melting points.

Structure of Crystalline Solids

In mineralogy and crystallography, crystal structure is a unique arrangement of atoms or molecules in a crystalline liquid or solid. A crystal structure is composed of a pattern, a set of atoms arranged in a particular way, and a lattice exhibiting long-range order and symmetry. Patterns are located upon the points of a lattice, which is an array of points repeating periodically in three dimensions. The points can be thought of as forming identical tiny boxes, called unit cells, that fill the space of the lattice. The lengths of the edges of a unit cell and the angles between them are called the *lattice parameters*. The symmetry properties of the crystal are embodied in its space group. A crystal's structure and symmetry play a role in determining many of its physical properties, such as cleavage, electronic band structure, and optical transparency

Unit cell

The crystal structure of a material (the arrangement of atoms within a given type of crystal) can be described in terms of its unit cell. The unit cell is a small box containing one or more atoms, a spatial arrangement of atoms. The unit cells stacked in three-dimensional space describe the bulk arrangement of atoms of the crystal. The crystal structure has a three-dimensional shape. The unit cell is given by its lattice parameters, which are the length of the cell edges and the angles between them, while the positions of the atoms inside the unit cell are described by the set of atomic positions (x_i , y_i , z_i) measured from a lattice point.

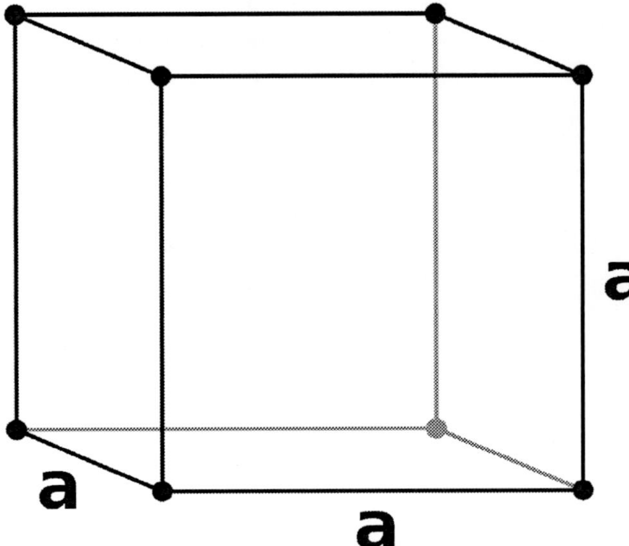

Figure: *Simple cubic (P)*

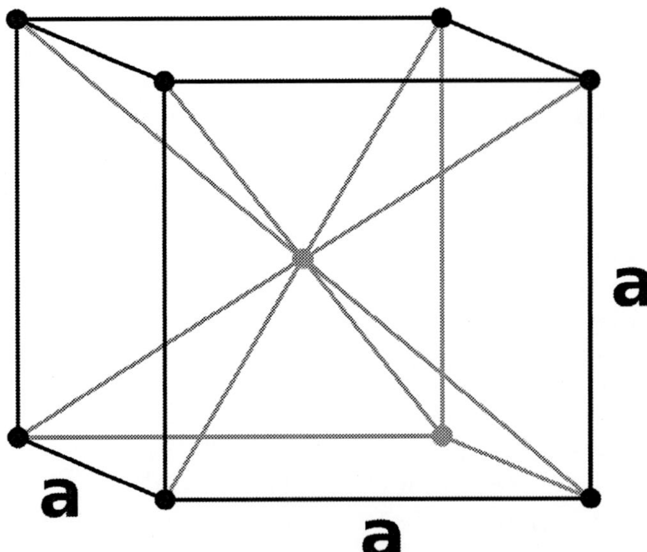

Figure: *Body-centred cubic (I)*

Within the unit cell is the asymmetric unit, smallest unit the crystal can be divided into using the crystallographic symmetry operations of the space group. The asymmetric unit is also what is generally solved when solving a structure of a molecule or protein by X-ray crystallography.

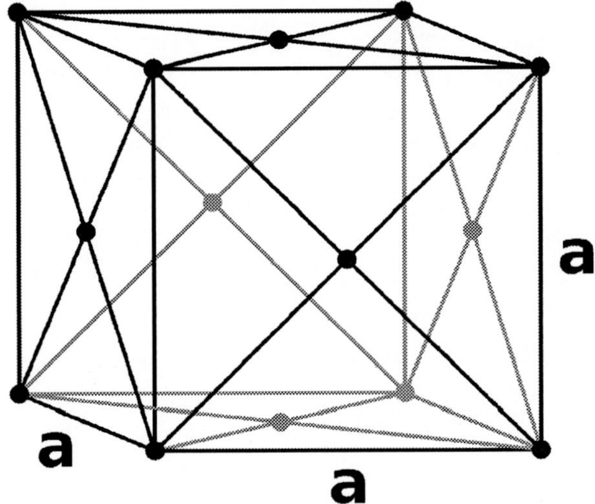

Figure: *Face-centred cubic (F)*

Miller Indices

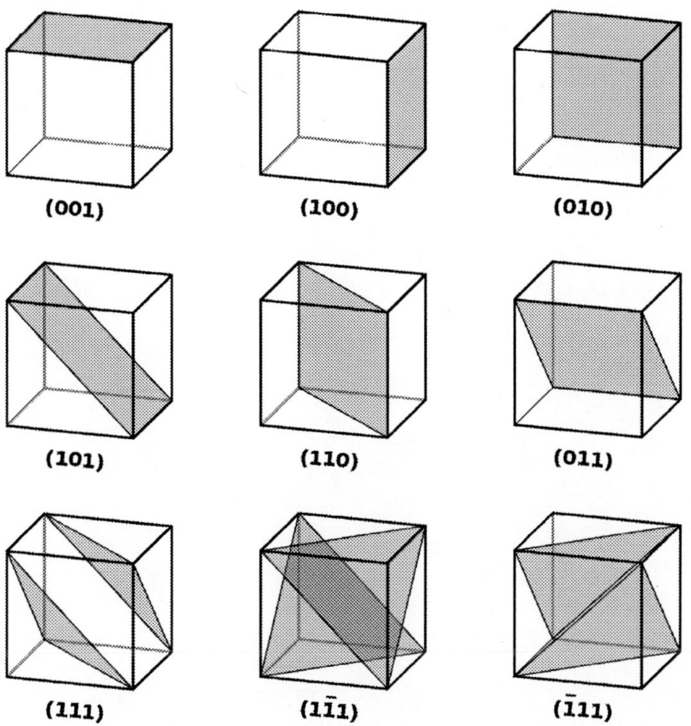

Figure: *Planes with different Miller indices in cubic crystals*

Miller indices form a notation system in crystallography for planes in crystal (Bravais) lattices. In particular, a family of lattice planes is determined by three integers *h*, *k*, and ℓ, the *Miller indices*. They are written (hkℓ), and each index denotes a plane orthogonal to a direction (h, k, ℓ) in the basis of the reciprocal lattice vectors. By convention, negative integers are written with a bar, as in 3 for –3. The integers are usually written in lowest terms, i.e. their greatest common divisor should be 1. Miller index 100 represents a plane orthogonal to direction *h*; index 010 represents a plane orthogonal to direction *k*, and index 001 represents a plane orthogonal to ℓ.

There are also several related notations:

- the notation {hkℓ} denotes the set of all planes that are equivalent to (hkℓ) by the symmetry of the lattice.

In the context of crystal directions (not planes), the corresponding notations are:

- [hkℓ], with square instead of round brackets, denotes a direction in the basis of the *direct* lattice vectors instead of the reciprocal lattice; and
- similarly, the notation $\langle hk\ell \rangle$ denotes the set of all directions that are equivalent to [hkℓ] by symmetry.

Miller indices were introduced in 1839 by the British mineralogist William Hallowes Miller. The method was also historically known as the Millerian system, and the indices as Millerian, although this is now rare. The Miller indices are defined with respect to any choice of unit cell and not only with respect to primitive basis vectors, as is sometimes stated.

Definition

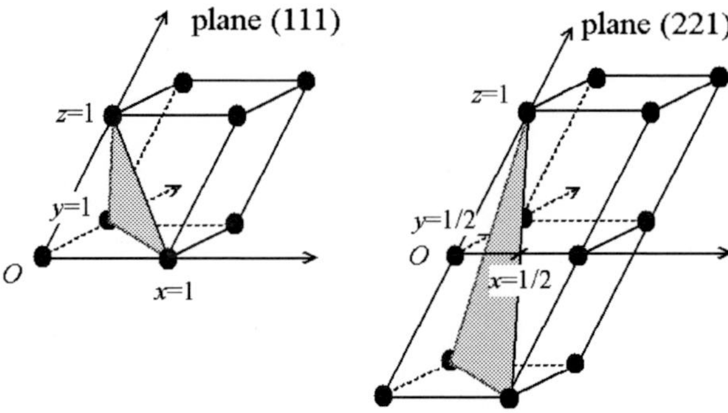

Figure: *Examples of determining indices for a plane using intercepts with axes; left (111), right (221)*

There are two equivalent ways to define the meaning of the Miller indices: via a point in the reciprocal lattice, or as the inverse intercepts along the lattice vectors.

Both definitions are given below. In either case, one needs to choose the three lattice vectors a_1, a_2, and a_3 that define the unit cell (note that the conventional unit cell may be larger than the primitive cell of the Bravais lattice, as the examples below illustrate). Given these, the three primitive reciprocal lattice vectors are also determined (denoted b_1, b_2, and b_3).

Then, given the three Miller indices h, k, ℓ, (hkℓ) denotes planes orthogonal to the reciprocal lattice vector:

$$g_{hk\ell} = h\mathrm{b}_1 + k\mathrm{b}_2 + \ell\mathrm{b}_3.$$

That is, (hkℓ) simply indicates a normal to the planes in the basis of the primitive reciprocal lattice vectors. Because the coordinates are integers, this normal is itself always a reciprocal lattice vector. The requirement of lowest terms means that it is the *shortest* reciprocal lattice vector in the given direction.

Equivalently, (hkℓ) denotes a plane that intercepts the three points a_1/h, a_2/k, and a_3/ℓ, or some multiple thereof. That is, the Miller indices are proportional to the *inverses* of the intercepts of the plane, in the basis of the lattice vectors. If one of the indices is zero, it means that the planes do not intersect that axis (the intercept is "at infinity").

Considering only (hkℓ) planes intersecting one or more lattice points (the *lattice planes*), the perpendicular distance *d* between adjacent lattice planes is related to the (shortest) reciprocal lattice vector orthogonal to the planes by the formula: $d = 2\pi / |g_{hk\ell}|$.

The related notation [hkℓ] denotes the *direction*:

$$h\mathrm{a}_1 + k\mathrm{a}_2 + \ell\mathrm{a}_3.$$

That is, it uses the direct lattice basis instead of the reciprocal lattice. Note that [hkℓ] is *not* generally normal to the (hkℓ) planes, except in a cubic lattice as described below.

Case of Cubic Structures

For the special case of simple cubic crystals, the lattice vectors are orthogonal and of equal length (usually denoted *a*); similar to the reciprocal lattice. Thus, in this common case, the Miller indices (hkℓ) and [hkℓ] both simply denote normals/directions in Cartesian coordinates.

For cubic crystals with lattice constant a, the spacing d between adjacent (hkℓ) lattice planes is (from above):

$$d_{hk\ell} = \frac{a}{\sqrt{h^2 + k^2 + \ell^2}}.$$

Because of the symmetry of cubic crystals, it is possible to change the place and sign of the integers and have equivalent directions and planes:

- Coordinates in *angle brackets* such as ⟨100⟩ denote a *family* of directions which are equivalent due to symmetry operations, such as [100], [010], [001] or the negative of any of those directions.
- Coordinates in *curly brackets* or *braces* such as {100} denote a family of plane normals which are equivalent due to symmetry operations, much the way angle brackets denote a family of directions.

For face-centred cubic and body-centred cubic lattices, the primitive lattice vectors are not orthogonal. However, in these cases the Miller indices are conventionally defined relative to the lattice vectors of the cubic supercell and hence are again simply the Cartesian directions.

Case of Hexagonal and Rhombohedral Structures

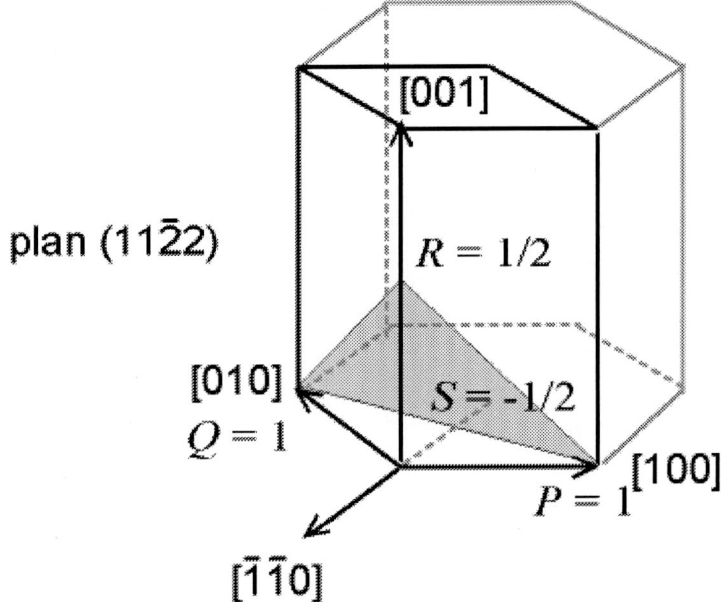

Figure: *Miller-Bravais indices*

With hexagonal and rhombohedral lattice systems, it is possible to use the Bravais-Miller index which has 4 numbers ($h\ k\ i\ \ell$)

$$i = -(h + k).$$

Here h, k and ℓ are identical to the Miller index, and i is a redundant index.

This four-index scheme for labelling planes in a hexagonal lattice makes permutation symmetries apparent. For example, the similarity between (110) ≡ (1120) and (120) ≡ (1210) is more obvious when the redundant index is shown.

In the figure at right, the (001) plane has a 3-fold symmetry: it remains unchanged by a rotation of 1/3 (2π/3 rad, 120°). The [100], [010] and the [110] directions are really similar. If S is the intercept of the plane with the [110] axis, then

$$i = 1/S.$$

There are also *ad hoc* schemes (e.g. in the transmission electron microscopy literature) for indexing hexagonal *lattice vectors* (rather than reciprocal lattice vectors or planes) with four indices. However they don't operate by similarly adding a redundant index to the regular three-index set.

For example, the reciprocal lattice vector (hkℓ) as suggested above can be written as ha*+kb*+ℓc*if the reciprocal-lattice basis-vectors are a*, b*, and c*. For hexagonal crystals this may be expressed in terms of direct-lattice basis-vectors a, b and c as

$$(hk\ell) = h\overrightarrow{a^*} + k\overrightarrow{b^*} + \ell\overrightarrow{c^*} = \frac{2}{3a^2}(2h+k)\vec{a} + \frac{2}{3a^2}(h+2k)\vec{b} + \frac{1}{c^2}(\ell)\vec{c}.$$

Hence zone indices of the direction perpendicular to plane (hkℓ) are, in suitably normalized triplet form, simply [2h+k,h+2k, ℓ(3/2)(a/c)^2]. When *four indices* are used for the zone normal to plane (hkℓ), however, the literature often uses [h,k,-h-k, ℓ(3/2)(a/c)^2] instead. Thus as you can see, four-index zone indices in square or angle brackets sometimes mix a single direct-lattice index on the right with reciprocal-lattice indices (normally in round or curly brackets) on the left.

The Crystallographic Planes and Directions

The crystallographic directions are fictitious lines linking nodes (atoms, ions or molecules) of a crystal. Similarly, the crystallographic planes are fictitious *planes* linking nodes. Some directions and planes have a higher density of nodes; these dense planes have an influence on the behaviour of the crystal:

- *optical properties:* in condensed matter, the light "jumps" from one atom to the other with the Rayleigh scattering; the velocity of light thus varies according to the directions, whether the atoms are close or far; this gives the birefringence
- *adsorption and reactivity:* the adsorption and the chemical reactions occur on atoms or molecules, these phenomena are thus sensitive to the density of nodes;
- *surface tension:* the condensation of a material means that the atoms, ions or molecules are more stable if they are surrounded by other similar species; the surface tension of an interface thus varies according to the density on the surface
  - o the pores and crystallites tend to have straight grain boundaries following dense planes
  - o cleavage
- dislocations (plastic deformation)
  - o the dislocation core tends to spread on dense planes (the elastic perturbation is "diluted"); this reduces the friction (Peierls–Nabarro force), the sliding occurs more frequently on dense planes;
  - o the perturbation carried by the dislocation (Burgers vector) is along a dense direction: the shift of one node in a dense direction is a lesser distortion;
  - o the dislocation line tends to follow a dense direction, the dislocation line is often a straight line, a dislocation loop is often a polygon.

For all these reasons, it is important to determine the planes and thus to have a notation system.

Integer vs. Irrational Miller indices: Lattice Planes and Quasicrystals

Ordinarily, Miller indices are always integers by definition, and this constraint is physically significant. To understand this, suppose that we allow a plane (abc) where the Miller "indices" *a*, *b* and *c* (defined as above) are not necessarily integers.

If *a*, *b* and *c* have rational ratios, then the same family of planes can be written in terms of integer indices (hk–!) by scaling *a*, *b* and *c* appropriately: divide by the largest of the three numbers, and then multiply by the least common denominator. Thus, integer Miller indices implicitly include indices with all rational ratios. The reason why planes where the components (in the reciprocal-lattice basis) have rational ratios are of special interest is that these are the lattice planes: they are the only planes whose intersections with the crystal are 2d-periodic.

For a plane (abc) where *a*, *b* and *c* have irrational ratios, on the other hand, the intersection of the plane with the crystal is *not* periodic. It forms an aperiodic pattern known as a quasicrystal. This construction corresponds precisely to the standard "cut-and-project" method of defining a quasicrystal, using a plane with irrational-ratio Miller indices. (Although many quasicrystals, such as the Penrose tiling, are formed by "cuts" of periodic lattices in more than three dimensions, involving the intersection of more than one such hyperplane.)

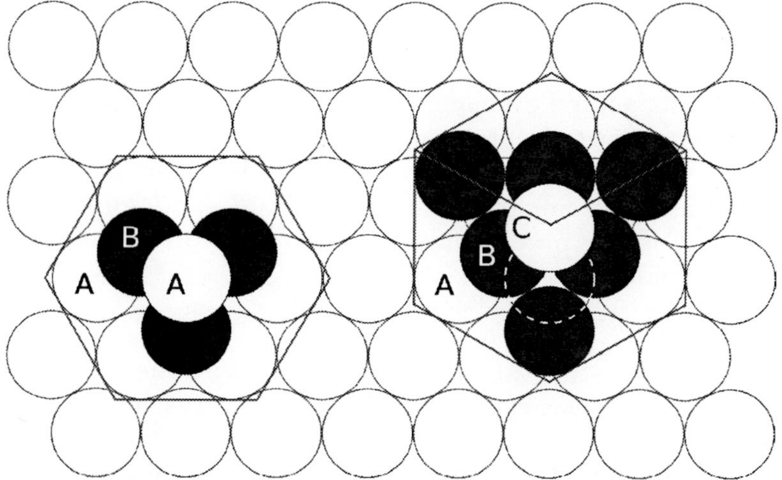

***Figure:** HCP lattice (left) and the fcc lattice (right)*

Close Packing

The principles involved can be understood by considering the most efficient way of packing together equal-sized spheres and stacking close-packed atomic planes in three dimensions. For example, if plane A lies beneath plane B, there are two possible ways of placing an additional atom on top of layer B. If an additional layer was placed directly over plane A, this would give rise to the following series :

...ABABABAB....

This type of crystal structure is known as hexagonal close packing (hcp).

If however, all three planes are staggered relative to each other and it is not until the fourth layer is positioned directly over plane A that the sequence is repeated, then the following sequence arises:

...ABCABCABC...

This type of crystal structure is known as cubic close packing (ccp).

The unit cell of the ccp arrangement is the face-centred cubic (fcc) unit cell. This is not immediately obvious as the closely packed layers are parallel to the {111} planes of the fcc unit cell. There are four different orientations of the close-packed layers.

The packing efficiency could be worked out by calculating the total volume of the spheres and dividing that by the volume of the cell as follows:

$$\frac{4 \times \frac{4}{3}\pi r^3}{16\sqrt{2}r^3} = \frac{\pi}{3\sqrt{2}} = 0.7405$$

The 74% packing efficiency is the maximum density possible in unit cells constructed of spheres of only one size. Most crystalline forms of metallic elements are hcp, fcc, or bcc (body-centred cubic). The coordination number of hcp and fcc is 12 and its atomic packing factor (APF) is the number mentioned above, 0.74. The APF of bcc is 0.68 for comparison.

Bravais Lattices

When the crystal systems are combined with the various possible lattice centerings, we arrive at the Bravais lattices. They describe the geometric arrangement of the lattice points, and thereby the translational symmetry of the crystal. In three dimensions, there are 14 unique Bravais lattices that are distinct from one another in the translational symmetry they contain. All crystalline materials recognized until now (not including quasicrystals) fit in one of these arrangements. The fourteen three-dimensional lattices, classified by crystal system, are shown above. The Bravais lattices are sometimes referred to as *space lattices*.

The crystal structure consists of the same group of atoms, the *basis*, positioned around each and every lattice point. This group of atoms therefore repeats indefinitely in three dimensions according to the arrangement of one of the 14 Bravais lattices. The characteristic rotation and mirror symmetries of the group of atoms, or unit cell, is described by its crystallographic point group.

Point Groups

The crystallographic point group or *crystal class* is the mathematical group comprising the symmetry operations that leave at least one point unmoved and that leave the appearance of the crystal structure unchanged. These symmetry operations include

- *Reflection*, which reflects the structure across a *reflection plane*
- *Rotation*, which rotates the structure a specified portion of a circle about a *rotation axis*
- *Inversion*, which changes the sign of the coordinate of each point with respect to a *centre of symmetry* or *inversion point*
- *Improper rotation*, which consists of a rotation about an axis followed by an inversion.

Rotation axes (proper and improper), reflection planes, and centres of symmetry are collectively called *symmetry elements*. There are 32 possible crystal classes. Each one can be classified into one of the seven crystal systems.

Space Groups

In addition to the operations of the point group, the space group of the crystal structure contains translational symmetry operations. These include:

- Pure *translations*, which move a point along a vector
- *Screw axes*, which rotate a point around an axis while translating parallel to the axis.
- *Glide planes*, which reflect a point through a plane while translating it parallel to the plane.

There are 230 distinct space groups.

Grain Boundaries

Grain boundaries are interfaces where crystals of different orientations meet. A grain boundary is a single-phase interface, with crystals on each side of the boundary being identical except in orientation. The term "crystallite boundary" is sometimes, though rarely, used. Grain boundary areas contain those atoms that have been perturbed from their original lattice sites, dislocations, and impurities that have migrated to the lower energy grain boundary.

Treating a grain boundary geometrically as an interface of a single crystal cut into two parts, one of which is rotated, we see that there are five variables required to define a grain boundary. The first two numbers come from the unit vector that specifies a rotation axis. The third number designates the angle of rotation of the grain. The final two numbers specify the plane of the grain boundary (or a unit vector that is normal to this plane).

Grain boundaries disrupt the motion of dislocations through a material, so reducing crystallite size is a common way to improve strength, as described by the Hall–Petch relationship. Since grain boundaries are defects in the crystal structure they tend to decrease the electrical and thermal conductivity of the material.

The high interfacial energy and relatively weak bonding in most grain boundaries often makes them preferred sites for the onset of corrosion and for the precipitation of new phases from the solid. They are also important to many of the mechanisms of creep.

Grain boundaries are in general only a few nanometers wide. In common materials, crystallites are large enough that grain boundaries account for a small fraction of the material. However, very small grain sizes are achievable.

In nanocrystalline solids, grain boundaries become a significant volume fraction of the material, with profound effects on such properties as diffusion and plasticity. In the limit of small crystallites, as the volume fraction of grain boundaries approaches 100%, the material ceases to have any crystalline character, and thus becomes an amorphous solid.

Defects and Impurities

Real crystals feature defects or irregularities in the ideal arrangements described above and it is these defects that critically determine many of the electrical and mechanical properties of real materials. When one atom substitutes for one of the principal atomic components within the crystal structure, alteration in the electrical and thermal properties of the material may ensue.

Impurities may also manifest as spin impurities in certain materials. Research on magnetic impurities demonstrates that substantial alteration of certain properties such as specific heat may be affected by small concentrations of an impurity, as for example impurities in semiconducting ferromagnetic alloys may lead to different properties as first predicted in the late 1960s. Dislocations in the crystal lattice allow shear at lower stress than that needed for a perfect crystal structure.

Prediction of Structure

Figure: *Crystal structure of sodium chloride (table salt)*

The difficulty of predicting stable crystal structures based on the knowledge of only the chemical composition has long been a stumbling block on the way to fully computational materials design. Now, with more powerful algorithms and high-performance computing, structures of medium complexity can be predicted using such approaches as evolutionary algorithms, random sampling, or metadynamics.

The crystal structures of simple ionic solids (e.g., NaCl or table salt) have long been rationalized in terms of Pauling's rules, first set out in 1929 by Linus Pauling, referred to by many since as the "father of the chemical bond". Pauling also considered the nature of the interatomic forces in metals, and concluded that about half of the five d-orbitals in the transition metals are involved in bonding, with the remaining nonbonding d-orbitals being responsible for the magnetic properties. He, therefore, was able to correlate the number of d-orbitals in bond formation with the bond length as well as many of the physical properties of the substance. He subsequently introduced the metallic orbital, an extra orbital necessary to permit uninhibited resonance of valence bonds among various electronic structures.

In the resonating valence bond theory, the factors that determine the choice of one from among alternative crystal structures of a metal or intermetallic compound revolve around the energy of resonance of bonds among interatomic positions. It is clear that some modes of resonance would make larger contributions (be more mechanically stable than others), and that in particular a simple ratio of number of bonds to number of positions would be exceptional. The resulting principle is that a special stability is associated with the simplest ratios or "bond numbers": 1/2, 1/3, 2/3, 1/4, 3/4, etc. The choice of structure and the value of the axial ratio (which determines the relative bond lengths) are thus a result of the effort of an atom to use its valency in the formation of stable bonds with simple fractional bond numbers.

After postulating a direct correlation between electron concentration and crystal structure in beta-phase alloys, Hume-Rothery analyzed the trends in melting points, compressibilities and bond lengths as a function of group number in the periodic table in order to establish a system of valencies of the transition elements in the metallic state. This treatment thus emphasized the increasing bond strength as a function of group number.

The operation of directional forces were emphasized in one article on the relation between bond hybrids and the metallic structures. The resulting correlation between electronic and crystalline structures is summarized by a single parameter, the weight of the d-electrons per hybridized metallic orbital. The "d-weight" calculates out to 0.5, 0.7 and 0.9 for the fcc, hcp and bcc structures respectively. The relationship between d-electrons and crystal structure thus becomes apparent.

Polymorphism

Quartz is one of the several thermodynamically stable crystalline forms of silica, SiO_2. The most important forms of silica include: α-quartz, β-quartz, tridymite, cristobalite, coesite, and stishovite.

Polymorphism refers to the ability of a solid to exist in more than one crystalline form or structure. According to Gibbs' rules of phase equilibria, these unique crystalline phases will be dependent on intensive variables such as pressure and temperature. Polymorphism can potentially be found in many crystalline materials including polymers, minerals, and metals, and is related to allotropy, which refers to elemental solids.

The complete morphology of a material is described by polymorphism and other variables such as crystal habit, amorphous fraction or crystallographic defects. Polymorphs have different stabilities and may spontaneously convert from a metastable form (or thermodynamically unstable form) to the stable form at a particular temperature. They also exhibit different melting points, solubilities, and X-ray diffraction patterns.

One good example of this is the quartz form of silicon dioxide, or SiO_2. In the vast majority of silicates, the Si atom shows tetrahedral coordination by 4 oxygens. All but one of the crystalline forms involve tetrahedral SiO_4 units linked together by shared vertices in different arrangements. In different minerals the tetrahedra show different degrees of networking and polymerization. For example, they occur singly, joined together in pairs, in larger finite clusters including rings, in chains, double chains, sheets, and three-dimensional frameworks. The minerals are classified into groups based on these structures. In each of its 7 thermodynamically stable crystalline forms or polymorphs of crystalline quartz, only 2 out of 4 of each the edges of the SiO_4 tetrahedra are shared with others, yielding the net chemical formula for silica: SiO_2.

Another example is elemental tin (Sn), which is malleable near ambient temperatures but is brittle when cooled. This change in mechanical properties due to existence of its two major allotropes, α- and β-tin. The two allotropes that are encountered at normal pressure and temperature, α-tin and β-tin, are more commonly known as *gray tin* and *white tin* respectively. Two more allotropes, γ and σ, exist at temperatures above 161 °C and pressures above several GPa. White tin is metallic, and is the stable crystalline form at or above room temperature. Below 13.2 °C, tin exists in the gray form, which has a diamond cubic crystal structure, similar to diamond, silicon or germanium. Gray tin has no metallic properties at all, is a dull-gray powdery material, and has few uses, other than a few specialized semiconductor applications. Although the α-β transformation temperature of tin is nominally 13.2 °C, impurities (e.g. Al, Zn, etc.) lower the transition temperature well below 0 °C, and upon addition of Sb or Bi the transformation may not occur at all.

Physical Properties

Twenty of the 32 crystal classes are piezoelectric, and crystals belonging to one of these classes (point groups) display piezoelectricity. All piezoelectric classes lack a centre of symmetry. Any material develops a dielectric polarization when an electric field is applied, but a substance that has such a natural charge separation even in the absence of a field is called a polar material.

Whether or not a material is polar is determined solely by its crystal structure. Only 10 of the 32 point groups are polar. All polar crystals are pyroelectric, so the 10 polar crystal classes are sometimes referred to as the pyroelectric classes.

There are a few crystal structures, notably the perovskite structure, which exhibit ferroelectric behaviour. This is analogous to ferromagnetism, in that, in the absence of an electric field during production, the ferroelectric crystal does not exhibit a polarization. Upon the application of an electric field of sufficient magnitude, the crystal becomes permanently polarized.

This polarization can be reversed by a sufficiently large counter-charge, in the same way that a ferromagnet can be reversed. However, although they are called ferroelectrics, the effect is due to the crystal structure (not the presence of a ferrous metal).

Coordination Number

In chemistry and crystallography, the coordination number of a central atom in a molecule or crystal is the number of its nearest neighbours. This number is determined somewhat differently for molecules than for crystals.

In chemistry, the emphasis is on bonding structure in molecules or ions and the coordination number of an atom is determined by simply counting the other atoms to which it is bonded (by either single or multiple bonds). For example, $[Cr(NH_3)_2Cl_2Br_2]^{1-}$ has Cr^{3+} as its central cation, which has a coordination number of 6.

However the solid-state structures of crystals often have less clearly defined bonds, so a simpler model is used, in which the atoms are represented by touching spheres.

In this model the coordination number of an atom is the number of other atoms that it touches. For an atom in the interior of a crystal lattice, the number of atoms touching the given atom is the bulk coordination number; for an atom at a surface of a crystal, this is the surface coordination number.

Chemistry Usage

In chemistry, coordination number (c.n.), defined originally in 1893 by Alfred Werner, is the total number of neighbours of a central atom in a molecule or ion. Although a carbon atom has four chemical bonds in most stable molecules, the coordination number of each carbon is four in methane (CH_4), three in ethylene ($H_2C{=}CH_2$, each C is bonded to 2H + 1C = 3 atoms), and two in acetylene. In effect we count the first bond (or sigma bond) to each neighbouring atom, but not the other bonds (pi bonds).

In inorganic chemistry also, only the first or sigma bond between each ligand and the central atom counts, but not any pi bonds to the same ligands. In tungsten hexacarbonyl, $W(CO)_6$, the coordination number of tungsten (W) is counted as six although pi as well as sigma bonding is important in such metal carbonyls.

Examples of high coordination number complexes are the ions formed by uranium and thorium with bidentate nitrate ion ligands, $U(NO_3)_6^{2-}$ and $Th(NO_3)_6^{2-}$. Here each nitrate ligand is bound to the metal by two oxygen atoms, so that the total coordination number of the U or Th atom is 12.

When the surrounding ligands are smaller than the central atom, even higher coordination numbers may be possible. One computational chemistry study predicted a particularly stable $PbHe_{15}^{2+}$ ion composed of a central lead ion coordinated with no fewer than 15 helium atoms.

Crystallography Usage

In materials science, the bulk coordination number of a given atom in the interior of a crystal lattice is the number of atoms touching the given atom. Iron at 20 °C has a body-centred cubic (BCC) crystal in which each interior iron atom occupies the centre of a cube formed by eight neighbouring iron atoms. The bulk coordination number for this structure is therefore 8.

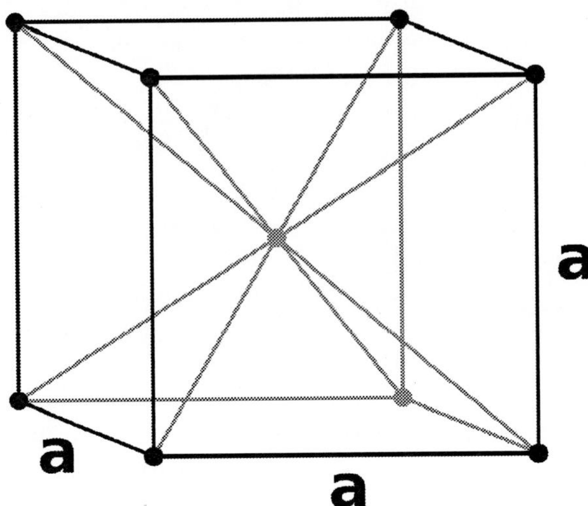

Figure: *BCC structure of iron. For greater visibility the atoms are indicated by points rather than touching spheres.*

The highest bulk coordination number is 12, found in both hexagonal close-packed (HCP) and cubic close-packed (CCP) (also known as face-centred cubic or FCC) structures. This value of 12 corresponds to the theoretical limit of the kissing number problem when all spheres are identical.

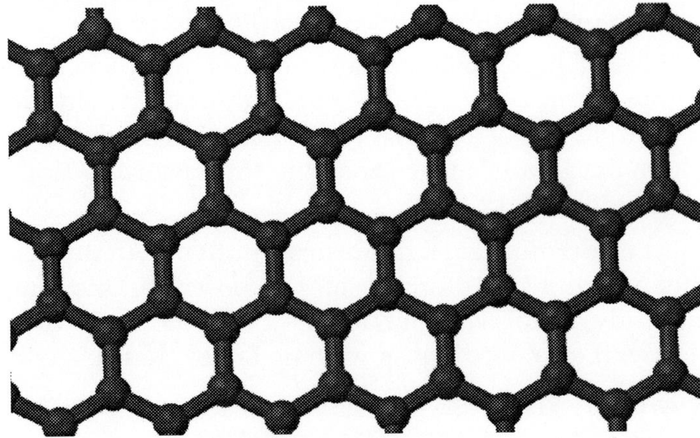

Figure: *One layer of a graphite crystal (graphene) with carbon atoms and C-C bonds shown in black.*

The two most common allotropes of carbon have different coordination numbers. In diamond, each carbon atom is at the centre of a tetrahedron formed by four other carbon atoms, so the coordination number is four, as for methane. Graphite is made of two-dimensional layers in which each carbon is covalently bonded to three other carbons. Atoms in other layers are much further away and are not nearest neighbours, so the coordination number of a carbon atom in graphite is 3 as in ethylene.

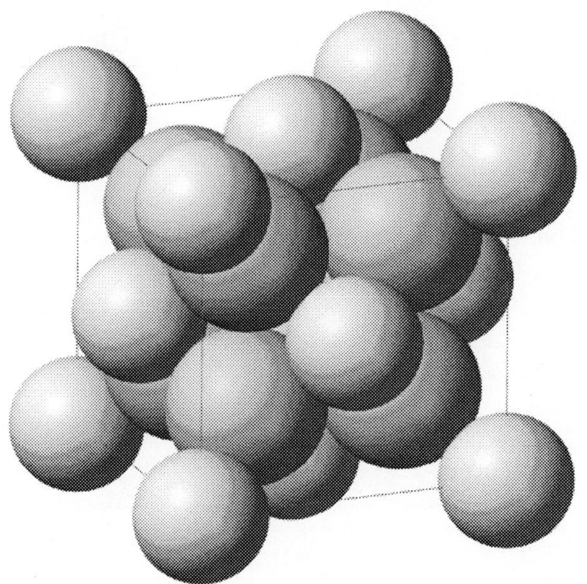

Figure: *Calcium fluoride unit cell. In this image the Ca^{2+}(gray) and F^{-} (green) ions are shown as touching spheres*

Simple ionic structures are described by two coordination numbers, one for each type of ion. Calcium fluoride (CaF_2) is an (8, 4) structure, meaning that each cation Ca^{2+} is surrounded by eight F^{-} anion neighbours, and each anion F^{-} by four Ca^{2+}. For sodium chloride (NaCl), the numbers of cations and anions are equal, and both coordination numbers are six so that the structure is (6, 6).

For an atom at a surface of a crystal, the surface coordination number is always less than the bulk coordination number. The surface coordination number is dependent on the Miller indices of the surface. In a body-centred cubic (BCC) crystal, the bulk coordination number is 8, whereas, for the (100) surface, the surface coordination number is 4.

Usage in Quasicrystal, Liquid and other Disordered Systems

The coordination number of systems with disorder cannot be precisely defined.

The first coordination number can be defined using the radial distribution function $g(r)$

$$n_1 = 4\pi\int_{r_0}^{r_1} r^2 g(r)\rho dr,$$

where r_0 is the rightmost position starting from $r=0$ whereon $g(r)$ is approximately zero, r_1 is the first minimum. Therefore, it is the area under the first peak of $g(r)$.

The second coordination number is defined similarly:

$$n_2 = 4\pi\int_{r_1}^{r_2} r^2 g(r)\rho dr.$$

Alternative definitions for the coordination number can be found in literature, but in essence the main idea is the same. One of those definition are as follows: Denote r_p as the position of the first peak,

$$n_{1'} = 4\pi \times 2\int_{r_0}^{r_p} r^2 g(r)\rho dr.$$

The first coordination shell is the spherical shell with radius between r_0 to r_1 around the central particle under investigation.

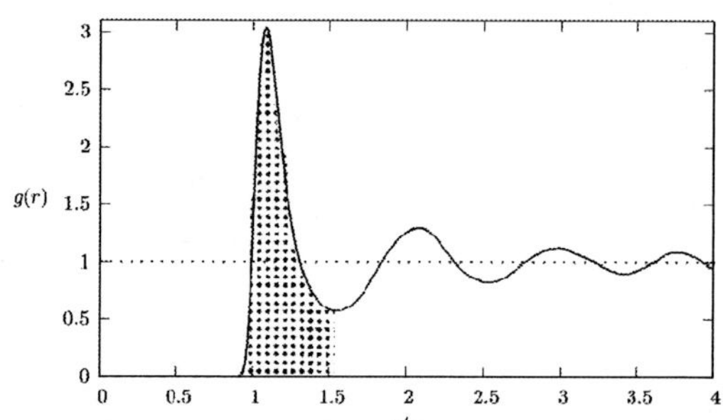

Figure: *First coordination number of Lennard-Jones fluid*

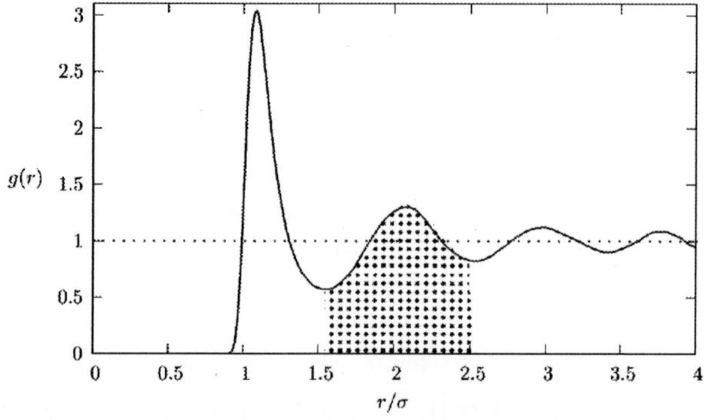

Figure: *Second coordination number of Lennard-Jones fluid*

Packing Factors

In crystallography, atomic packing factor (APF) or packing fraction is the fraction of volume in a crystal structure that is occupied by atoms. It is dimensionless and always less than unity. For practical purposes, the APF of a crystal structure is determined by assuming that atoms are rigid spheres. The radius of the spheres is taken to be the maximal value such that the atoms do not overlap. For one-component crystals (those that contain only one type of atom), the APF is represented mathematically by

$$\mathrm{APF} = \frac{N_{\mathrm{atoms}} V_{\mathrm{atom}}}{V_{\mathrm{unitcell}}}$$

where N_{atoms} is the number of atoms in the unit cell, V_{atom} is the volume of an atom, and $V_{unit\ cell}$ is the volume occupied by the unit cell. It can be proven mathematically that for one-component structures, the most dense arrangement of atoms has an APF of about 0.74. In reality, this number can be higher due to specific intermolecular factors. For multiple-component structures, the APF can exceed 0.74.

Worked out Examples

Body-centred cubic crystal structure:

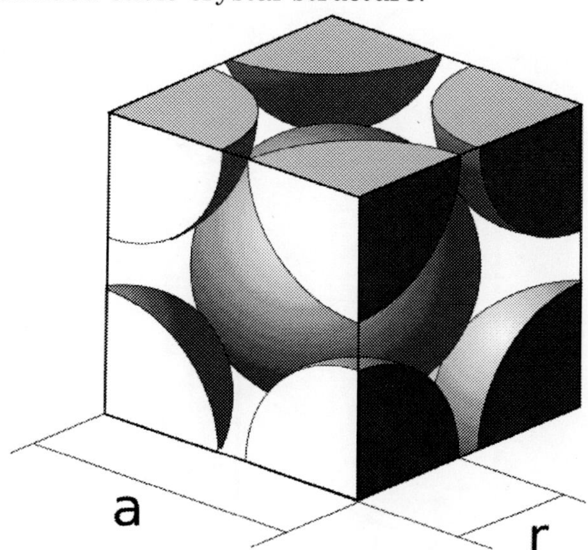

Figure: *BCC structure*

The primitive unit cell for the body-centred cubic (BCC) crystal structure contains several fractions taken from nine atoms: one on each corner of the cube and one atom in the centre. Because the volume of each corner atom is shared between adjacent cells, each BCC cell contains two atoms.

Each corner atom touches the centre atom. A line that is drawn from one corner of the cube through the centre and to the other corner passes through $4r$, where r is the radius of an atom. By geometry, the length of the diagonal is $a\sqrt{3}$. Therefore, the length of each side of the BCC structure can be related to the radius of the atom by

$$a = \frac{4r}{\sqrt{3}}.$$

Knowing this and the formula for the volume of a sphere, it becomes possible to calculate the APF as follows:

$$\mathrm{APF} = \frac{N_{\mathrm{atoms}} V_{\mathrm{atom}}}{V_{\mathrm{crystal}}} = \frac{2(4/3)\pi r^3}{(4r/\sqrt{3})^3}$$

$$= \frac{\pi\sqrt{3}}{8} \approx 0.68.$$

Hexagonal close-packed crystal structure

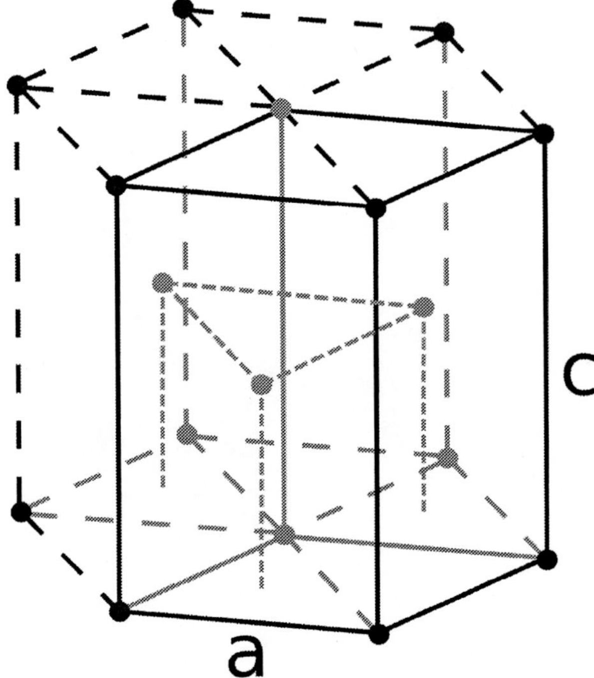

Figure: HCP structure

For the hexagonal close-packed (HCP) structure the derivation is similar. Here the unit cell is a hexagonal prism containing six atoms. Let a be the side length of its base and c be its height. Then:

$$a = 2r$$

$$c = \sqrt{\frac{2}{3}}(4r).$$

It is then possible to calculate the APF as follows:

$$\mathrm{APF} = \frac{N_{\mathrm{atoms}} \cdot V_{\mathrm{atom}}}{V_{\mathrm{crystal}}} = \frac{6 \cdot (4/3)\pi r^3}{[(3\sqrt{3})/2](a^2)(c)}$$

$$= \frac{6(4/3)\pi r^3}{[(3\sqrt{3})/2](2r)^2(\sqrt{\frac{2}{3}})(4r)} = \frac{6(4/3)\pi r^3}{[(3\sqrt{3})/2](\sqrt{\frac{2}{3}})(16r^3)}$$

$$= \frac{\pi}{\sqrt{18}} \approx 0.74.$$

APF of Common Structures

By similar procedures, the ideal atomic packing factors of all crystal structures can be found. The common ones are collected here as reference, rounded to the nearest hundredth.

gσ

Hexagonal Crystal System

In crystallography, the hexagonal crystal system is one of the 7 crystal systems, the hexagonal lattice system is one of the 7 lattice systems, and the hexagonal crystal family is one of the 6 crystal families. They are closely related and often confused with each other, but they are not the same.

The hexagonal lattice system consists of just one Bravais lattice type: the hexagonal one. The hexagonal crystal system consists of the 7 point groups such that all their space groups have the hexagonal lattice as underlying lattice. The hexagonal crystal family consists of the 12 point groups such that at least one of their space groups has the hexagonal lattice as underlying lattice, and is the union of the hexagonal crystal system and the trigonal crystal system.

In some cases, it is useful or instructive to redraw a hexagonal structure with orthohexagonal axes, wherein the b axis is redrawn at 90° to the a and c axes.

Graphite is an example of a crystal that crystallizes in the hexagonal crystal system.

Hexagonal Lattice System

The hexagonal lattice system is one of the seven lattice systems, consisting of the hexagonal Bravais lattice. It is associated with 45 space groups whose underlying lattice has point group of order 24. It is often confused with the smaller hexagonal crystal system, which consists of the 27 space groups such that all space groups with the same point group are in the hexagonal lattice system, or with the larger hexagonal crystal family, consisting of the 52 space groups in either the hexagonal or rhombohedral lattice systems.

Hexagonal Crystal Family

The hexagonal crystal family consists of the 12 point groups such that at least one of their space groups has the hexagonal lattice as underlying lattice, and is the union of the hexagonal crystal system and the trigonal crystal system. There are 50 space groups associated with it, which are exactly those whose Bravais lattice is either hexagonal or rhombohedral.

Cubic Crystal System

In crystallography, the cubic (or isometric) crystal system is a crystal system where the unit cell is in the shape of a cube. This is one of the most common and simplest shapes found in crystals and minerals.

There are three main varieties of these crystals:

- Primitive cubic (abbreviated cP and alternatively called simple cubic)
- Body-centred cubic (abbreviated cI or bcc),
- Face-centred cubic (abbreviated cF or fcc, and alternatively called cubic close-packed or ccp)

Each is subdivided into other variants listed below. Note that although the *unit cell* in these crystals is conventionally taken to be a cube, the *primitive* unit cell often is not. This is related to the fact that in most cubic crystal systems, there is more than one atom per cubic unit cell.

Voids in the Unit Cell

A simple cubic unit cell has a single cubic void in the centre. A body-centred cubic unit cell has six octahedral voids located at the centre of each face of the unit cell, for a total of three net octahedral voids. Additionally, there are 36 tetrahedral voids located in an octahedral spacing around each octahedral void, for a total of eighteen net tetrahedral voids. These tetrahedral voids are not local maxima and are not technically voids, but they do occasionally appear in multi-atom unit cells.

A face-centred cubic unit cell has eight tetrahedral voids located slightly towards the centre from each corner of the unit cell, for a total of eight net tetrahedral voids. Additionally, there are twelve octahedral voids located at the centre of edge of the unit cell as well as one octahedral hole in the very centre, for a total of four net octahedral voids.

Atomic Packing Factors and Examples

One important characteristic of a crystalline structure is its atomic packing factor. This is calculated by assuming that all the atoms are identical spheres, with a radius large enough that each sphere abuts the next. The atomic packing factor is the proportion of space filled by these spheres.

Assuming one atom per lattice point, in a primitive cubic lattice with cube side length a, the sphere radius would be $a/2$ and the atomic packing factor turns out to be about 0.524 (which is quite low). Similarly, in a *bcc* lattice, the atomic packing factor is 0.680, and in *fcc* it is 0.740. The *fcc* value is the highest theoretically possible value for any lattice, although there are other lattices which also achieve the same value, such as hexagonal close packed and one version of tetrahedral bcc.

As a rule, since atoms in a solid attract each other, the more tightly packed arrangements of atoms tend to be more common. (Loosely packed arrangements do occur, though, for example if the orbital hybridization demands certain bond angles.) Accordingly, the primitive-cubic structure, with especially low atomic packing factor, is rare in nature, but is found in polonium. The *bcc* and fcc, with their higher densities, are both quite common in nature. Examples of *bcc* include iron, chromium, tungsten, and niobium. Examples of *fcc* include aluminium, copper, gold and silver.

Multi-element Compounds

Compounds that consist of more than one element (e.g. binary compounds) often have crystal structures based on a cubic crystal system. Some of the more common ones are listed here.

Interpenetrating Primitive Cubic (Caesium Chloride) Structure

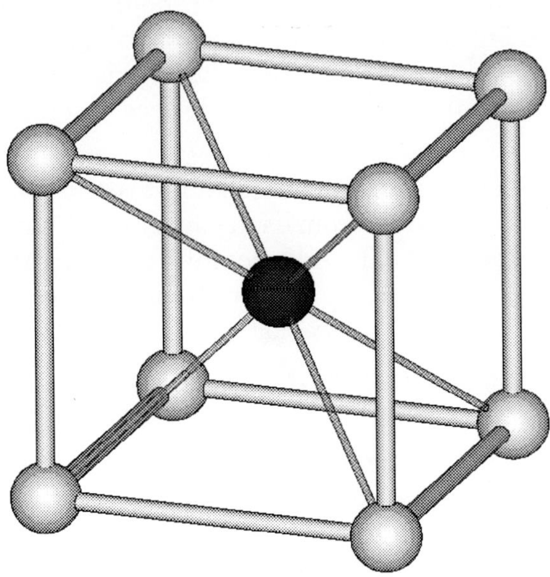

Figure: *A caesium chloride unit cell. The two colours of spheres represent the two types of atoms.*

One structure is the "interpenetrating primitive cubic" structure, also called the "caesium chloride" structure. Each of the two atom types forms a separate primitive cubic lattice, with an atom of one type at the centre of each cube of the other type. Altogether, the arrangement of atoms is the same as body-centred cubic, but with alternating types of atoms at the different lattice sites. Alternately, one could view this lattice as a simple cubic structure with a secondary atom in its cubic void. Examples of compounds with this structure include caesium chloride itself, as well as certain other alkali halides when prepared at low temperatures or high pressures. More generally, this structure is more likely to be formed from two elements whose ions are of roughly the same size (for example, ionic radius of Cs^+ = 167 pm, and Cl^- = 181 pm) .

The space group of this structure is called Pm3m (in Hermann–Mauguin notation), or "221" (in the International Tables for Crystallography). The Strukturbericht designation is "B2".

The coordination number of each atom in the structure is 8: the central cation is coordinated to 8 anions on the corners of a cube as shown, and similarly, the central anion is coordinated to 8 cations on the corners of a cube.

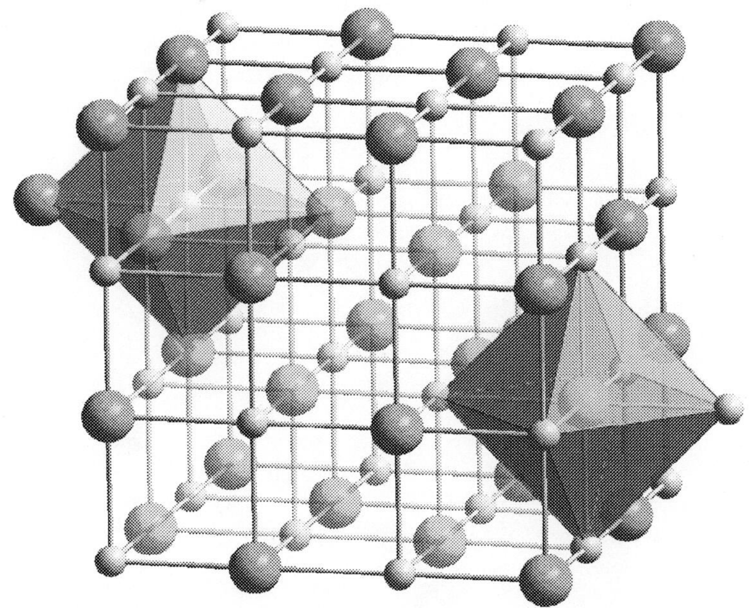

Figure: *The rock-salt crystal structure. Each atom has six nearest neighbours, with octahedral geometry.*

Rock-salt Structure

Another structure is the "rock salt" or "sodium chloride" (halite) structure. In this, each of the two atom types forms a separate face-centred cubic lattice, with the two lattices interpenetrating so as to form a 3D checkerboard pattern. Alternately, one could view this structure as a face-centred cubic structure with secondary atoms in its octahedral holes.

Examples of compounds with this structure include sodium chloride itself, along with almost all other alkali halides, and "many divalent metal oxides, sulphides, selenides, and tellurides". More generally, this structure is more likely to be formed if the cation is slightly smaller than the anion (a cation/anion radius ratio of 0.414 to 0.732).

The space group of this structure is called Fm3m (in Hermann–Mauguin notation), or "225" (in the International Tables for Crystallography). The Strukturbericht designation is "B1".

The coordination number of each atom in this structure is 6: each cation is coordinated to 6 anions at the vertices of an octahedron, and similarly, each anion is coordinated to 6 cations at the vertices of an octahedron.

The interatomic distance (distance between cation and anion, or half the unit cell length a) in some rock-salt-structure crystals are: 2.3 Å (2.3×10^{-10} m) for NaF, 2.8 Å for NaCl, and 3.2 Å for SnTe.

Zincblende Structure

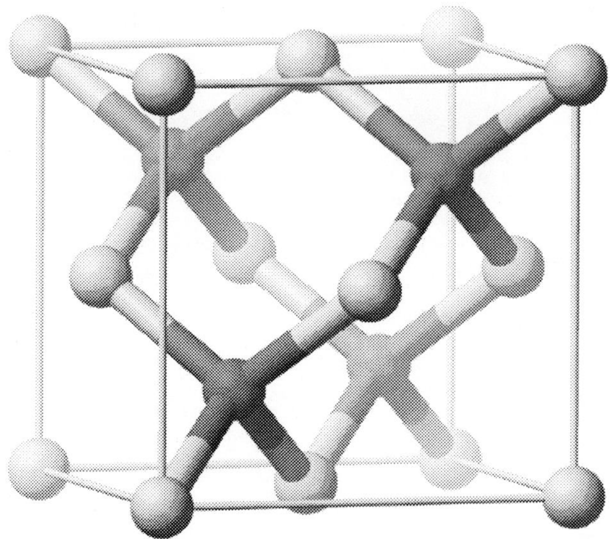

Figure: *A zincblende unit cell*

Another common structure is the "zincblende" structure (also spelled "zinc blende"), named after the mineral zincblende (sphalerite). As in the rock-salt structure, the two atom types form two interpenetrating face-centred cubic lattices. However, it differs from rock-salt structure in how the two lattices are positioned relative to one another. The zincblende structure has tetrahedral coordination: Each atom's nearest neighbours consist of four atoms of the opposite type, positioned like the four vertices of a regular tetrahedron. Altogether, the arrangement of atoms in zincblende structure is the same as diamond cubic structure, but with alternating types of atoms at the different lattice sites.

Examples of compounds with this structure include zincblende itself, lead(II) nitrate, many compound semiconductors (such as gallium arsenide and cadmium telluride), and a wide array of other binary compounds.

Additionally, if one were to take the two atoms to both be carbon, the zincblende structure is equivalent to the structure of diamond.

The space group of this structure is called F43m (in Hermann–Mauguin notation), or "#216" in the International Tables for Crystallography. The Strukturbericht designation is "B3".

Diamond Cubic

The diamond cubic crystal structure is a repeating pattern of 8 atoms that certain materials may adopt as they solidify. While the first known example was diamond, other elements in group 14 also adopt this structure, including α-tin, the semiconductors silicon and germanium, and silicon/germanium alloys in any proportion.

Crystallographic Structure

Diamond cubic is in the Fd3m space group, which follows the face-centred cubic bravais lattice. The lattice describes the repeat pattern; for diamond cubic crystals this lattice is "decorated" with a *motif* of two tetrahedrally bonded atoms in each primitive cell, separated by 1/4 of the width of the unit cell in each dimension.

Many compound semiconductors such as gallium arsenide, β-silicon carbide and indium antimonide adopt the analogous zincblende structure, where each atom has nearest neighbours of an unlike element. Zincblende's space group is F43m, but many of its structural properties are quite similar to the diamond structure.

The atomic packing factor of the diamond cubic structure (the proportion of space that would be filled by spheres that are centred on the vertices of the structure and are as large as possible without overlapping) is $\frac{\pi\sqrt{3}}{16} \approx 0.34$, significantly smaller (indicating a less dense structure) than the packing factors for the face-centred and body-centred cubic lattices. Zinc blende structures have higher packing factors than 0.34 depending on the relative sizes of their two component atoms.

The first, second, third and fourth nearest-neighbour distances in units of the cubic lattice constant are $\frac{\sqrt{3}}{4}, \frac{\sqrt{2}}{2}, \frac{\sqrt{11}}{4}$, and 1, respectively.

Mathematical Structure

Mathematically, the points of the diamond cubic structure can be given coordinates as a subset of a three-dimensional integer lattice by using a cubical unit cell four units across. With these coordinates, the points of the structure have coordinates (x, y, z) satisfying the equations

$$x = y = z \pmod 2), \text{ and}$$

$$x + y + z = 0 \text{ or } 1 \pmod 4).$$

There are eight points (modulo 4) that satisfy these conditions:

(0,0,0), (0,2,2), (2,0,2), (2,2,0),

(3,3,3), (3,1,1), (1,3,1), (1,1,3)

All of the other points in the structure may be obtained by adding multiples of four to the x, y, and z coordinates of these eight points. Adjacent points in this structure are at distance $\sqrt{3}$ apart in the integer lattice; the edges of the diamond structure lie along the body diagonals of the integer grid cubes. This structure may be scaled to a cubical unit cell that is some number a of units across by multiplying all coordinates by $a/4$.

Alternatively, each point of the diamond cubic structure may be given by four-dimensional integer coordinates such that sum to either zero or one.

Two points are adjacent in the diamond structure if and only if their four-dimensional coordinates differ by one in a single coordinate. The total difference in coordinate values between any two points (their four-dimensional Manhattan distance) gives the number of edges in the shortest path between them in the diamond structure. The four nearest neighbours of each point may be obtained, in this coordinate system, by adding one to each of the four coordinates, or by subtracting one from each of the four coordinates, accordingly as the coordinate sum is zero or one. These four-dimensional coordinates may be transformed into three-dimensional coordinates by the formula

$$(a, b, c, d) \rightarrow (a + b - c - d, a - b + c - d, -a + b + c - d).$$

Because the diamond structure forms a distance-preserving subset of the four-dimensional integer lattice, it is a partial cube.

Yet another coordinatization of the diamond cubic involves the removal of some of the edges from a three-dimensional grid graph. In this coordinatization, which has a distorted geometry from the standard diamond cubic structure but has the same topological structure, the vertices of the diamond cubic are represented by all possible 3d grid points and the edges of the diamond cubic are represented by a subset of the 3d grid edges.

The diamond cubic is sometimes called the "diamond lattice" but it is not, mathematically, a lattice: there is no translational symmetry that takes the point (0,0,0) into the point (3,3,3), for instance. However, it is still a highly symmetric structure: any incident pair of a vertex and edge can be transformed into any other incident pair by a congruence of Euclidean space.

Manufacturing Considerations

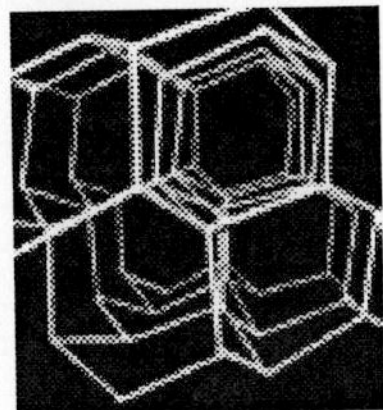

Figure: *A diamond cubic crystal viewed from a <110> direction.*

Since materials with diamond or zinc blende structures are important for formation of wafers used in the fabrication of modern electronics, it is important to know that they present open, hexagonal ion channels when ion implantation is carried out from any of the <110> directions (that is, 45° from one of the cube edges). Their open structure also results in a volume reduction upon melting or amorphization, as is also seen in ice.

They display octahedral cleavage, which means that they have four planes—directions following the faces of the octahedron where there are fewer bonds and therefore points of structural weakness—along which single crystals can easily split, leaving smooth surfaces. Similarly, this lack of bonds can guide chemical etching of the right chemistry (i.e., potassium hydroxide solutions for Si) to produce pyramidal structures such as mesas, points, or etch pits, a useful technique for microelectromechanical systems.

Chapter 2

Defects in Crystals

Crystalline solids exhibit a periodic crystal structure. The positions of atoms or molecules occur on repeating fixed distances, determined by the unit cell parameters. However, the arrangement of atoms or molecules in most crystalline materials is not perfect. The regular patterns are interrupted by crystallographic defects.

Point Defects

Point defects are defects that occur only at or around a single lattice point. They are not extended in space in any dimension. Strict limits for how small a point defect is are generally not defined explicitly, typically, however, these defects involve at most a few extra or missing atoms. Larger defects in an ordered structure are usually considered dislocation loops. For historical reasons, many point defects, especially in ionic crystals, are called *centres*: for example a vacancy in many ionic solids is called a luminescence centre, a colour centre, or F-centre. These dislocations permit ionic transport through crystals leading to electrochemical reactions. These are frequently specified using Kröger–Vink Notation.

Vacancy Defect

In crystallography, a vacancy is a type of point defect in a crystal. Crystals inherently possess imperfections, sometimes referred to as crystalline defects. A defect in which an atom, such as silicon, is missing from one of the lattice sites is known as a 'vacancy' defect. It is also known as a Schottky defect, although in ionic crystals the concepts are not identical.

Vacancies occur naturally in all crystalline materials. At any given temperature, up to the melting point of the material, there is an equilibrium concentration (ratio of vacant lattice sites to those containing atoms). At the melting point of some metals the ratio can be approximately 1:1000.

It is the simplest point defect. In this system, an atom is missing from its regular atomic site. Vacancies are formed during solidification due to vibration of atoms, local rearrangement of atoms, plastic deformation and ionic bombardments.

The creation of a vacancy can be simply modelled by considering the energy required to break the bonds between an atom inside the crystal and its nearest neighbour atoms. Once that atom is removed from the lattice site, it is put back on the surface of the crystal and some energy is retrieved because new bonds are established with other atoms on the surface. However, there is a net input of energy because there are fewer bonds between surface atoms than between atoms in the interior of the crystal.

At any given temperature, the amount of energy needed to create a vacancy is diminished because creating a vacancy disorders the interior of the crystal. The measure of this disorder is called the entropy of the system. Adding vacancies to the material increases the entropy, which tends to reduce the total energy required to create the vacancy. This energy is called the free energy and this is the energy that is required to create an equilibrium concentration of vacancies at a given temperature.

Interstitial Defect

Interstitials are a variety of crystallographic defects, i.e. atoms which occupy a site in the crystal structure at which there is usually not an atom, or two or more atoms sharing one or more lattice sites such that the number of atoms is larger than the number of lattice sites. They are generally high energy configurations. Small atoms in some crystals can occupy interstitial sites in an energetically favourable configuration, such as hydrogen in palladium. Interstitials can be produced for instance by particle irradiation above the threshold displacement energy, but may also exist in small concentrations in thermodynamic equilibrium.

Self-interstitials

Self-interstitial defects are interstitial defects which contain only atoms which are the same as those already present in the lattice.

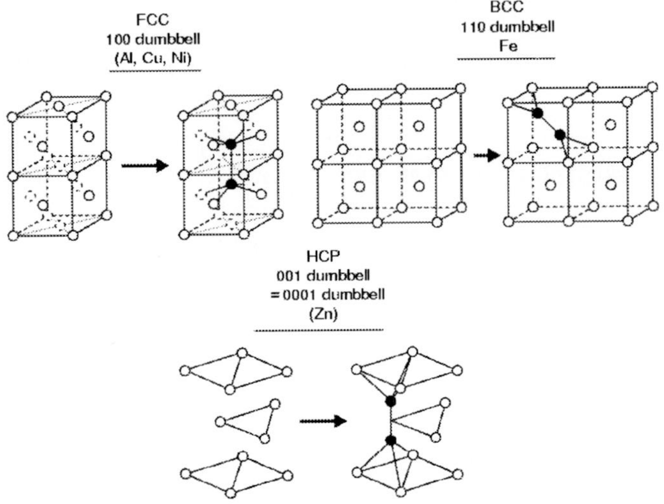

Figure: *Structure of self-interstitial in some common metals. The left-hand side of each crystal type shows the perfect crystal and the right-hand side the one with a defect.*

The structure of interstitial defects has been experimentally determined in some metals and semiconductors. Contrary to what one might intuitively expect, most self-interstitials in metals with a known structure have a 'split' structure, in which two atoms share the same lattice site. Typically the centre of mass of the two atoms is at the lattice site, and they are displaced symmetrically from it along one of the principal lattice directions. For instance, in several common FCC metals such as copper, nickel and platinum, the ground state structure of the self-interstitial is the split interstitial structure, where two atoms are displaced in a positive and negative direction from the lattice site. In BCC iron the ground state interstitial structure is similarly a split interstitial. These split interstitials are often called dumbbell interstitials, because plotting the two atoms forming the interstitial with two large spheres and a thick line joining them makes the structure resemble a dumbbell weight-lifting device. In other BCC metals than iron, the ground state structure is believed based on recent Density-functional theory calculations to be the crowdion interstitial, which can be understood as a long chain (typically some 10-20) of atoms along the lattice direction, compressed compared to the perfect lattice such that the chain contains one extra atom.

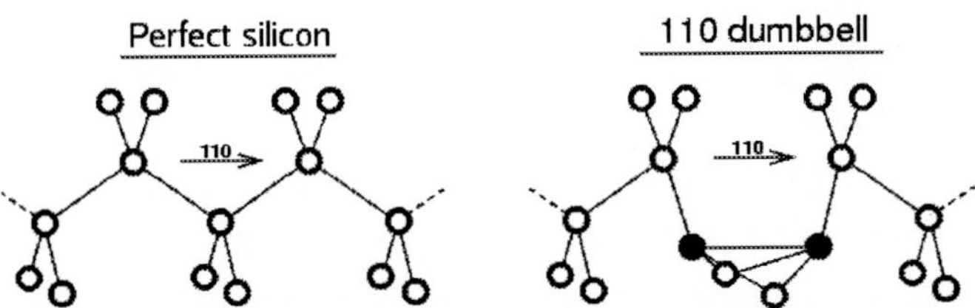

Figure: *Structure of dumbbell self-interstitial in silicon. Note that the structure of the interstitial in silicon may depend on charge state and doping level of the material.*

In semiconductors the situation is more complex, since defects may be charged and different charge states may have different structures. For instance, in silicon, the interstitial may either have a split structure or a tetrahedral truly interstitial one.

Impurity Interstitials

Small impurity interstitial atoms are usually on true off-lattice sites between the lattice atoms. Such sites can be characterized by the symmetry of the interstitial atom position with respect to its nearest lattice atoms. For instance, an impurity atom I with 4 nearest lattice atom A neighbours (at equal distances) in a FCC lattice is in a tetrahedral symmetry position, and thus can be called a tetrahedral interstitial. Large impurity interstitials can also be in split interstitial configurations together with a lattice atom, similar to those of the self-interstitial atom.

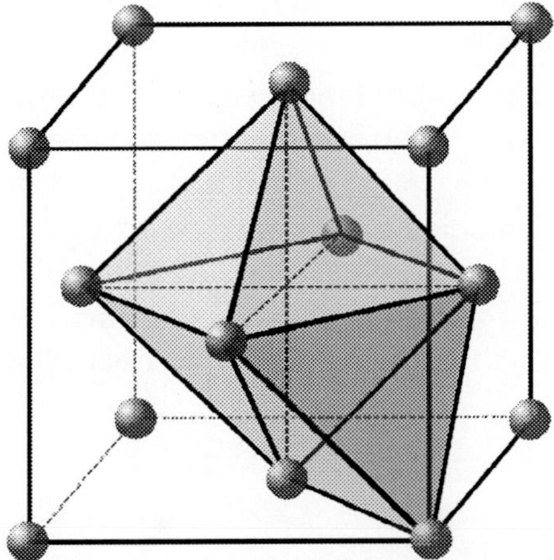

Figure: Octahedral *(red) and* tetrahedral *(blue) interstitial symmetry polyhedra in a face-centred cubic lattice. The actual interstitial atom would ideally be in the middle of one of the polyhedra.*

Effects of Interstitials

Interstitials modify the physical and chemical properties of materials.

- Interstitial carbon atoms have a crucial role for the properties and processing of steels, in particular carbon steels.
- Impurity interstitials can be used e.g. for storage of hydrogen in metals.
- The amorphization of semiconductors such as silicon during ion irradiation is often explained by the buildup of a high concentration of interstitials leading eventually to the collapse of the lattice as it becomes unstable
- Creation of large amounts of interstitials in a solid can lead to a significant energy buildup, which on release can even lead to severe accidents in certain old types of nuclear reactors (Wigner effect). The high-energy states can be released by annealing.
- At least in face-centred cubic (FCC) lattice, interstitials have a large diaelastic softening effect on the material
- It has been proposed that interstitials are related to the onset of melting and the glass transition.

Schottky Defect

A Schottky defect is a type of point defect in a crystal lattice named after Walter H. Schottky. In non-ionic crystals it means a lattice vacancy defect.

In ionic crystals, the defect forms when oppositely charged ions leave their lattice sites, creating vacancies. These vacancies are formed in stoichiometric units, to maintain an overall neutral charge in the ionic solid. The vacancies are then free to move about as their own entities. Normally these defects will lead to a decrease in the density of the crystal. The following are the chemical equations in Kröger-Vink Notation for the formation of Schottky defects in TiO_2 and $BaTiO_3$.

$$\varnothing \Leftrightarrow V_{Ti''''} + 2V_O^{\bullet\bullet}$$

$$\varnothing \Leftrightarrow V_{Ba''} + V_{Ti''''} + 3V_O^{\bullet\bullet}$$

This can be illustrated schematically with a two-dimensional diagram of a sodium chloride crystal lattice:

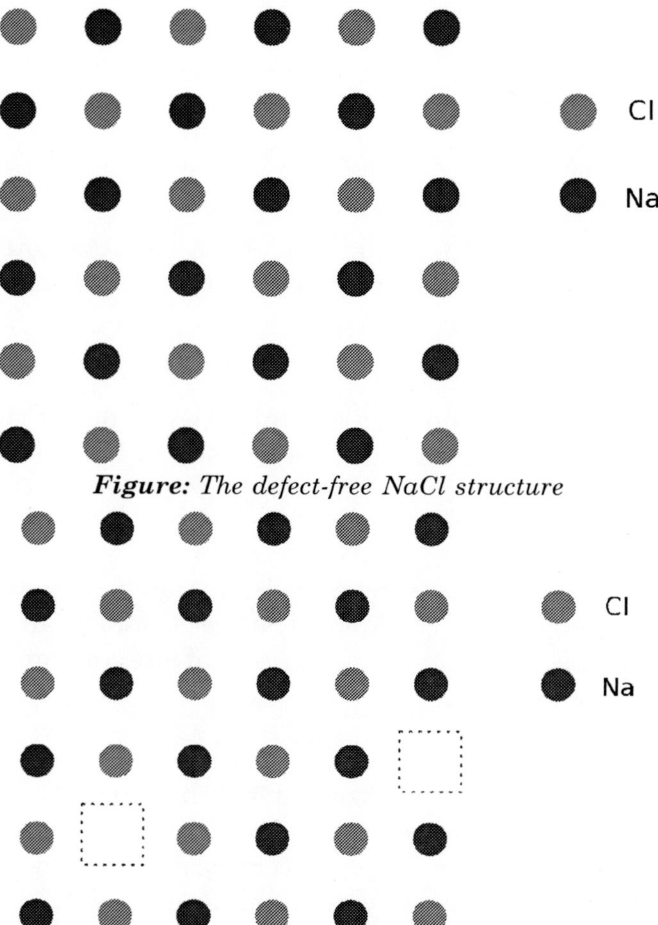

Figure: *The defect-free NaCl structure*

Figure: *Schottky defects within the NaCl structure*

Definition

If in an ionic crystal of type A^+B^- an equal number of cations and anions are missing from their lattice sites so that electrical neutrality as well as stoichiometry is maintained this is called a Schottky Defect.

It is a vacancy defect (due to missing ions) and also a stoichiometric defect, as the ratio of the number of cations and anions remains the same. It occurs only when there is small difference in size between cations and anions. This is produced as the result of the thermal incorporation of unoccupied lattice sites from the exterior of the crystal. The lattice undergoes thermal vibration and thermal expansion when the temperature is raised above 0K. When it happens the pair of vacancies are incorporated in the crystal.

Examples

This type of defect is shown in compounds with :

- highly ionic compounds
- high co-ordination number
- small difference in sizes of cations and anions

Examples : NaCl, KCl, CsCl, KBr, AgCl. etc. A Schottky defect is when 2 oppositely charged ions leave their normal lattice positions forming 2 vacancies in the lattice structure.

Effect on Density

The total number of ions in a crystal with this defect is less than the theoretical value of ions, thus, the density of the solid crystal is less than normal.

Frenkel Defect

The Frenkel Defect is shown by ionic solids. The smaller ion (usually the cation) is displaced from its lattice position to an interstitial site. It creates a *vacancy defect* at its original site and an *interstitial defect* at its new location.

Definition

A Frenkel defect, Frenkel point defect in a crystal lattice. The defect forms when an atom or cation leaves its place in the lattice, creating a vacancy, and becomes an interstitial by lodging in a nearby location not usually occupied by an atom. Frenkel defects occur due to thermal vibrations, and it is theorized that there will be no defects in a crystal at 0 K. The phenomenon is named after the Soviet physicist Yakov Frenkel, who discovered it in 1926.

Effect on Density

This defect does not change the density of the solid as it involves only the migration of the ions within the crystal, thus preserving both the volume as well as mass.

Examples

It is shown in ionic solids with large size difference between the anion and cation (with the cation usually smaller due to an increased effective nuclear charge) Some solids which display this defect - ZnS, AgCl, AgBr, AgI (due to the comparatively smaller size of Zn^{2+} and Ag^{+} ions)

To be noted : AgBr shows both Frenkel as well as Schottky defects. For example, consider a lattice formed by X and M ions. Suppose an M ion leaves the M sublattice, leaving the X sublattice unchanged. The number of interstitials formed will equal the number of vacancies formed. One form of a Frenkel defect reaction in MgO with the oxygen ion leaving the lattice and going into the interstitial site written in Kröger–Vink notation:

$$Mg_{Mg}^{\times} + O_O^{\times} \rightarrow O_i'' + V_O^{\bullet\bullet} + Mg_{Mg}^{\times}$$

This can be illustrated with the example of the sodium chloride crystal structure. The diagrams below are schematic two-dimensional representations.

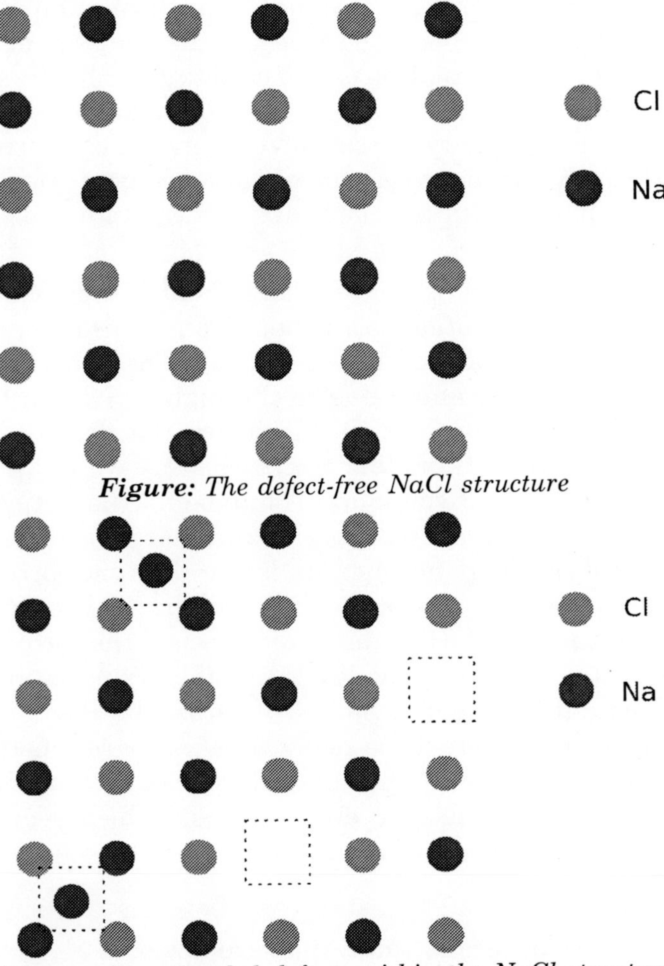

Figure: *The defect-free NaCl structure*

Figure: *Two Frenkel defects within the NaCl structure*

Stone–Wales Defect

A Stone–Wales defect is a crystallographic defect that occurs on carbon nanotubes and graphene and is thought to have important implications for the mechanical properties of carbon nanotubes. The defect is named after Anthony Stone and David Wales of Cambridge University, who described it in a 1986 paper on the isomerization of fullerenes. However, a rather similar defect was described much earlier by Peter Thrower in a paper on defects in graphite. For this reason the term Stone–Thrower–Wales defect is sometimes used. The defect is thought to be responsible for nanoscale plasticity and the brittle–ductile transitions in carbon nanotubes.

The Stone–Wales, or "pyracylene", transformation is the 90° rotation of two carbon atoms with respect to the midpoint of the bond. The Stone–Wales transformation is also used to describe the structural changes of sp^2-bonded carbon nanosystems. For example, it has been proposed that the coalescence process of fullerenes or carbon nanotubes may occur through a sequence of such a rearrangement. By the Stone–Wales transformation, four hexagons are changed into two pentagons and two heptagons.

A Stone–Wales defect is the rearrangement of the six-membered rings of graphene into pentagons and heptagons. This rearrangement is a result of $\pi/2$ (90°) rotation of a C–C bond. The density of Stone–Wales defects is usually small due to the high activation barrier of several eV for the bond rotation. However, the defects has been imaged using scanning tunneling microscopy as well as resonant–vibrational spectroscopy techniques. A number of theoretical studies have shown that the absorption energies and charge transfer energies for single-walled carbon nanotubes are larger than the corresponding values on pristine carbon nanotubes. The diminished resonance and higher strain energy of this defect increased the probability of nucleophilic attack which provides a possible explanation for the enhanced reactivity. Research has also shown the incorporation of defects along a carbon-nanotube network can programme a carbon-nanotube circuit to enhance the conductance along a specific path. These defects lead to a charge delocalization which redirects and incoming electron down a given trajectory.

Impurities

Impurities occur because materials are never 100% pure. In the case of an impurity, the atom is often incorporated at a regular atomic site in the crystal structure. This is neither a vacant site nor is the atom on an interstitial site and it is called a *substitutional* defect. The atom is not supposed to be anywhere in the crystal, and is thus an impurity. In some cases where the radius of the substitutional atom (ion) is substantially smaller than that of the atom (ion) it is replacing, its equilibrium position can be shifted away from the lattice site. These types of substitutional defects are often referred to as off-centre ions. There are two different types of substitutional defects: Isovalent substitution and aliovalent substitution. Isovalent substitution is where the ion that is substituting the original ion is of the same oxidation state as the ion it is replacing. Aliovalent substitution is where the ion that is substituting the original ion is of a different oxidation state as the ion it is replacing. Aliovalent substitutions change the overall charge within the ionic compound, but the ionic compound must be neutral. Therefore a charge compensation mechanism is required. Hence either one of the metals is partially or fully oxidised or reduced, or ion vacancies are created.

Antisite Defects

Antisite defects occur in an ordered alloy or compound when atoms of different type exchange positions. For example, some alloys have a regular structure in which every other atom is a different species; for illustration assume that type A atoms sit on the corners of a cubic lattice, and type B atoms sit in the centre of the cubes. If one cube has an A atom at its centre, the atom is on a site usually occupied by a B atom, and is thus an antisite defect. This is neither a vacancy nor an interstitial, nor an impurity.

Topological defects

Topological defects are regions in a crystal where the normal chemical bonding environment is topologically different from the surroundings. For instance, in a perfect sheet of graphite (graphene) all atoms are in rings containing six atoms. If the sheet contains regions where the number of atoms in a ring is different from six, while the total number of atoms remains the same, a topological defect has formed. An example is the Stone Wales defect in nanotubes, which consists of two adjacent 5-membered and two 7-membered atom rings.

Other Defects

Also amorphous solids may contain defects. These are naturally somewhat hard to define, but sometimes their nature can be quite easily understood. For instance, in ideally bonded amorphous silica all Si atoms have 4 bonds to O atoms and all O atoms have 2 bonds to Si atom.

Thus e.g. an O atom with only one Si bond (a dangling bond) can be considered a defect in silica. Moreover, defects can also be defined in amorphous solids based on empty or densely packed local atomic neighbourhoods, and the properties of such 'defects' can be shown to be similar to normal vacancies and interstitials in crystals,

Complexes can form between different kinds of point defects. For example, if a vacancy encounters an impurity, the two may bind together if the impurity is too large for the lattice. Interstitials can form 'split interstitial' or 'dumbbell' structures where two atoms effectively share an atomic site, resulting in neither atom actually occupying the site.

Dislocation

In materials science, a dislocation is a crystallographic defect, or irregularity, within a crystal structure. The presence of dislocations strongly influences many of the properties of materials. The theory describing the elastic fields of the defects was originally developed by Vito Volterra in 1905, but the term 'dislocation' to refer to a defect on the atomic scale was coined by G. I. Taylor in 1934 . Some types of dislocations can be visualized as being caused by the termination of a plane of atoms in the middle of a crystal. In such a case, the surrounding planes are not straight, but they instead bend around the edge of the terminating plane so that the crystal structure is perfectly ordered on either side.

The analogy with a stack of paper is apt: if a half a piece of paper is inserted in a stack of paper, the defect in the stack is only noticeable at the edge of the half sheet.

There are two primary types: *edge dislocations* and *screw dislocations. Mixed dislocations* are intermediate between these.

Mathematically, dislocations are a type of topological defect, sometimes called a soliton. The mathematical theory explains why dislocations behave as stable particles: they can be moved about, but they maintain their identity as they move. Two dislocations of opposite orientation, when brought together, can cancel each other (this is the process of annihilation), but a single dislocation typically cannot "disappear" on its own.

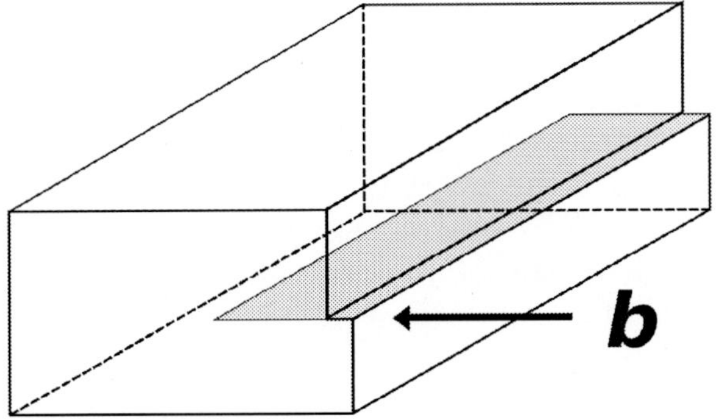

Figure: *An edge-dislocation (b = Burgers vector)*

Dislocation Geometry

Figure: *Crystal lattice showing atoms and lattice planes*

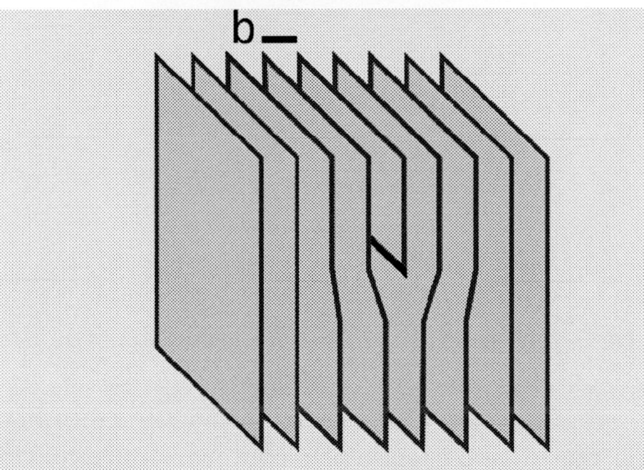

Figure: *Schematic diagram (lattice planes) showing an edge dislocation. Burgers vector in black, dislocation line in blue.*

Two main types of dislocation exist: edge and screw. Dislocations found in real materials typically are *mixed*, meaning that they have characteristics of both. A crystalline material consists of a regular array of atoms, arranged into lattice planes (imagine stacking oranges in a grocery, each of the trays of oranges are the lattice planes). One approach is to begin by considering a 3D representation of a perfect crystal lattice, with the atoms represented by spheres. The viewer may then start to simplify the representation by visualising planes of atoms instead of the atoms themselves.

Edge Dislocations

An edge dislocation is a defect where an extra half-plane of atoms is introduced mid way through the crystal, distorting nearby planes of atoms. When enough force is applied from one side of the crystal structure, this extra plane passes through planes of atoms breaking and joining bonds with them until it reaches the grain boundary. A simple schematic diagram of such atomic planes can be used to illustrate lattice defects such as dislocations. (Figure B represents the "extra half-plane" concept of an edge type dislocation). The dislocation has two properties, a line direction, which is the direction running along the bottom of the extra half plane, and the Burgers vector which describes the magnitude and direction of distortion to the lattice. In an edge dislocation, the Burgers vector is perpendicular to the line direction.

The stresses caused by an edge dislocation are complex due to its inherent asymmetry. These stresses are described by three equations:

$$\sigma_{xx} = \frac{-\mu b}{2\pi(1-\nu)} \frac{y(3x^2 + y^2)}{(x^2 + y^2)^2}$$

$$\sigma_{yy} = \frac{\mu b}{2\pi(1-\nu)} \frac{y(x^2 - y^2)}{(x^2 + y^2)^2}$$

$$\tau_{xy} = \frac{\mu b}{2\pi(1-\nu)} \frac{x(x^2 - y^2)}{(x^2 + y^2)^2}$$

where μ is the shear modulus of the material, b is the Burgers vector, ν is Poisson's ratio and x and y are coordinates.

These equations suggest a vertically oriented dumbbell of stresses surrounding the dislocation, with compression experienced by the atoms near the "extra" plane, and tension experienced by those atoms near the "missing" plane.

Screw Dislocations

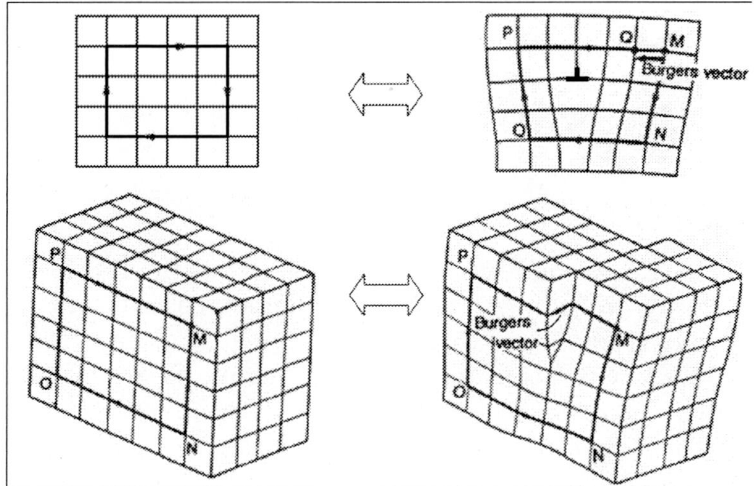

Figure: *Top right: edge dislocation. Bottom right: screw dislocation.*

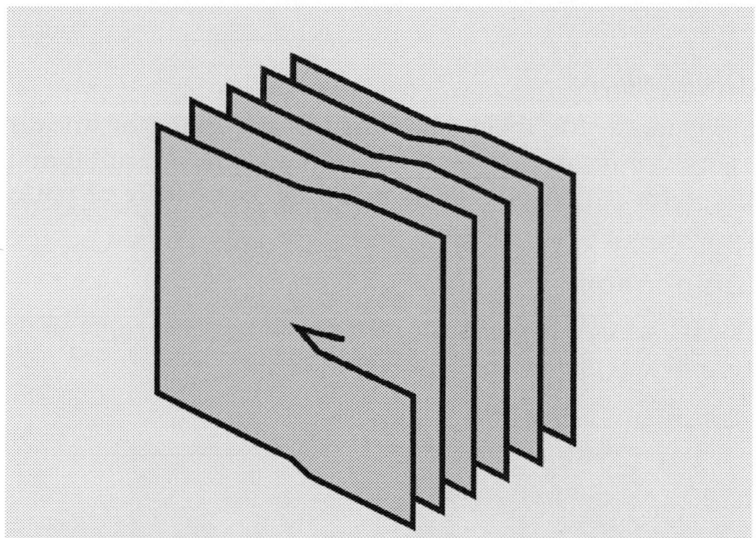

Figure: *C Schematic diagram (lattice planes) showing a screw dislocation.*

A *screw dislocation* is much harder to visualize. Imagine cutting a crystal along a plane and slipping one half across the other by a lattice vector, the halves fitting back together without leaving a defect. If the cut only goes part way through the crystal, and then slipped, the boundary of the cut is a screw dislocation. It comprises a structure in which a helical path is traced around the linear defect (dislocation line) by the atomic planes in the crystal lattice (Figure

C). Perhaps the closest analogy is a spiral-sliced ham. In pure screw dislocations, the Burgers vector is parallel to the line direction. Despite the difficulty in visualization, the stresses caused by a screw dislocation are less complex than those of an edge dislocation. These stresses need only one equation, as symmetry allows only one radial coordinate to be used:

$$\tau_r = \frac{-\mu b}{2\pi r}$$

where μ is the shear modulus of the material, b is the Burgers vector, and r is a radial coordinate. This equation suggests a long cylinder of stress radiating outward from the cylinder and decreasing with distance. Please note, this simple model results in an infinite value for the core of the dislocation at r=0 and so it is only valid for stresses outside of the core of the dislocation.

If the Burgers vector is very large, the core may actually be empty resulting in a micropipe, as commonly observed in silicon carbide.

Mixed Dislocations

In many materials, dislocations are found where the line direction and Burgers vector are neither perpendicular nor parallel and these dislocations are called *mixed dislocations*, consisting of both screw and edge character.

Partial Dislocations

Dislocations can decompose into partial dislocations in order to facilitate movement through a crystal lattice.

Observation of Dislocations

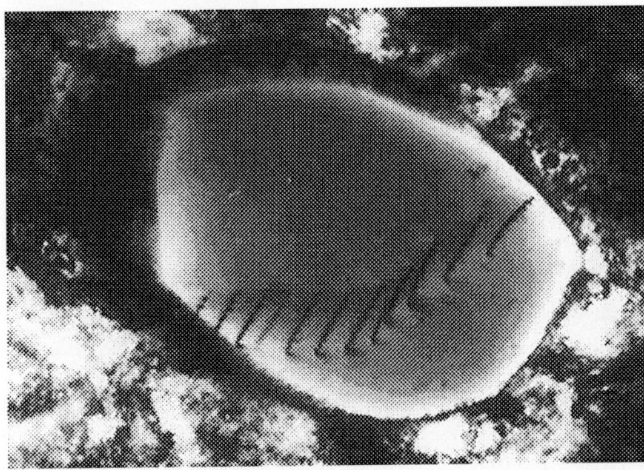

Figure: *Transmission Electron Micrograph of Dislocations*

When a dislocation line intersects the surface of a metallic material, the associated strain field locally increases the relative susceptibility of the material to acidic etching and an etch pit of regular geometrical format results. If the material is strained (deformed) and repeatedly re-etched, a series of etch pits can be produced which effectively trace the movement of the dislocation in question.

Transmission electron microscopy can be used to observe dislocations within the microstructure of the material. Thin foils of metallic samples are prepared to render them transparent to the electron beam of the microscope. The electron beam suffers diffraction by the regular crystal lattice planes of the metal atoms and the differing relative angles between the beam and the lattice planes of each grain in the metal's microstructure result in image contrast (between grains of different crystallographic orientation). The less regular atomic structures of the grain boundaries and in the strain fields around dislocation lines have different diffractive properties than the regular lattice within the grains, and therefore present different contrast effects in the electron micrographs. (The dislocations are seen as dark lines in the lighter, central region of the micrographs on the right). Transmission electron micrographs of dislocations typically utilize magnifications of 50,000 to 300,000 times (though the equipment itself offers a wider range of magnifications than this).

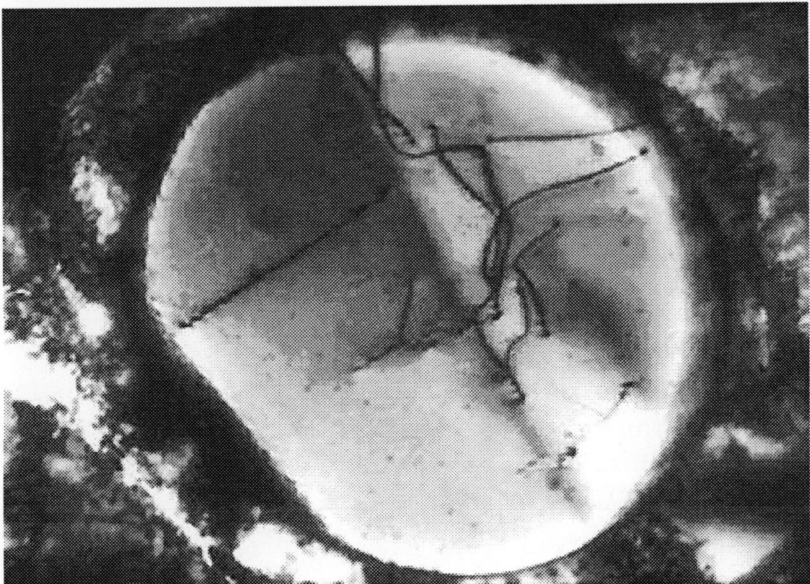

Figure: *Transmission Electron Micrograph of Dislocations*

Some microscopes also permit the in-situ heating and/or deformation of samples, thereby permitting the direct observation of dislocation movement and their interactions. Note the characteristic 'wiggly' contrast of the dislocation lines as they pass through the thickness of the material. Note also that a dislocation cannot end within a crystal; the dislocation lines in these images end at the sample surface. A dislocation can only be contained within a crystal as a complete loop.

Field ion microscopy and atom probe techniques offer methods of producing much higher magnifications (typically 3 million times and above) and permit the observation of dislocations at an atomic level. Where surface relief can be resolved to the level of an atomic step, screw dislocations appear as distinctive spiral features - thus revealing an important mechanism of crystal growth: where there is a surface step, atoms can more easily add to the crystal, and the surface step associated with a screw dislocation is never destroyed no matter how many atoms are added to it.

Figure: *Dislocations in silicon, orientation 111*

(By contrast, traditional optical microscopy, which is not appropriate for the *direct* observation of dislocations, typically offers magnifications up to a maximum of only around 2000 times).

After chemical etching, small pits are formed where the etching solution preferentially attacks the sample surface around the dislocations intercepting this surface, due to the more highly strained state of the material . Thus, the image features indicate points at which dislocations intercept the sample surface. In this way, dislocations in silicon, for example, can be observed *indirectly* using an interference microscope. Crystal orientation can be determined by the shape of the etch pits associated with the dislocations (in the case of the illustration below; 100 elliptical, 111 - triangular/pyramidal).

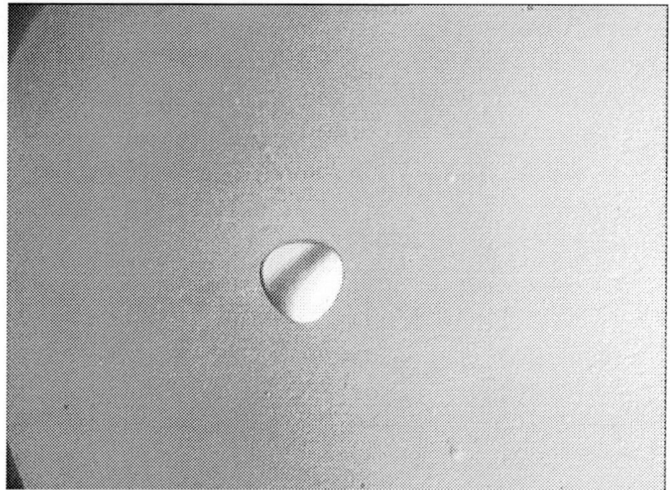

Figure: *Dislocation in silicon, orientation 111*

Sources of Dislocations

Dislocation density in a material can be increased by plastic deformation by the following relationship: $\tau \propto \rho^{1/2}$. Since the dislocation density increases with plastic deformation, a mechanism for the creation of dislocations must be activated in the material. Three mechanisms for dislocation formation are homogeneous nucleation, grain boundary initiation, and interfaces between the lattice and the surface, precipitates, dispersed phases, or reinforcing fibres.

The creation of a dislocation by homogeneous nucleation is a result of the rupture of the atomic bonds along a line in the lattice. A plane in the lattice is sheared, resulting in 2 oppositely faced half planes or dislocations. These dislocations move away from each other through the lattice. Since homogeneous nucleation forms dislocations from perfect crystals and requires the simultaneous breaking of many bonds, the energy required for homogeneous nucleation is high. For instance the stress required for homogeneous nucleation in copper has been shown to be $\frac{\tau_{hom}}{G} = 7.4 \times 10^{-2}$, where G is the shear modulus of copper (46 GPa). Solving for τ_{hom}, we see that the required stress is

3.4 GPa, which is very close to the theoretical strength of the crystal. Therefore, in conventional deformation homogeneous nucleation requires a concentrated stress, and is very unlikely. Grain boundary initiation and interface interaction are more common sources of dislocations.

Irregularities at the grain boundaries in materials can produce dislocations which propagate into the grain. The steps and ledges at the grain boundary are an important source of dislocations in the early stages of plastic deformation.

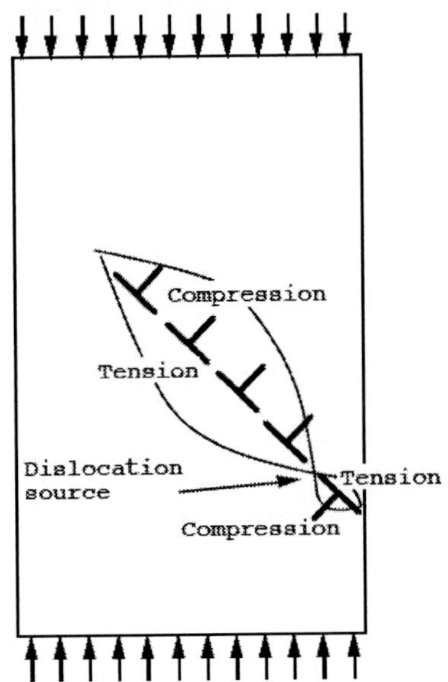

Figure: *Dislocation source (schematic)*

The surface of a crystal can produce dislocations in the crystal. Due to the small steps on the surface of most crystals, stress in certain regions on the surface is much larger than the average stress in the lattice. The dislocations are then propagated into the lattice in the same manner as in grain boundary initiation.

In monocrystals, the majority of dislocations are formed at the surface. The dislocation density 200 micrometres into the surface of a material has been shown to be six times higher than the density in the bulk.

However, in polycrystalline materials the surface sources cannot have a major effect because most grains are not in contact with the surface. The interface between a metal and an oxide can greatly increase the number of dislocations created. The oxide layer puts the surface of the metal in tension because the oxygen atoms squeeze into the lattice, and the oxygen atoms are under compression. This greatly increases the stress on the surface of the metal and consequently the amount of dislocations formed at the surface. The increased amount of stress on the surface steps results in an increase of dislocations.

- The stresses produced by a dislocation source may be visualized by photoelasticity in a gamma-irradiated LiF single crystal. The tensile stress along the glide plane is red. The compressive stress is dark green:

Figure: *Stresses in a dislocation pile-up*

Dislocations, Slip and Plasticity

Until the 1930s, one of the enduring challenges of materials science was to explain plasticity in microscopic terms. A simplistic attempt to calculate the shear stress at which neighbouring atomic planes *slip* over each other in a perfect crystal suggests that, for a material with shear modulus G, shear strength τ_m is given approximately by:

$$\tau_m = \frac{G}{2\pi}.$$

As shear modulus in metals is typically within the range 20 000 to 150 000 MPa, this is difficult to reconcile with shear stresses in the range 0.5 to 10 MPa observed to produce plastic deformation in experiments.

In 1934, Egon Orowan, Michael Polanyi and G. I. Taylor, roughly simultaneously, realized that plastic deformation could be explained in terms of the theory of dislocations. Dislocations can move if the atoms from one of the surrounding planes break their bonds and rebond with the atoms at the terminating edge. In effect, a half plane of atoms is moved in response to shear stress by breaking and reforming a line of bonds, one (or a few) at a time. The energy required to break a single bond is far less than that required to break all the bonds on an entire plane of atoms at once. Even this simple model of the force required to move a dislocation shows that plasticity is possible at much lower stresses than in a perfect crystal. In many materials, particularly ductile materials, dislocations are the "carrier" of plastic deformation, and the energy required to move them is less than the energy required to fracture the material. Dislocations give rise to the characteristic malleability of metals.

When metals are subjected to "cold working" (deformation at temperatures which are relatively low as compared to the material's absolute melting temperature, T_m, i.e., typically less than 0.3 T_m) the dislocation density increases due to the formation of new dislocations and dislocation multiplication. The consequent increasing overlap between the strain fields of adjacent dislocations gradually increases the resistance to further dislocation motion. This causes a hardening of the metal as deformation progresses. This effect is known as strain hardening (also "work hardening"). Tangles of dislocations are found at the early stage of deformation and appear as non well-defined boundaries; the process of dynamic recovery leads eventually to the formation of a cellular structure containing boundaries with misorientation lower than 15° (low angle grain boundaries). In addition, adding pinning points that inhibit the motion of dislocations, such as alloying elements, can introduce stress fields that ultimately strengthen the material by requiring a higher applied stress to overcome the pinning stress and continue dislocation motion.

The effects of strain hardening by accumulation of dislocations and the grain structure formed at high strain can be removed by appropriate heat treatment (annealing) which promotes the recovery and subsequent recrystallisation of the material.

The combined processing techniques of work hardening and annealing allow for control over dislocation density, the degree of dislocation entanglement, and ultimately the yield strength of the material.

Dislocation Climb

Dislocations can slip in planes containing both the dislocation and the Burgers Vector. For a screw dislocation, the dislocation and the Burgers vector are parallel, so the dislocation may slip in any plane containing the dislocation. For an edge dislocation, the dislocation and the Burgers vector are perpendicular, so there is only one plane in which the dislocation can slip. There is an alternative mechanism of dislocation motion, fundamentally different from slip, that allows an edge dislocation to move out of its slip plane, known as dislocation climb. Dislocation climb allows an edge dislocation to move perpendicular to its slip plane.

The driving force for dislocation climb is the movement of vacancies through a crystal lattice. If a vacancy moves next to the boundary of the extra half plane of atoms that forms an edge dislocation, the atom in the half plane closest to the vacancy can "jump" and fill the vacancy. This atom shift "moves" the vacancy in line with the half plane of atoms, causing a shift, or positive climb, of the dislocation. The process of a vacancy being absorbed at the boundary of a half plane of atoms, rather than created, is known as negative climb. Since dislocation climb results from individual atoms "jumping" into vacancies, climb occurs in single atom diameter increments.

During positive climb, the crystal shrinks in the direction perpendicular to the extra half plane of atoms because atoms are being removed from the half plane. Since negative climb involves an addition of atoms to the half plane, the crystal grows in the direction perpendicular to the half plane.

Therefore, compressive stress in the direction perpendicular to the half plane promotes positive climb, while tensile stress promotes negative climb. This is one main difference between slip and climb, since slip is caused by only shear stress.

One additional difference between dislocation slip and climb is the temperature dependence. Climb occurs much more rapidly at high temperatures than low temperatures due to an increase in vacancy motion. Slip, on the other hand, has only a small dependence on temperature.

Topological Defect

In mathematics and physics, a topological soliton or a topological defect is a solution of a system of partial differential equations or of a quantum field theory homotopically distinct from the vacuum solution; it can be proven to exist because the boundary conditions entail the existence of homotopically distinct solutions. Typically, this occurs because the boundary on which the boundary conditions are specified has a non-trivial homotopy group which is preserved in differential equations; the solutions to the differential equations are then topologically distinct, and are classified by their homotopy class. Topological defects are not only stable against small perturbations, but cannot decay or be undone or be de-tangled, precisely because there is no continuous transformation that will map them (homotopically) to a uniform or "trivial" solution.

Examples include the soliton or solitary wave which occurs in many exactly solvable models, the screw dislocations in crystalline materials, the skyrmion and the Wess–Zumino–Witten model in quantum field theory.

Topological defects are believed to drive phase transitions in condensed matter physics. Notable examples of topological defects are observed in Lambda transition universality class systems including: screw/edge-dislocations in liquid crystals, magnetic flux tubes in superconductors, vortices in superfluids.

Cosmology

Certain grand unified theories predict topological defects to have formed in the early universe. According to the Big Bang theory, the universe cooled from an initial hot, dense state triggering a series of phase transitions much like what happens in condensed-matter systems.

In physical cosmology, a topological defect is an (often) stable configuration of matter predicted by some theories to form at phase transitions in the very early universe.

Symmetry breakdown

Depending on the nature of symmetry breakdown, various solitons are believed to have formed in the early universe according to the Higgs–Kibble mechanism. The well-known topological defects are magnetic monopoles, cosmic strings, domain walls, Skyrmions and textures.

As the universe expanded and cooled, symmetries in the laws of physics began breaking down in regions that spread at the speed of light; topological defects occur where different regions came into contact with each other. The matter in these defects is in the original symmetric phase, which persists after a phase transition to the new asymmetric phase is completed.

Types of Topological Defects

Various different types of topological defects are possible, with the type of defect formed being determined by the symmetry properties of the matter and the nature of the phase transition. They include:

- Domain walls, two-dimensional membranes that form when a discrete symmetry is broken at a phase transition. These walls resemble the walls of a closed-cell foam, dividing the universe into discrete cells.
- Cosmic strings are one dimensional lines that form when an axial or cylindrical symmetry is broken.
- Monopoles, cube-like defects that form when a spherical symmetry is broken, are predicted to have magnetic charge, either north or east (and so are commonly called "magnetic monopoles").
- Textures form when larger, more complicated symmetry groups are completely broken. They are not as localized as the other defects, and are unstable. Other more complex hybrids of these defect types are also possible.
- Extra-dimensions and higher dimensions.

Observation

Topological defects, of the cosmological type, are extremely high-energy phenomena and are likely impossible to produce in artificial Earth-bound physics experiments, but topological defects that formed during the universe's formation could theoretically be observed.

No topological defects of any type have yet been observed by astronomers, however, and certain types are not compatible with current observations; in particular, if domain walls and monopoles were present in the observable universe, they would result in significant deviations from what astronomers can see. Theories that predict the formation of these structures *within the observable universe* can therefore be largely ruled out. On the other hand, cosmic strings have been suggested as providing the initial 'seed'-gravity around which the large-scale structure of the cosmos of matter has condensed. Textures are similarly benign. In late 2007, a cold spot in the cosmic microwave background was interpreted as possibly being a sign of a texture lying in that direction.

Condensed Matter

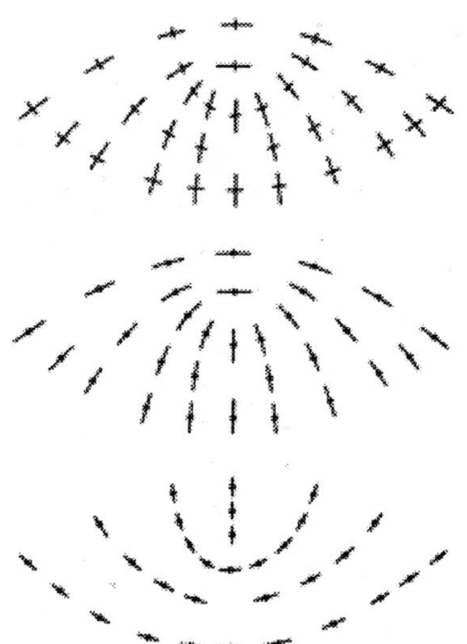

***Figure:** Classes of stable defects in Biaxial nematics*

In condensed matter physics, the theory of homotopy groups provides a natural setting for description and classification of defects in ordered systems. Topological methods have been used in several problems of condensed matter theory. Poénaru and Toulouse used topological methods to obtain a condition for line (string) defects in liquid crystals can cross each other without entanglement.

It was a non-trivial application of topology that first led to the discovery of peculiar hydrodynamic behaviour in the *A*-phase of superfluid Helium-3.

Classification

An *ordered medium* is defined as a region of space described by a function *f(r)* that assigns to every point in the region an *order parameter*, and the possible values of the order parameter space constitute an *order parameter space*. The homotopy theory of defects uses the fundamental group of the order parameter space of a medium to discuss the existence, stability and classifications of topological defects in that medium.

Suppose *R* is the order parameter space for a medium, and let *G* be a Lie group of transformations on *R*. Let *H* be the symmetry subgroup of *G* for the medium. Then, the order parameter space can be written as the Lie group quotient $R=G/H$.

If *G* is a universal cover for G/H then, it can be shown that $\pi_n (G/H)=\pi_{n-1} (H)$, where π_i denotes the *i*-th homotopy group.

Various types of defects in the medium can be characterized by elements of various homotopy groups of the order parameter space. For example, (in three dimensions), line defects correspond to elements of $\pi_1 (R)$, point defects correspond to elements of $\pi_2 (R)$, textures correspond to elements of $\pi_3 (R)$. However, defects which belong to the same conjugacy class of $\pi_1 (R)$ can be deformed continuously to each other, and hence, distinct defects correspond to distinct conjugacy classes. Poénaru and Toulouse showed that crossing defects get entangled if and only if they are members of separate conjugacy classes of $\pi_1 (R)$.

Stable Defects

The homotopy theory is deeply related to the stability of topological defects. In the case of line defect, if the closed path can be continuously deformed into one point, the defect is not stable, and otherwise, it is stable. Unlike in cosmology and field theory, topological defects in condensed matter can be experimentally observed. Ferromagnetic materials have regions of magnetic alignment separated by domain walls. Nematic and bi-axial nematic liquid crystals display a variety of defects including monopoles, strings, textures etc. Defects can also been found in biochemistry, notably in the process of protein folding.

Partial Dislocations

Partial Dislocations are a decomposed form of dislocations that occur within a material.

Reaction Favourability

A dislocation will decompose into partial dislocations if the energy state of the sum of the partials is less than the energy state of the original dislocation. This is summarized by *Frank's Energy Criterion*:

$$|b_1|^2 > |b_2|^2 + |b_3|^2 \quad \text{(favorable, will decompose)}$$

$$|b_1|^2 < |b_2|^2 + |b_3|^2 \quad \text{(not favorable, will not decompose)}$$

$$|b_1|^2 = |b_2|^2 + |b_3|^2 \quad \text{(will remain in original state)}$$

Shockley Partial Dislocations

Shockley Partial Dislocations generally refer to a pair of dislocations which can lead to the presence of stacking faults. This pair of partial dislocations can enable dislocation motion by allowing an alternate path for atomic motion.

$$b_1 \rightarrow b_2 + b_3$$

In FCC systems, an example of Shockley decomposition is:

$$\frac{a}{2}[10\bar{1}] \rightarrow \frac{a}{6}[2\bar{1}\bar{1}] + \frac{a}{6}[11\bar{2}]$$

Which is energetically favourable:

$$|\frac{a}{2}\sqrt{1^2+0^2+(-1)^2}|^2 > |\frac{a}{6}\sqrt{2^2+(-1)^2+(-1)^2}|^2 + |\frac{a}{6}\sqrt{1^2+1^2+(-2)^2}|^2$$

$$\frac{a^2}{2} > \frac{a^2}{6} + \frac{a^2}{6}$$

The components of the *Shockley Partials* must add up to the original vector that is being decomposed:

$$\frac{a}{2}(1) = \frac{a}{6}(2) + \frac{a}{6}(1)$$

$$\frac{a}{2}(0) = \frac{a}{6}(-1) + \frac{a}{6}(1)$$

$$\frac{a}{2}(-1) = \frac{a}{6}(-1) + \frac{a}{6}(-2)$$

Frank Partial Dislocations

Frank Partial Dislocations are sessile, or immobile. In FCC systems, Frank partials are given by:

$$b_{\text{frank}} = \frac{a}{3}[1\ 1\ 1]$$

Thompson Tetrahedron

Shockley Partials and Frank Partials can combine to form a *Thompson Tetrahedron*, or a *stacking fault tetrahedron.*

Lomer-Cottrell Lock

The Lomer-Cottrell Lock is formed by partial dislocations and is sessile.

Bulk Defects

- Voids are small regions where there are no atoms, and can be thought of as clusters of vacancies.

- Impurities can cluster together to form small regions of a different phase. These are often called precipitates.

Mathematical Classification Methods

A successful mathematical classification method for physical lattice defects, which works not only with the theory of dislocations and other defects in crystals but also, e.g., for disclinations in liquid crystals and for excitations in superfluid ^{3}He, is the topological homotopy theory.

Computer Simulation Methods

Density-functional theory, classical molecular dynamics and kinetic Monte Carlo simulations are widely used to study the properties of defects in solids with computer simulations. Simulating jamming of hard spheres of different sizes and/or in containers with non-commeasurable sizes using the Lubachevsky-Stillinger algorithm can be an effective technique for demonstrating some types of crystallographic defects.

Jog (Dislocations)

Jog describes the turns of a dislocation line inside a crystal structure. A dislocation line is rarely uniformly straight as in the figure, often containing many curves and/or steps to facilitate movement through the crystal in incremental amounts, rather than shifting the entire line at once. One of these step types is a jog, the other is a kink. However, both are typically referred to as jogs, which can be a source of confusion.

Segments of dislocation line that have a component of their sense vector normal to the glide plane are termed jogs. Segments of dislocation line that do not leave the original glide plane are termed kinks.

Jogs are often very immobile compared to kinks, and require diffusion of crystallographic defects like vacancies or interstitial atoms to climb. They are not capable of glide (movement in the glide plane) because the direction of motion is in the plane normal direction, not on the plane itself as with kinks.

Dislocation Creep

Dislocation creep is a deformation mechanism in crystalline materials. Dislocation creep involves the movement of dislocations through the crystal lattice of the material. It causes plastic deformation of the individual crystals and in the end the material itself.

Dislocation creep is highly sensitive to the differential stress on the material. At relatively low temperatures it is the dominant deformation mechanism in most crystalline materials.

Principles

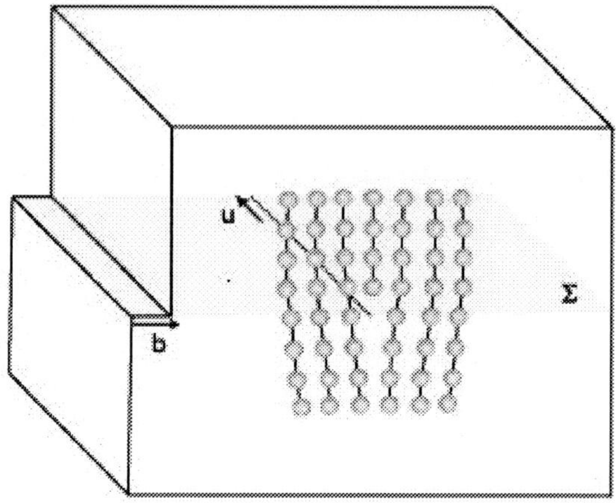

Figure: *Schematic representation of an edge dislocation in a crystal lattice. The yellow plane is the glide plane, the vector u represents the dislocation, b is the Burgers vector. When the dislocation moves from left to right through the crystal, the lower half of the crystal has moved one Burgers vector length to the left, relative to the upper half.*

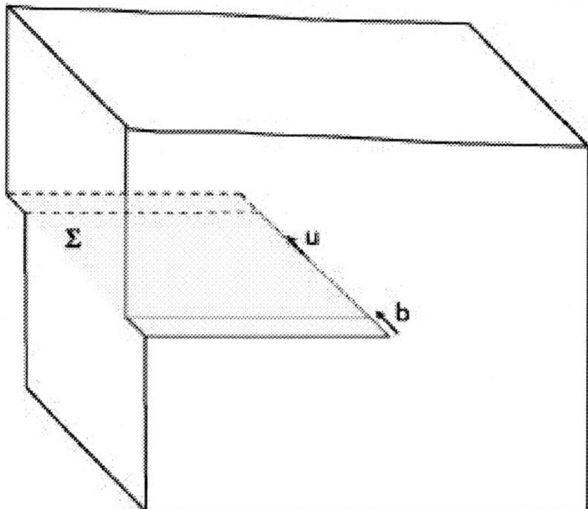

Figure: *Schematic representation of a screw dislocation in a crystal lattice. The yellow plane (Σ) is again the glide plane, u the dislocation and b the Burgers vector. When the dislocation moves from the back to the front of the crystal, the lower half moves one Burgers vector length to the front, relative to the upper half.*

Dislocations and Glide Planes in Crystals

Dislocation creep takes place due to the movement of dislocations through a crystal lattice. Each time a dislocation moves through a crystal, part of the crystal moves one lattice point along a plane, relative to the rest of the crystal. The plane that separates both parts and along which the movement takes place is called a glide plane. To allow the movement, all ionic bonds along the plane have to be broken. If all bonds were broken at once, this would require so much energy that dislocation creep would only in theory be possible. When it is assumed that the movement takes place step by step, the breaking of bonds is immediately followed by the creation of new ones and the energy required is much lower. Calculations of molecular dynamics and analysis of deformed materials have shown that deformation creep can be an important factor in deformation processes, under certain circumstances.

By moving a dislocation step by step through a crystal lattice a linear lattice defect is created between parts of the crystal lattice, which is called a dislocation. Two types of dislocations exist. Edge dislocations form the edge of an extra layer of atoms inside the crystal lattice. Screw dislocations form a line along which the crystal lattice jumps one lattice point. In both cases the dislocation line forms a linear defect through the crystal lattice, the crystal can be perfect on all sides of the line.

Edge dislocation move in a direction perpendicular to the dislocation line, screw dislocations move parallel to the dislocation line. In both cases this causes a part of the crystal to move relative to other parts. Meanwhile the dislocation itself moves further on along a glide plane. The crystal system of the material (mineral or metal) determines how many glide planes are possible, and in which orientations. The orientation of the differential stress then determines which glide planes are active and which are not. The Von Mises criterion states that to deform a material, movement along at least five different glide planes is required. A dislocation will not always be a straight line and can thus move along more than one glide plane. Where the orientation of the dislocation line changes, a screw dislocation can continue as an edge dislocation and vice versa.

The length of the displacement in the crystal caused by the movement of the dislocation is called the Burgers vector. It equals the distance between two atoms or ions in the crystal lattice. Therefore each material has its own characteristic Burgers vectors for each glide plane.

Dislocation Movement

A dislocation can ideally move through a crystal until it reaches a grain boundary (the boundary between two crystals). When it reaches a grain boundary, the dislocation will disappear. In that case the whole crystal is sheared a little.

There are however different ways in which the movement of a dislocation can be slowed or stopped. When a dislocation moves along several different glide planes, it can have different velocities in these different planes, due to the anisotropy of some materials. Dislocations can also encounter other defects in the crystal on their ways, such as other dislocations or point defects. In such cases a part of the dislocation could slow down or even stop moving altogether.

In alloy design, this effect is used to a great extent. On adding a dissimilar atom or phase, such as a small amount of carbon to iron, it is hardened, meaning deformation of the material will be more difficult (the material becomes stronger). The carbon atoms act as interstitial particles (point defects) in the crystal lattice of the iron, and dislocations will not be able to move as easily as before.

Dislocation Recovery

Dislocations are imperfections in a crystal lattice, that from a thermodynamic point of view reduce the amount of free energy in the system. Therefore, parts of a crystal that have more dislocations will be relatively unstable. By recrystallisation the crystal can heal itself. Recovery of the crystal structure can also take place when two dislocations with opposite displacement meet each other.

A dislocation that has been brought to a halt by an obstacle (a point defect) can overcome the obstacle and start moving again by a process called dislocation climb. For dislocation climb to occur, vacancies have to be able to move through the crystal.

When a vacancy arrives at the place where the dislocation is stuck it can cause the dislocation to climb out of its glide plane, after which the point defect is no longer in its way.

Dislocation climb is therefore dependent from the velocity of vacancy diffusion. As with all diffusion processes, this is highly dependent on the temperature. At higher temperatures dislocations will more easily be able to move around obstacles. For this reason, many hardened materials become exponentially weaker at higher temperatures.

To increase the free energy in the system, dislocations can tend to concentrate themselves in certain zones, so that other regions will stay free of dislocations. This leads to the formation of 'dislocation walls', planes in a crystal where dislocations localise. Edge dislocations form so called tilt walls, while screw dislocations form twist walls. In both cases the increasing localisation of dislocations in the wall will increase the angle between the orientation of the crystal lattice on both sides of the wall. This leads to the formation of subgrains. The process is called subgrain rotation (SGR) and can eventually lead to the formation of new grains when the dislocation wall becomes a new grain boundary.

Origin of Dislocations

When a crystalline material is put under differential stress, new dislocations form at the grain boundaries, and begin moving through the crystal.

Another way in which new dislocations can form are so called Frank-Read sources. These form when a dislocation is stopped at two places. The part of the dislocation in between will move along, causing the dislocation line to curve. This curving can continue until the dislocation curves over itself to form a circle. In the centre of such a circle the source will produce a new dislocation, and this process will produce a sequence of dislocations on top of each other. Frank-Read sources are also created when screw dislocations double cross-slip (change glide planes twice), as the jogs in the dislocation line pin the dislocation in the 3rd plane.

Peierls Stress

Peierls stress is the force (first discovered by Rudolph Peierls and modified by Frank Nabarro) needed to move a dislocation within a plane of atoms in the unit cell. The magnitude varies periodically as the dislocation moves within the plane. Peierls stress depends on the size and width of a dislocation and the distance between planes. Because of this, Peierls stress decreases with increasing distance between atomic planes. Yet since the distance between planes increases with planar atomic density, slip of the dislocation is preferred on closely packed planes.

Peierls–Nabarro stress proportionality

$$\tau_{PN} = \frac{2G}{1-\nu} e^{-2\pi W/b}$$

Where:

$W = \frac{a}{1-v}$ = the dislocation width

G= shear modulus

V= Poisson's ratio

b= interatomic spacing

a= interplanar spacing

The Peierls Stress and Yield Strength Temperature Sensitivity

The Peierls stress also relates to the temperature sensitivity of the yield strength of a material because it too depends on both short range atomic order and atomic bond strength. As temperature increases, the vibration of atoms increases and thus both peierls stress and yield strength decrease as a result of weaker atomic bond strength at high temperatures.

Kröger–Vink Notation

Kröger–Vink notation is set of conventions used to describe electric charge and lattice position for point defect species in crystals. It is primarily used for ionic crystals and is particularly useful for describing various defect reactions. It was proposed by F. A. Kröger and H. J. Vink.

Schottky and Frenkel Defects

Schottky Pairs: A Schottky defect is an intrinsic point defect that creates vacancies on both the cation and anion sub-lattices. This defect occurs in ionic crystals when one positively charged cation and one negatively charged anion leave the lattice simultaneously, resulting in vacant lattice sites. Because mass, site, and charge numbers must remain balanced, these vacancies always occur in stoichiometric ratios. However, due to the loss of ions within the crystal lattice, these Schottky defects tend to lead to a decrease in the density of the material because vacancies have been created.

Frenkel Pairs

Similarly to Schottky defect, a Frenkel defect is an intrinsic point defect that produces a vacancy site on either the cation or anion sub-lattice along with an interstitial site on that same lattice. In an ionic crystal, this occurs when a cation or anion leaves its site in the sub-lattice, creating the vacancy site, and moves to another location to create a cation/anion interstitial. These Frenkel defect pairs maintain a balanced mass, site, and charge ratio throughout the relocation of the ions. Since the movement remains within the material's single lattice, the density remains the same.

Notation

$$M_S^C$$

M corresponds to the species. These can be

- atoms – e.g., Si, Ni, O, Cl,
- vacancies – V or v (since V is also the symbol for vanadium)
- interstitials – i
- electrons – e
- electron holes – h

S indicates the lattice site that the species occupies. For instance, Ni might occupy a Cu site. In this case, M would be replaced by Ni and S would be replaced by Cu. The site may also be a lattice interstice. In this case the symbol 'i' is used. A cation site can be represented by the symbols C or M (for metal), and an anion site can be represented by either an A or X.

C corresponds to the electronic charge of the species relative to the site that it occupies. The charge of the species is calculated by the charge on the current site minus the charge on the original site. To continue the previous example, Ni often has the same valency as Cu, so the relative charge is zero. To indicate null charge, × is used. A single • indicates a single positive charge, while two would represent two positive charges. Finally, 2 signifies a single negative charge, so two would indicate a double negative charge.

Examples

$Al_{Al}^{\times}$ = an aluminum ion sitting on an aluminum lattice site, with neutral charge.

$Ni_{Cu}^{\times}$ = a nickel ion sitting on a copper lattice site, with neutral charge.

$V_{Cl}^{\bullet}$ = a chlorine vacancy, with singular positive charge.

$Ca_{i}^{\bullet\bullet}$ = a calcium interstitial ion, with double positive charge.

$Cl_{i}^{'}$ = a chlorine anion on an interstitial site, with singular negative charge.

O_i'' = an oxygen anion on an interstitial site, with double negative charge.

e' = an electron. A site isn't normally specified.

Procedure

When using Kroger-Vink notation for both intrinsic and extrinsic defects, it is imperative to keep all masses, sites, and charges balanced in each reaction. If any piece is unbalanced, the reactants and the products do not equal the same entity and therefore all quantities are not being conserved as they should. The first step in this process is determining the correct type of defect and reaction that comes along with it; Schottky and Frenkel defects begin with a null or Ø reactant and produce either cation and anion vacancies (Schottky) or cation/anion vacancies and interstitials (Frenkel). Otherwise, a compound is broken down into its respective cation and anion parts for the process to begin on each lattice. From here, depending on the required steps for the desired outcome, several possibilities occur. For example, the defect may result in an ion on its own ion site or a vacancy on the cation site. To complete the reactions, the proper number of each ion must be present (mass balance), an equal number of sites must exist (site balance), and the charges of the reactants and products must also be equivalent (charge balance).

Example Usage

$\emptyset \Leftrightarrow V_{Ti''''} + 2V_O^{\bullet\bullet}$ = A Kröger–Vink representation of Schottky defect formation in TiO_2.

$\emptyset \Leftrightarrow V_{Ba''} + V_{Ti''''} + 3V_O^{\bullet\bullet}$ = A Kröger–Vink representation of Schottky defect formation in $BaTiO_3$.

$Mg_{Mg}^{\times} + O_O^{\times} \Leftrightarrow O_i'' i + V_O^{\bullet\bullet} + Mg_{Mg}^{\times}$ = A Kröger-Vink representation of Frenkel defect formation in MgO.

$Mg_{Mg}^{\times} + O_O^{\times} \Leftrightarrow V''_{Mg} + V_O^{\bullet\bullet} + Mg_{surface}^{\times} + O_{surface}^{\times}$ = A Kröger-Vink representation of Schottky defect formation in MgO.

Basic Types of Defect Reactions

- Assume that the cation has +1 charge and anion has -1 charge.

1. *Schottky Defect* – Forming vacancy pair on both anion and cation sites

$\emptyset \Leftrightarrow V'_C + V_A^{\bullet} \Leftrightarrow V'_M + V_X^{\bullet}$

2. *Schottky Defect (Charged)* – Forming electron-hole pair

$\emptyset \Leftrightarrow e' + h^{\bullet}$

3. *Frenkel Defect* – Forming interstitial and vacancy pair on anion or cation site

$\emptyset \Leftrightarrow V'_C + i_C^{\bullet} \Leftrightarrow V'_M + i_M^{\bullet}$ (Cation Frenkel defect)

$\emptyset \Leftrightarrow V'_A + i_A^{\bullet} \Leftrightarrow V'_X + i_X^{\bullet}$ (Anion Frenkel defect)

4. Associates – Forming entropically favoured site, usually depending on temperature. For the two equations shown below, the right side, it is usually at high temperature as they allow for more movement of electrons. On the left side, it is usually at low temperature as the electrons lose its mobility due to loss in kinetic energy.

$M_M^{\times} + e' = M'_M$ (Metal Site Reduced)

$B_M^{\bullet} + e' = B_M^{\times}$ (Metal Site Oxidized, where B is an arbitrary cation having one extra positive charge than the original site.)

Oxidation/Reduction Tree

The following oxidation/reduction tree of an ionic species shows the various ways that a substance can be broken down. Depending on the cation to anion ratio, the species can either be reduced and therefore classified as n-type, or if the converse is true, the ionic species is classified as p-type. Below, the tree is shown for a further explanation of the pathways and results of each breakdown of the substance.

A = Cation

X = Anion

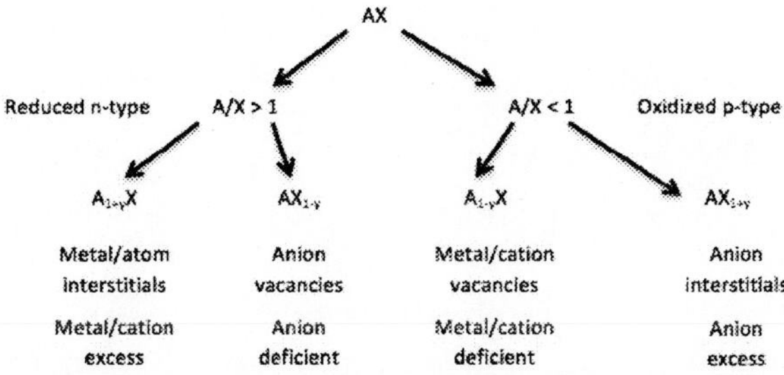

Figure: *Tree diagram representation of the oxidation and reduction of intrinsic defects in ionic substances.*

Schematic Examples

From the chart above, there are total of four possible chemical reactions using Kroger-Vink Notation depending on the intrinsic deficiency of atoms within the material. Assume the chemical composition to be AX with A being the cation and X being the anion. (The following assumed that X is a diatomic gas, namely, oxygen and therefore A has a +2 charge as a cation.)

- Note that this could be used to make an oxygen sensor.

1. For the reduced n-type, with cation in excess on the interstitial site:

$$A_A^{\times} + X_X^{\times} \Leftrightarrow A_i^{\bullet\bullet} + \frac{1}{2}X_2(g) + 2e'$$

2. For the reduced n-type, with anion in deficiency on the lattice site:

$$A(s) \Leftrightarrow A_A^{\times} + V_X^{\bullet\bullet} + 2e'$$

3. For the oxidized p-type, with cation in deficiency on the lattice site:

$$\frac{1}{2}X_2(g) \Leftrightarrow V''_A + X_X^{\times} + 2h^{\bullet}$$

4. For the oxidized p-type, with anion in excess on the interstitial site:

$$A_A^{\times} + X_X^{\times} \Leftrightarrow A(s) + X''i + 2h^{\bullet}$$

Relating Chemical Reactions to the Equilibrium Constant

Using Law of Mass Action, defect concentration can be related to its Gibbs free energy of formation, and the energy terms (enthalpy of formation) can be calculated given the defect concentration or vice versa.

Examples

For Schottky Reaction for MgO, the Kroger-Vink defect reaction can be written as follows:

$$\varnothing \Leftrightarrow V''_{\mathrm{Mg}} + V_{\mathrm{O}}^{\bullet\bullet} \ (1)$$

Note that the vacancy on the Mg sub-lattice site has a -2 charge and the vacancy on the oxygen sub-lattice site has a +2 charge, detailed explanation of how charge is calculated is shown above. Using the Law of mass action, the reaction equilibrium constant can be written as,

$$\mathrm{k} = [V''_{\mathrm{Mg}}][V_{\mathrm{O}}^{\bullet\bullet}] \ (2)$$

According to the reaction, the stoichiometric relation is as follows,

$$[V''_{Mg}] = [V_O^{\bullet\bullet}]\ (3)$$

Also, the equilibrium constant can be related to the Gibbs free energy according to the following relations,

$k = e^{-\frac{\Delta G_F}{k_B T}}$, where k_B is the Boltzmann's Constant. (4)

$$\Delta G = \Delta H - T\Delta S\ (5)$$

Relating equation 2 and 4, it yields

$$e^{-\frac{\Delta G_F}{k_B T}} = [V''_{MG}]^2$$

Using equation 5, the formula can be simplified into the following form where the enthalpy of formation can be directly calculated,

$[V''_{\mathrm{Mg}}] = e^{-\frac{\Delta H_F}{2k_B T} + \frac{\Delta S}{2k_B}} = Ae^{-\frac{\Delta H_F}{2k_B T}}$, where A is a constant containing the entropic term.

Therefore, given a temperature and the formation energy of Schottky defect, the intrinsic Schottky defect concentration can be calculated from the above equation.

Burgers vector

The Burgers vector, named after Dutch physicist Jan Burgers, is a vector, often denoted b, that represents the magnitude and direction of the lattice distortion of dislocation in a crystal lattice.

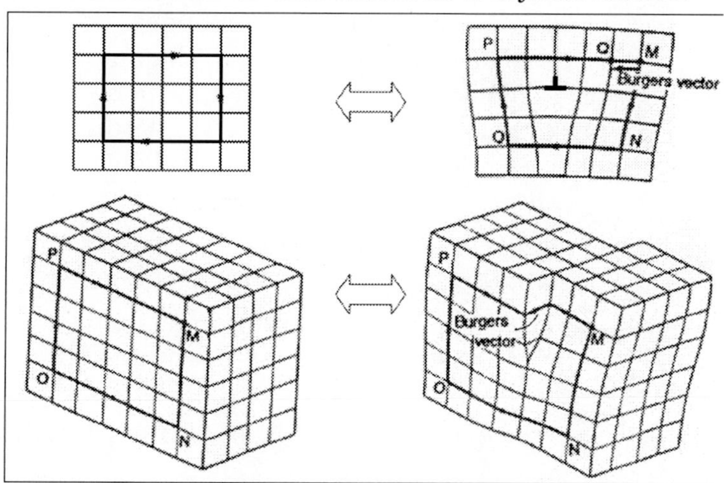

Figure: *Burgers vectors*

The vector's magnitude and direction is best understood when the dislocation-bearing crystal structure is first visualized *without* the dislocation, that is, the *perfect* crystal structure. In this perfect crystal structure, a rectangle whose lengths and widths are integer multiples of "a" (the unit cell length) is drawn *encompassing* the site of the original dislocation's origin. Once thi</ref>s encompassing rectangle is drawn, the dislocation can be introduced. This dislocation will have the effect of deforming, not only the perfect crystal structure, but the rectangle as well. The said rectangle could have one of its sides disjoined from the perpendicular side, severing the connection of the length and width line segments of the rectangle at one of the rectangle's corners, and displacing each line segment from each other. What was once a rectangle before the dislocation was introduced is now an open geometric figure, whose opening defines the direction and magnitude of the Burgers vector. Specifically, the breadth of the opening defines the magnitude of the Burgers vector, and, when a set of fixed coordinates is introduced, an angle between the termini of the dislocated rectangle's length line segment and width line segment may be specified.

The direction of the vector depends on the plane of dislocation, which is usually on the closest-packed plane of unit cell. The magnitude is usually represented by the equation:

$$\| b \| = \frac{a}{2}\sqrt{h^2 + k^2 + l^2}$$

where a is the unit cell length of the crystal, $\|b\|$ is the magnitude of Burgers vector and h, k, and l are the components of the Burgers vector, b = <h k l>. In most metallic materials, the magnitude of the Burgers vector for a dislocation is of a magnitude equal to the interatomic spacing of the material, since a single dislocation will offset the crystal lattice by one close-packed crystallographic spacing unit.

In edge dislocations, the Burgers vector and dislocation line are at right angles to one another. In screw dislocations, they are parallel.

The Burgers vector is significant in determining the yield strength of a material by affecting solute hardening, precipitation hardening and work hardening.

In materials science/engineering it is often useful to know the magnitude of the Burger's vector in metres. This is easily done for BCC and FCC lattice materials using the previously mentioned equation as *only the slip system and unit cell length 'a' need to be known. So for a FCC lattice where a = 2 R (2)^1/2 and the slip system is <110> the length of the Burger's vector is simply b = 2 R, where R is the atomic radius. The HCP system is even simpler with the Burger's vector simply equal to the unit cell length a, i.e. b =a, where a = 2 R for the HCP system. Hence the Burger's vector of titanium would be 0.29 nm long.*

Planar Defects

- Grain boundaries occur where the crystallographic direction of the lattice abruptly changes. This usually occurs when two crystals begin growing separately and then meet.
- Antiphase boundaries occur in ordered alloys: in this case, the crystallographic direction remains the same, but each side of the boundary has an opposite phase: For example, if the ordering is usually ABABABAB, an antiphase boundary takes the form of ABABBABA.
- Stacking faults occur in a number of crystal structures, but the common example is in close-packed structures. Face-centred cubic (fcc) structures differ from hexagonal close packed (hcp) structures only in stacking order: both structures have close packed atomic planes with sixfold symmetry—the atoms form equilateral triangles. When stacking one of these layers on top of another, the atoms are not directly on top of one another—the first two layers are identical for hcp and fcc, and labelled AB. If the third layer is placed so that its atoms are directly above those of the first layer, the stacking will be ABA—this is the hcp structure, and it continues ABABABAB. However, there is another possible location for the third layer, such that its atoms are not above the first layer. Instead, it is the atoms in the fourth layer that are directly above the first layer. This produces the stacking ABCABCABC, and is actually a cubic arrangement of the atoms. A stacking fault is a one or two layer interruption in the stacking sequence, for example, if the sequence ABCABABCAB were found in an fcc structure.

Stacking Faults

The stacking-fault energy (SFE) is a material property on a very small scale. It is noted as γ_{SFE} in units of joules per square metre.

A stacking fault is a one or two layer interruption in the stacking sequence of the crystal structure. These interruptions carry a certain stacking fault energy. The width of stacking fault is a consequence of the balance between the repulsive force between two partial dislocations on one hand and the attractive force due to the surface tension of the stacking fault on the other hand. The equilibrium width is determined by the stacking fault energy. When the SFE is high, the dissociation of a perfect dislocation into two partial is unlikely and the material deforms only by dislocation glide. Lower SFE materials display wider stacking faults and have more difficulties for cross-slip and climb. The SFE modifies the ability of a dislocation in a crystal to glide onto an intersecting slip plane. When the SFE is low, the mobility of dislocations in a material decreases.

Material	*Ag (Silver)*	*Si (Silicon)*	*Ni (Nickel)*	*Cu (Copper)*	*Mg (Magnesium)*	*Al (Aluminum)*
SFE (mJ m?2)	20-30	>42	90	70 -78	125	160-200

Stacking Faults and Stacking Fault Energy

A stacking fault is an irregularity in the planar stacking sequence of atoms in a crystal – in FCC metals the normal stacking sequence is ABCABC etc., but if a stacking fault is introduced it may change to ABCBC.

Influences on Stacking Fault Energy

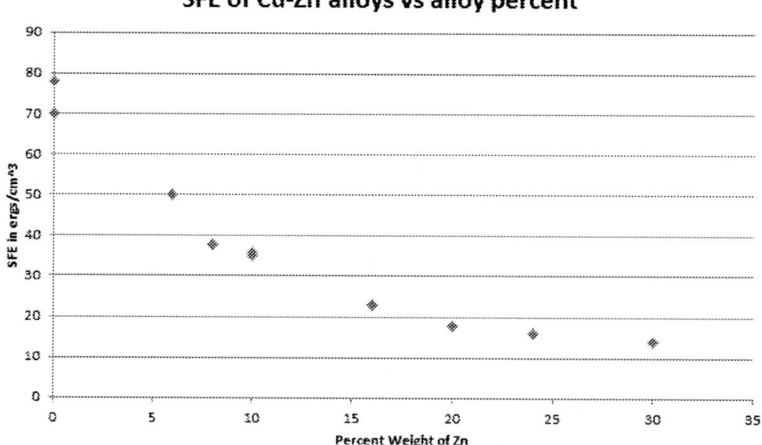

Figure: *A graph of how SFE quickly diminishes with zinc alloy content. Data taken from.*

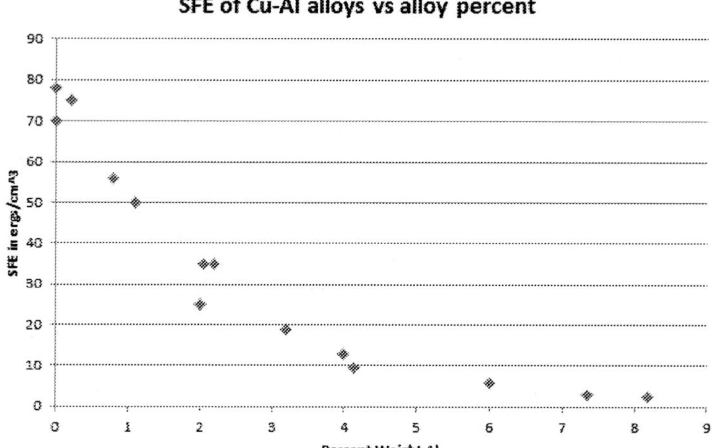

Figure: *A graph of how SFE quickly diminishes with aluminum alloy content. Data taken from.*

Stacking fault energy is heavily influenced by a few major factors, specifically base metal, alloying metals, percent of alloy metals, and valence-electron to atom ratio.

Alloying Elements Effects on SFE

It has long been established that the addition of alloying elements significantly lowers the SFE of most metals. Which element and how much is added dramatically affects the SFE of a material. The figures on the right show how the SFE of copper lowers with the addition or two different alloying elements, zinc and aluminum. In both cases, the SFE of the brass decreases with increasing alloy content. However, the SFE of the Cu-Al alloy decreases faster and reaches a low minimum.

e/a ratio

Another factor that has a significant effect on the SFE of a material and is very interrelated with alloy content is the e/a ratio, or the ratio of valence electrons to atoms. Thornton showed this in 1962 by plotting the e/a ratio vs SFE for a few Cu based alloys. He found that the valence-electron to atom ratio is a good predictor of stacking fault energy, even when the alloying element is changed. This directly support to graphs to the right. Zinc is a heavier element and only has two valence electrons, whereas aluminum is lighter and has three valence electrons. Thus each weight percent of aluminum has a much greater impact on the SFE of the Cu based alloy than does zinc.

Effects of Stacking Fault Energy on Deformation and Texture

The two primary methods of deformation in metals are slip and twinning. Slip occurs by dislocation glide of either screw or edge dislocations within a slip plane. Slip is by far the most common mechanism. Twinning is less common but readily occurs under some circumstances.

A twin is a very large stacking fault. Twinning occurs when there are not enough slip systems to accommodate deformation and/or when the material has a very low SFE. Twins are abundant in many low SFE metals like copper alloys, but are rarely seen in high SFE metals like aluminum.

In order to accommodate large strains without fracturing, there must be at least five independent and active slip systems. When cross-slip frequently occurs and certain other criteria are met, sometimes only three independent slip systems are needed for accommodating large deformations.

Because of the different deformation mechanisms in high and low SFE materials, they develop different textures.

High SFE materials

High SFE materials deform by glide of full dislocations. Because there are no stacking faults, the screw dislocations may cross-slip. Smallman found that cross-slip happens under low stress for high SFE materials like aluminum (1964).This gives a metal extra ductility because with cross-slip it needs only three other active slip systems to undergo large strains. This is true even when the crystal is not ideally oriented.

High SFE materials therefore do not need to change orientation in order to accommodate large deformations because of cross-slip. Some reorientation and texture development will occur as the grains move during deformation. Extensive cross-slip due to large deformation also causes some grain rotation. However, this re-orientation of grains in high SFE materials is much less prevalent than in low SFE materials.

Low SFE materials

Low SFE materials twin and create partial dislocations. Partials form instead of screw dislocations. Screws which do exist cannot cross-slip across stacking faults, even under high stresses. Five or more slip systems must be active for large deformations to occur because of the absence of cross-slip. For both the <111> and <100> directions there are six and eight different slip systems, respectively. If loading is not applied near one of those directions, five slip systems might be active. In this case, other mechanisms must also be in place to accommodate large strains.

Low SFE materials also twin when strained. If deformation twinning is combined with regular shear deformation, the grains eventually align towards a more preferred orientation. When many different grains align a highly anisotropic texture is created.

Twins

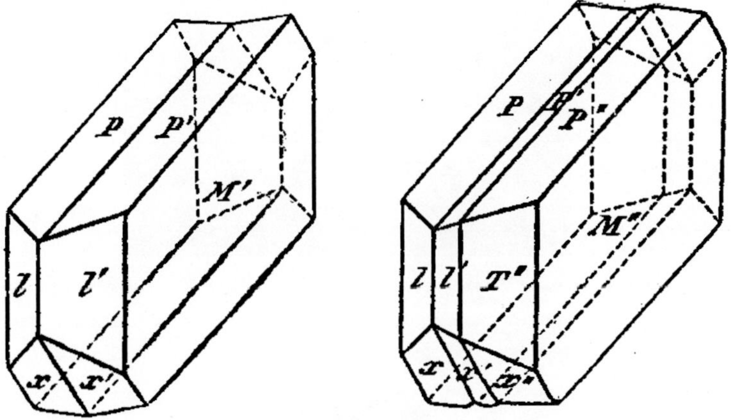

Figure: *Diagram of twinned crystals of Albite. On the more perfect cleavage, which is parallel to the basal plane (P), is a system of fine striations, parallel to the second cleavage (M).*

Crystal twinning occurs when two separate crystals share some of the same crystal lattice points in a symmetrical manner. The result is an intergrowth of two separate crystals in a variety of specific configurations. A twin boundary or composition surface separates the two crystals. Crystallographers classify twinned crystals by a number of twin laws. These twin laws are specific to the crystal system. The type of twinning can be a diagnostic tool in mineral identification.

Twinning can often be a problem in X-ray crystallography, as a twinned crystal does not produce a simple diffraction pattern.

Types of Twinning

Simple twinned crystals may be contact twins or penetration twins. Contact twins share a single composition surface often appearing as mirror images across the boundary. Plagioclase, quartz, gypsum, and spinel often exhibit contact twinning. Merohedral twinning occurs when the lattices of the contact twins superimpose in three dimensions, such as by relative rotation of one twin from the other. An example is metazeunerite. In penetration twins the individual crystals have the appearance of *passing through* each other in a symmetrical manner. Orthoclase, staurolite, pyrite, and fluorite often show penetration twinning.

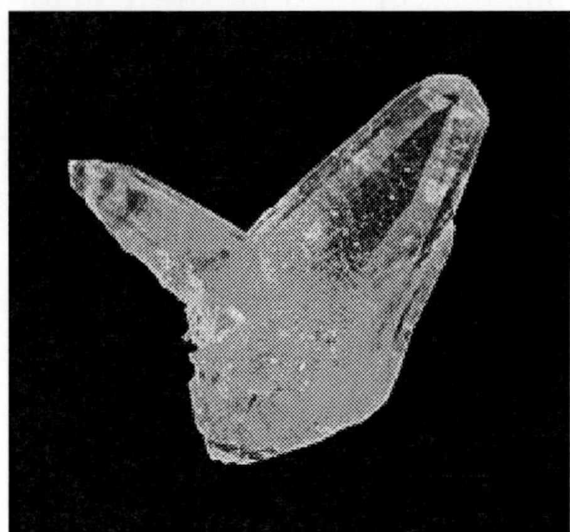

Figure: *Quartz - Japan twin*

Figure: *Twinned pyrite crystal group*

If several twin crystal parts are aligned by the same twin law they are referred to as multiple or repeated twins. If these multiple twins are aligned in parallel they are called polysynthetic twins. When the multiple twins are not parallel they are cyclic twins.

Albite, calcite, and pyrite often show polysynthetic twinning. Closely spaced polysynthetic twinning is often observed as striations or fine parallel lines on the crystal face. Rutile, aragonite, cerussite, and chrysoberyl often exhibit cyclic twinning, typically in a radiating pattern.

Figure: *Galvanized surface with macroscopic crystalline features. Twin boundaries are visible as striations within each crystallite, most prominently in the bottom-left and top-right.*

Modes of Formation

There are three modes of formation of twinned crystals. Growth twins are the result of an interruption or change in the lattice during formation or growth due to a possible deformation from a larger substituting ion. Annealing or transformation twins are the result of a change in crystal system during cooling as one *form* becomes unstable and the crystal structure must re-organize or *transform* into another more stable form. Deformation or gliding twins are the result of stress on the crystal after the crystal has formed. If a FCC metal like aluminum experiences extreme stresses, it will experience twinning as seen in the case of explosions. Deformation twinning is a common result of regional metamorphism.

Crystals that grow adjacent to each other may be aligned to resemble twinning. This parallel growth simply reduces system energy and is not twinning.

Twin Boundaries

Twin boundaries occur when two crystals of the same type intergrow, so that only a slight misorientation exists between them. It is a highly symmetrical interface, often with one crystal the mirror image of the other; also, atoms are shared by the two crystals at regular intervals. This is also a much lower-energy interface than the grain boundaries that form when crystals of arbitrary orientation grow together.

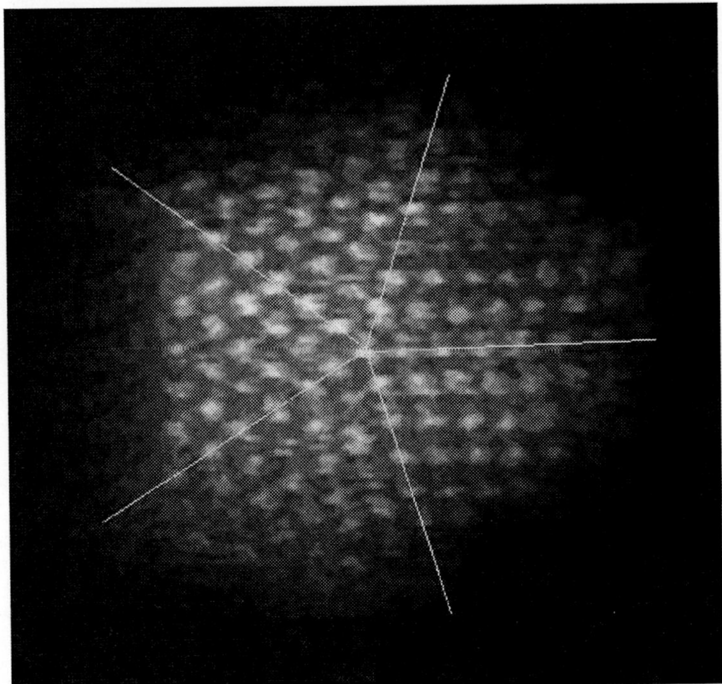

Figure: *Fivefold twinning in a gold nanoparticle (electron microscope image).*

Twin boundaries are partly responsible for shock hardening and for many of the changes that occur in cold work of metals with limited slip systems or at very low temperatures.

They also occur due to martensitic transformations: the motion of twin boundaries is responsible for the pseudoelastic and shape-memory behaviour of nitinol, and their presence is partly responsible for the hardness due to quenching of steel.

In certain types of high strength steels, very fine deformation twins act as primary obstacles against dislocation motion. These steels are referred to as 'TWIP' steels, where TWIP stands for TWinning Induced Plasticity.

Deformation Twins

Of the three common crystalline structures BCC, FCC, and HCP, the HCP structure is the most likely to form deformation twins when strained, because they rarely have a sufficient number of slip systems for an arbitrary shape change. High strain rates, low Stacking-fault energy and low temperatures facilitate deformation twinning.

Grain Boundaries

A grain boundary is the interface between two grains, or crystallites, in a polycrystalline material. Grain boundaries are defects in the crystal structure, and tend to decrease the electrical and thermal conductivity of the material. Most grain boundaries are preferred sites for the onset of corrosion and for the precipitation of new phases from the solid. They are also important to many of the mechanisms of creep. On the other hand, grain boundaries disrupt the motion of dislocations through a material, so reducing crystallite size is a common way to improve strength, as described by the Hall–Petch relationship.

High and Low Angle Boundaries

It is convenient to separate grain boundaries by the extent of the mis-orientation between the two grains. *Low angle grain boundaries* (LAGBs) are those with a misorientation less than about 11 degrees. Generally speaking they are composed of an array of dislocations and their properties and structure are a function of the misorientation. In contrast the properties of *high angle grain boundaries* (HAGBs) whose misorientation is greater than about 11 degrees (the transition angle varies from 10–15 degrees depending on the material) are normally found to be independent of the misorientation. However, there are 'special boundaries' at particular orientations whose interfacial energies are notably lower than those of general HAGBs.

The most simple boundary is that of a tilt boundary where the rotation axis is parallel to the boundary plane. This boundary can be conceived as forming from a single, contiguous crystallite or grain which is gradually bent by some external force.

The energy associated with the elastic bending of the lattice can be reduced by inserting a dislocation, which is essentially a half-plane of atoms that act like a wedge, that creates a permanent misorientation between the two sides. As the grain is bent further, more and more dislocations must be introduced to accommodate the deformation resulting in a growing wall of dislocations – a low-angle boundary. The grain can now be considered to have split into two sub-grains of related crystallography but notably different orientations.

An alternative is a twist boundary where the mis-orientation occurs around an axis that is perpendicular to the boundary plane. This type of boundary incorporates two sets of screw dislocations. If the Burgers vectors of the dislocations are orthogonal then the dislocations do not strongly interact and form a square network. In other cases the dislocations may interact to form a more complex hexagonal structure.

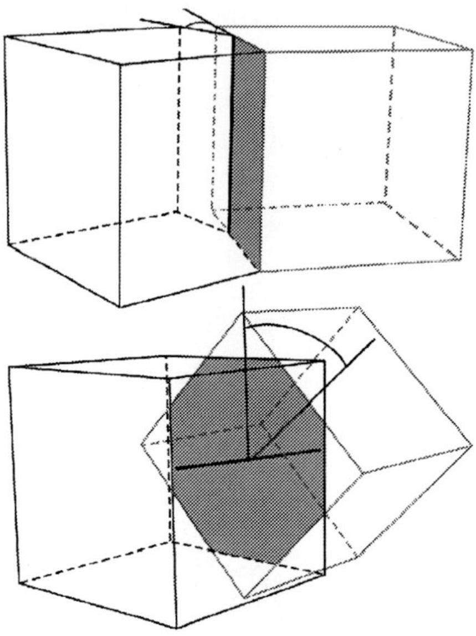

Figure: *Schematic representations of a tilt boundary (top) and a twist boundary between two idealised grains.*

These concepts of tilt and twist boundaries represent somewhat idealized cases. The majority of boundaries are of a mixed-type, containing dislocations of different types and burgers vector, in order to create the best fit between the neighbouring grains.

While the dislocations in the boundary remain isolated and distinct the boundary can be considered to be low-angle. If deformation continues the density of dislocations will increase and so reduce the spacing between neighbouring dislocations. Eventually, the cores of the dislocations will begin to overlap and the ordered nature of the boundary will begin to break down. At this point the boundary can be considered to be high-angle and the original grain to have separated into two entirely separate grains.

In comparison to LAGBs, high-angle boundaries are considerably more disordered with large areas of poor fit and a comparatively open structure. Indeed, they were originally thought to be some form of amorphous or even liquid layer between the grain. However, this model could not explain the observed strength of grain boundaries and, after the invention of electron microscopy, direct evidence of the grain structure meant the hypothesis had to be discarded. It is now accepted that a boundary consists of structural units which depend on both the misorientation of the two grains and the plane of the interface. The types of structural unit that exist can be related to the concept of the *coincidence site lattice*, in which repeated units are formed from points where the two misoriented lattices happen to coincide.

In coincident site lattice (CSL) theory, the degree of fit (Σ) between the structures of the two grains is described by the reciprocal of the ratio of coincidence sites to the total number of sites. Thus a boundary with high Σ might be expected to have a higher energy than one with low Σ. Low-angle boundaries, where the distortion is entirely accommodated by dislocations, are $\Sigma 1$. Some other low Σ boundaries have special properties especially when the boundary plane is one that contains a high density of coincident sites. Examples include coherent twin boundaries ($\Sigma 3$) and high-mobility boundaries in FCC materials ($\Sigma 7$). Deviations from the ideal CSL orientation may be accommodated by local atomic relaxation or the inclusion of dislocations into the boundary.

Describing a Boundary

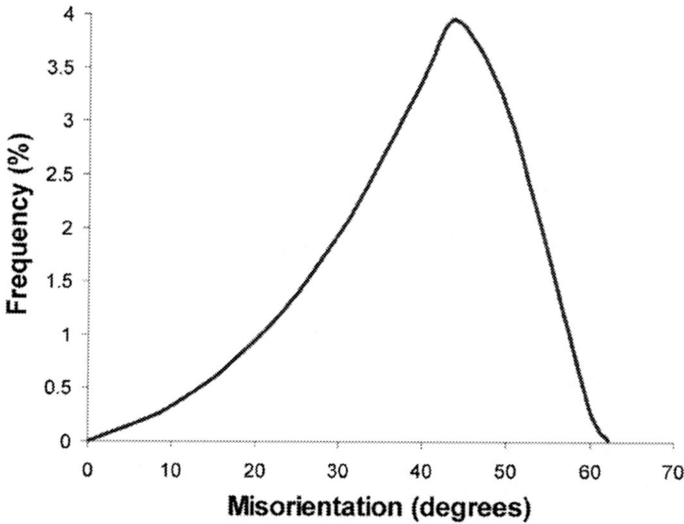

Figure: *The characteristic distribution of boundary misorientations in a completely randomly oriented set of grains for cubic symmetry materials.*

A boundary can be described by the orientation of the boundary to the two grains and the 3-D rotation required to bring the grains into coincidence. Thus a boundary has 5 macroscopic degrees of freedom.

However, it is common to describe a boundary only as the orientation relationship of the neighbouring grains. Generally, the convenience of ignoring the boundary plane orientation, which is very difficult to determine, outweighs the reduced information. The relative orientation of the two grains is described using the rotation matrix:

$$R = \begin{bmatrix} a_{11} & a_{12} & a_{13} \\ a_{21} & a_{22} & a_{23} \\ a_{31} & a_{32} & a_{33} \end{bmatrix}$$

Using this system the rotation angle θ is:

$$2\cos\theta + 1 = a_{11} + a_{22} + a_{33}$$

while the direction [uvw] of the rotation axis is:

$$[(a_{32} - a_{23}), (a_{13} - a_{31}), (a_{21} - a_{12})]$$

The nature of the crystallography involved limits the misorientation of the boundary. A completely random polycrystal, with no texture, thus has a characteristic distribution of boundary misorientations. However, such cases are rare and most materials will deviate from this ideal to a greater or lesser degree.

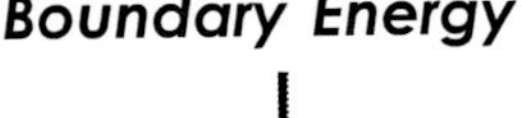

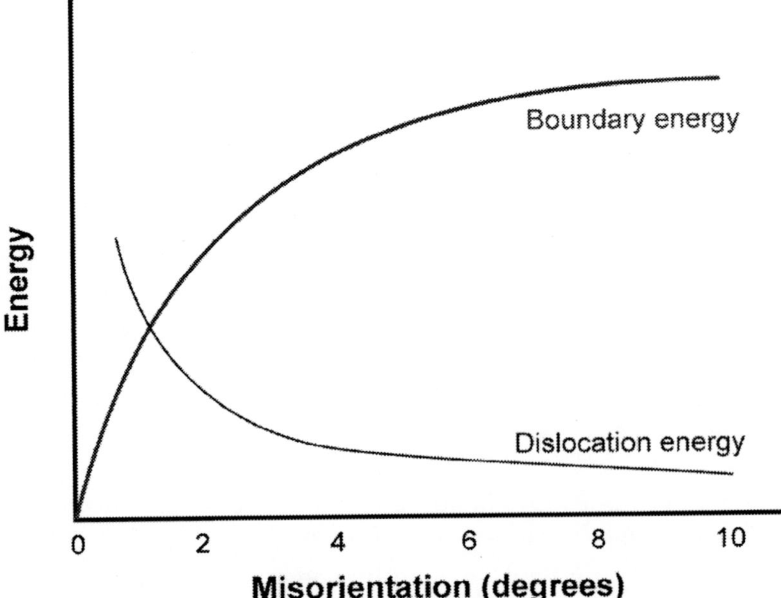

Figure: *The energy of a tilt boundary and the energy per dislocation as the misorientation of the boundary increases.*

The energy of a low-angle boundary is dependent on the degree of misorientation between the neighbouring grains up to the transition to high-angle status. In the case of simple *tilt boundaries* the energy of a boundary made up of dislocations with Burgers vector b and spacing h is predicted by the Read-Shockley equation:

$$\gamma_s = \gamma_0 \theta (A - \ln \theta)$$

where $\theta = b/h$, $\gamma_0 = Gb/4\,\pi(1 - \nu)$, $A = 1 + \ln(b/2\,\pi r_0)$, G is the shear modulus, ν is poisson's ratio, and r_0 is the radius of the dislocation core. It can be seen that as the energy of the boundary increases the energy per dislocation decreases. Thus there is a driving force to produce fewer, more misoriented boundaries (i.e., grain growth).

The situation in high-angle boundaries is more complex. Although theory predicts that the energy will be a minimum for ideal CSL configurations, with deviations requiring dislocations and other energetic features, empirical measurements suggest the relationship is more complicated. Some predicted troughs in energy are found as expected while others missing or substantially reduced. Surveys of the available experimental data have indicated that simple relationships such as low Σ are misleading:

It is concluded that no general and useful criterion for low energy can be enshrined in a simple geometric framework. Any understanding of the variations of interfacial energy must take account of the atomic structure and the details of the bonding at the interface.

Boundary Migration

The movement of grain boundaries (HAGB) has implications for recrystallization and grain growth while subgrain boundary (LAGB) movement strongly influences recovery and the nucleation of recrystallization.

A boundary moves due to a pressure acting on it. It is generally assumed that the velocity is directly proportional to the pressure with the constant of proportionality being the mobility of the boundary. The mobility is strongly temperature dependent and often follows an Arrhenius type relationship:

$$M = M_0 \exp\left(-\frac{Q}{RT}\right)$$

The apparent activation energy (Q) may be related to the thermally activated atomistic processes that occur during boundary movement. However, there are several proposed mechanisms where the mobility will depend on the driving pressure and the assumed proportionality may break down. It is generally accepted that the mobility of low-angle boundaries is much lower than that of high-angle boundaries. The following observations appear to hold true over a range of conditions:

- The mobility of low-angle boundaries is proportional to the pressure acting on it.
- The rate controlling process is that of bulk diffusion
- The boundary mobility increases with misorientation.

Since low-angle boundaries are composed of arrays of dislocations and their movement may be related to dislocation theory. The most likely mechanism, given the experimental data, is that of dislocation climb, rate limited by the diffusion of solute in the bulk.

The movement of high-angle boundaries occurs by the transfer of atoms between the neighbouring grains.

The ease with which this can occur will depend on the structure of the boundary, itself dependent on the crystallography of the grains involved, impurity atoms and the temperature.

It is possible that some form of diffusionless mechanism (akin to diffusionless phase transformations such as martensite) may operate in certain conditions. Some defects in the boundary, such as steps and ledges, may also offer alternative mechanisms for atomic transfer.

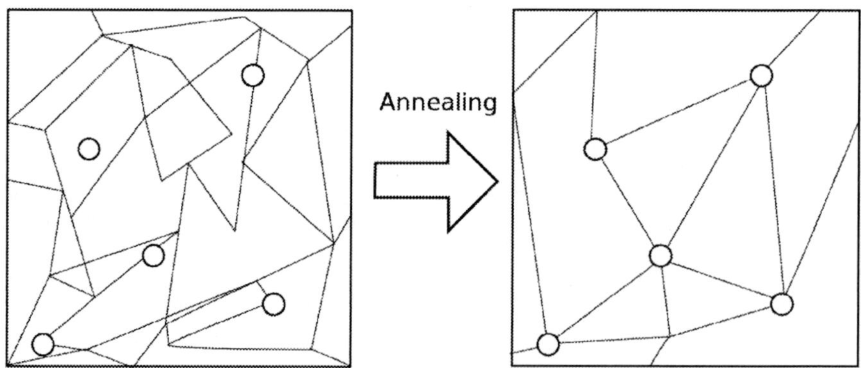

Figure: *Grain growth can be inhibited by second phase particles via Zener pinning.*

Since a high-angle boundary is imperfectly packed compared to the normal lattice it has some amount of *free space* or *free volume* where solute atoms may possess a lower energy. As a result a boundary may be associated with a *solute atmosphere* that will retard its movement. Only at higher velocities will the boundary be able to break free of its atmosphere and resume normal motion.

Both low- and high-angle boundaries are retarded by the presence of particles via the so-called Zener pinning effect. This effect is often exploited in commercial alloys to minimise or prevent recrystallization or grain growth during heat-treatment.

Chapter 3

Ceramic Materials

Introduction to Ceramic Materials

Ceramic materials are inorganic, non-metallic materials made from compounds of a metal and a non metal. Ceramic materials may be crystalline or partly crystalline. They are formed by the action of heat and subsequent cooling. Clay was one of the earliest materials used to produce ceramics, as pottery, but many different ceramic materials are now used in domestic, industrial and building products. Ceramic materials tend to be strong, stiff, brittle, chemically inert, and non-conductors of heat and electricity, but their properties vary widely. For example, porcelain is widely used to make electrical insulators, but some ceramic compounds are superconductors.

Types of Ceramic Materials

A ceramic material may be defined as any inorganic crystalline material, compounded of a metal and a non-metal. It is solid and inert. Ceramic materials are brittle, hard, strong in compression, weak in shearing and tension. They withstand chemical erosion that occurs in an acidic or caustic environment. In many cases withstanding erosion from the acid and bases applied to it. Ceramics generally can withstand very high temperatures such as temperatures that range from 1,000 °C to 1,600 °C (1,800 °F to 3,000 °F). Exceptions include inorganic materials that do not have oxygen such as silicon carbide. Glass by definition is not a ceramic because it is an amorphous solid (non-crystalline). However, glass involves several steps of the ceramic process and its mechanical properties behave similarly to ceramic materials.

Traditional ceramic raw materials include clay minerals such as kaolinite, more recent materials include aluminium oxide, more commonly known as alumina. The modern ceramic materials, which are classified as advanced ceramics, include silicon carbide and tungsten carbide. Both are valued for their abrasion resistance, and hence find use in applications such as the wear plates of crushing equipment in mining operations. Advanced ceramics are also used in the medicine, electrical and electronics industries.

Crystalline Ceramics

Crystalline ceramic materials are not amendable to a great range of processing. Methods for dealing with them tend to fall into one of two categories - either make the ceramic in the desired shape, by reaction in situ, or by "forming" powders into the desired shape, and then sintering to form a solid body. Ceramic forming techniques include shaping by hand (sometimes including a rotation process called "throwing"), slip casting, tape casting (used for making very thin ceramic capacitors, etc.), injection moulding, dry pressing, and other variations. A few methods use a hybrid between the two approaches of butohing and pukiying.

Non-crystalline Ceramics

Non-crystalline ceramics, being glasses, tend to be formed from melts. The glass is shaped when either fully molten, by casting, or when in a state of toffee-like viscosity, by methods such as blowing to a mold. If later heat-treatments cause this glass to become partly crystalline, the resulting material is known as a glass-ceramic.

Properties of Ceramics

The physical properties of any ceramic substance are a direct result of its crystalline structure and chemical composition. Solid state chemistry reveals the fundamental connection between microstructure and properties such as localized density variations, grain size distribution, type of porosity and second-phase content, which can all be correlated with ceramic properties such as mechanical strength σ by the Hall-Petch equation, hardness, toughness, dielectric constant, and the optical properties exhibited by transparent materials.

Physical properties of chemical compounds which provide evidence of chemical composition include odor, colour, volume, density (mass / volume), melting point, boiling point, heat capacity, physical form at room temperature (solid, liquid or gas), hardness, porosity, and index of refraction.

Ceramography is the art and science of preparation, examination and evaluation of ceramic microstructures. Evaluation and characterization of ceramic microstructures is often implemented on similar spatial scales to that used commonly in the emerging field of nanotechnology: from tens of angstroms (A) to tens of micrometers (μm). This is typically somewhere between the minimum wavelength of visible light and the resolution limit of the naked eye.

The microstructure includes most grains, secondary phases, grain boundaries, pores, micro-cracks, structural defects and hardness microindentions. Most bulk mechanical, optical, thermal, electrical and magnetic properties are significantly affected by the observed microstructure. The fabrication method and process conditions are generally indicated by the microstructure. The root cause of many ceramic failures is evident in the cleaved and polished microstructure. Physical properties which constitute the field of materials science and engineering include the following:

Mechanical Properties

Mechanical properties are important in structural and building materials as well as textile fabrics. They include the many properties used to describe the strength of materials such as: elasticity / plasticity, tensile strength, compressive strength, shear strength, fracture toughness & ductility (low in brittle materials), and indentation hardness.

Fracture mechanics is the field of mechanics concerned with the study of the formation and subsequent propagation of microcracks in materials. It uses methods of analytical solid mechanics to calculate the thermodynamic driving force on a crack and the methods of experimental solid mechanics to characterize the material's resistance to fracture and catastrophic failure.

In modern materials science, fracture mechanics is an important tool in improving the mechanical performance of materials and components. It applies the physics of stress and strain, in particular the theories of elasticity and plasticity, to the microscopic crystallographic defects found in real materials in order to predict the macroscopic mechanical failure of bodies. Fractography is widely used with fracture mechanics to understand the causes of failures and also verify the theoretical failure predictions with real life failures.

Thus, since cracks and other microstructural defects can lower the strength of a structure beyond that which might be predicted by the theory of crystalline objects, a different property of the material—above and beyond conventional strength—is needed to describe the fracture resistance of engineering materials. This is the reason for the need for fracture mechanics: the evaluation of the strength of flawed structures.

In this context, Fracture toughness is a property which describes the ability of a material containing a crack to resist fracture, and is one of the most important properties of any material for virtually all design applications. Fracture toughness is a quantitative way of expressing a material's resistance to brittle fracture when a crack is present. If a material has a large value of fracture toughness it will probably undergo ductile fracture.

Brittle fracture is very characteristic of materials with a low fracture toughness value. Fracture mechanics, which leads to the concept of fracture toughness, was largely based on the work of A. A. Griffith who, amongst other things, studied the behaviour of cracks in brittle materials. This describes "ceramics are strong in compression and weak in tension"

Ceramic materials are usually ionic or covalent bonded materials, and can be crystalline or amorphous. A material held together by either type of bond will tend to fracture before any plastic deformation takes place, which results in poor toughness in these materials. Additionally, because these materials tend to be porous, the pores and other microscopic imperfections act as stress concentrators, decreasing the toughness further, and reducing the tensile strength. These combine to give catastrophic failures, as opposed to the normally much more gentle failure modes of metals.

These materials do show plastic deformation. However, due to the rigid structure of the crystalline materials, there are very few available slip systems for dislocations to move, and so they deform very slowly. With the non-crystalline (glassy) materials, viscous flow is the dominant source of plastic deformation, and is also very slow. It is therefore neglected in many applications of ceramic materials. To overcome the brittle behaviour, ceramic material development has introduced the class of ceramic matrix composite materials, in which ceramic fibres are embedded and with specific coatings are forming fibre bridges across any crack. This mechanism substantially increases the fracture toughness of such ceramics. The ceramic disc brakes are, for example using a ceramic matrix composite material manufactured with a specific process.

Electrical Properties

Semiconductors: Some ceramics are semiconductors. Most of these are transition metal oxides that are II-VI semiconductors, such as zinc oxide.

While there are prospects of mass-producing blue LEDs from zinc oxide, ceramicists are most interested in the electrical properties that show grain boundary effects.

One of the most widely used of these is the varistor. These are devices that exhibit the property that resistance drops sharply at a certain threshold voltage. Once the voltage across the device reaches the threshold, there is a breakdown of the electrical structure in the vicinity of the grain boundaries, which results in its electrical resistance dropping from several megohms down to a few hundred ohms. The major advantage of these is that they can dissipate a lot of energy, and they self-reset – after the voltage across the device drops below the threshold, its resistance returns to being high.

This makes them ideal for surge-protection applications; as there is control over the threshold voltage and energy tolerance, they find use in all sorts of applications. The best demonstration of their ability can be found in electrical substations, where they are employed to protect the infrastructure from lightning strikes. They have rapid response, are low maintenance, and do not appreciably degrade from use, making them virtually ideal devices for this application.

Semiconducting ceramics are also employed as gas sensors. When various gases are passed over a polycrystalline ceramic, its electrical resistance changes. With tuning to the possible gas mixtures, very inexpensive devices can be produced.

Superconductivity

Under some conditions, such as extremely low temperature, some ceramics exhibit high temperature superconductivity. The exact reason for this is not known, but there are two major families of superconducting ceramics.

Ferroelectricity and Supersets

Piezoelectricity, a link between electrical and mechanical response, is exhibited by a large number of ceramic materials, including the quartz used to measure time in watches and other electronics. Such devices use both properties of piezoelectrics, using electricity to produce a mechanical motion (powering the device) and then using this mechanical motion to produce electricity (generating a signal). The unit of time measured is the natural interval required for electricity to be converted into mechanical energy and back again.

The piezoelectric effect is generally stronger in materials that also exhibit pyroelectricity, and all pyroelectric materials are also piezoelectric. These materials can be used to inter convert between thermal, mechanical, or electrical energy; for instance, after synthesis in a furnace, a pyroelectric crystal allowed to cool under no applied stress generally builds up a static charge of thousands of volts. Such materials are used in motion sensors, where the tiny rise in temperature from a warm body entering the room is enough to produce a measurable voltage in the crystal.

In turn, pyroelectricity is seen most strongly in materials which also display the ferroelectric effect, in which a stable electric dipole can be oriented or reversed by applying an electrostatic field. Pyroelectricity is also a necessary consequence of ferroelectricity. This can be used to store information in ferroelectric capacitors, elements of ferroelectric RAM.

The most common such materials are lead zirconate titanate and barium titanate. Aside from the uses mentioned above, their strong piezoelectric response is exploited in the design of high-frequency loudspeakers, transducers for sonar, and actuators for atomic force and scanning tunneling microscopes.

Positive Thermal Coefficient

Increases in temperature can cause grain boundaries to suddenly become insulating in some semiconducting ceramic materials, mostly mixtures of heavy metal titanates. The critical transition temperature can be adjusted over a wide range by variations in chemistry. In such materials, current will pass through the material until joule heating brings it to the transition temperature, at which point the circuit will be broken and current flow will cease. Such ceramics are used as self-controlled heating elements in, for example, the rear-window defrost circuits of automobiles.

At the transition temperature, the material's dielectric response becomes theoretically infinite. While a lack of temperature control would rule out any practical use of the material near its critical temperature, the dielectric effect remains exceptionally strong even at much higher temperatures. Titanates with critical temperatures far below room temperature have become synonymous with "ceramic" in the context of ceramic capacitors for just this reason.

Optical Properties

Optically transparent materials focus on the response of a material to incoming lightwaves of a range of wavelengths. Frequency selective optical filters can be utilized to alter or enhance the brightness and contrast of a digital image. Guided lightwave transmission via frequency selective waveguides involves the emerging field of fibre optics and the ability of certain glassy compositions as a transmission medium for a range of frequencies simultaneously (multi-mode optical fibre) with little or no interference between competing wavelengths or frequencies. This resonant mode of energy and data transmission via electromagnetic (light) wave propagation, though low powered, is virtually lossless. Optical waveguides are used as components in Integrated optical circuits (e.g. light-emitting diodes, LEDs) or as the transmission medium in local and long haul optical communication systems. Also of value to the emerging materials scientist is the sensitivity of materials to radiation in the thermal infrared (IR) portion of the electromagnetic spectrum. This heat-seeking ability is responsible for such diverse optical phenomena as Night-vision and IR luminescence.

Thus, there is an increasing need in the military sector for high-strength, robust materials which have the capability to transmit light (electromagnetic waves) in the visible (0.4 – 0.7 micrometers) and mid-infrared (1 – 5 micrometers) regions of the spectrum. These materials are needed for applications requiring transparent armor, including next-generation high-speed missiles and pods, as well as protection against improvised explosive devices (IED).

In the 1960s, scientists at General Electric (GE) discovered that under the right manufacturing conditions, some ceramics, especially aluminium oxide (alumina), could be made translucent. These translucent materials were transparent enough to be used for containing the electrical plasma generated in high-pressure sodium street lamps. During the past two decades, additional types of transparent ceramics have been developed for applications such as nose cones for heat-seeking missiles, windows for fighter aircraft, and scintillation counters for computed tomography scanners.

In the early 1970s, Thomas Soules pioneered computer modelling of light transmission through translucent ceramic alumina. His model showed that microscopic pores in ceramic, mainly trapped at the junctions of microcrystalline grains, caused light to scatter and prevented true transparency. The volume fraction of these microscopic pores had to be less than 1% for high-quality optical transmission.

This is basically a particle size effect. Opacity results from the incoherent scattering of light at surfaces and interfaces. In addition to pores, most of the interfaces in a typical metal or ceramic object are in the form of grain boundaries which separate tiny regions of crystalline order. When the size of the scattering centre (or grain boundary) is reduced below the size of the wavelength of the light being scattered, the scattering no longer occurs to any significant extent.

In the formation of polycrystalline materials (metals and ceramics) the size of the crystalline grains is determined largely by the size of the crystalline particles present in the raw material during formation (or pressing) of the object. Moreover, the size of the grain boundaries scales directly with particle size. Thus a reduction of the original particle size below the wavelength of visible light (~ 0.5 micrometers for shortwave violet) eliminates any light scattering, resulting in a transparent material.

Recently, Japanese scientists have developed techniques to produce ceramic parts that rival the transparency of traditional crystals (grown from a single seed) and exceed the fracture toughness of a single crystal. In particular, scientists at the Japanese firm Konoshima Ltd., a producer of ceramic construction materials and industrial chemicals, have been looking for markets for their transparent ceramics.

Livermore researchers realized that these ceramics might greatly benefit high-powered lasers used in the National Ignition Facility (NIF) Programmes Directorate. In particular, a Livermore research team began to acquire advanced transparent ceramics from Konoshima to determine if they could meet the optical requirements needed for Livermore's Solid-State Heat Capacity Laser (SSHCL). Livermore researchers have also been testing applications of these materials for applications such as advanced drivers for laser-driven fusion power plants.

Examples of Ceramics Materials

Until the 1950s, the most important ceramic materials were (1) pottery, bricks and tiles, (2) cements and (3) glass. A composite material of ceramic and metal is known as cermet.

- Barium titanate (often mixed with strontium titanate) displays ferroelectricity, meaning that its mechanical, electrical, and thermal responses are coupled to one another and also history-dependent. It is widely used in electromechanical transducers, ceramic capacitors, and data storage elements. Grain boundary conditions can create PTC effects in heating elements.
- Bismuth strontium calcium copper oxide, a high-temperature superconductor
- Boron nitride is structurally isoelectronic to carbon and takes on similar physical forms: a graphite-like one used as a lubricant, and a diamond-like one used as an abrasive.
- Earthenware used for domestic ware such as plates and mugs.
- Ferrite is used in the magnetic cores of electrical transformers and magnetic core memory.
- Lead zirconate titanate (PZT) was developed at the United States National Bureau of Standards in 1954. PZT is used as an ultrasonic transducer, as its piezoelectric properties greatly exceed those of Rochelle salt.
- Magnesium diboride (MgB_2) is an unconventional superconductor.
- Porcelain is used for a wide range of household and industrial products.
- Sialon (Silicon Aluminium Oxynitride) has high strength; high thermal, shock, chemical and wear resistance, and low density. These ceramics are used in non-ferrous molten metal handling, weld pins and the chemical industry.
- Silicon carbide (SiC) is used as a susceptor in microwave furnaces, a commonly used abrasive, and as a refractory material.
- Silicon nitride (Si_3N_4) is used as an abrasive powder.
- Steatite (magnesium silicates) is used as an electrical insulator.
- Titanium carbide Used in space shuttle re-entry shields and scratchproof watches.

- Uranium oxide (UO_2), used as fuel in nuclear reactors.
- Yttrium barium copper oxide ($YBa_2Cu_3O_{7-x}$), another high temperature superconductor.
- Zinc oxide (ZnO), which is a semiconductor, and used in the construction of varistors.
- Zirconium dioxide (zirconia), which in pure form undergoes many phase changes between room temperature and practical sintering temperatures, can be chemically "stabilized" in several different forms. Its high oxygen ion conductivity recommends it for use in fuel cells and automotive oxygen sensors. In another variant, metastable structures can impart transformation toughening for mechanical applications; most ceramic knife blades are made of this material.
- Partially stabilised zirconia (PSZ) is much less brittle than other ceramics and is used for metal forming tools, valves and liners, abrasive slurries, kitchen knives and bearings subject to severe abrasion.

Ceramic Structures

Ceramics and related materials cover a wide range of objects. Ceramics are a little more complex than metallic structures, which is why metals were covered first. A ceramic has traditionally been defined as "an inorganic, nonmetallic solid that is prepared from powdered materials and is fabricated into products through the application of heat. Most ceramics are made up of two or more elements. This is called a compound. For example, alumina (Al2O3) is a compound made up of aluminum atoms and oxygen atoms.

The two most common chemical bonds for ceramic materials are covalent and ionic. The bonding of atoms together is much stronger in covalent and ionic bonding than in metallic. This is why ceramics generally have the following properties: high hardness, high compressive strength, and chemical inertness. This strong bonding also accounts for the less attractive properties of ceramics, such as low ductility and low tensile strength. The absence of free electrons is responsible for making most ceramics poor conductors of electricity and heat.

However, it should be noted that the crystal structures of ceramics are many and varied and this results in a very wide range of properties. For example, while ceramics are perceived as electrical and thermal insulators, ceramic oxide (initially based on Y-Ba-Cu-O) is the basis for high temperature superconductivity. Diamond and silicon carbide have a higher thermal conductivity than aluminum or copper. Control of the microstructure can overcome inherent stiffness to allow the production of ceramic springs, and ceramic composites which have been produced with a fracture toughness about half that of steel. Also, the atomic structures are often of low symmetry that gives some ceramics interesting electromechanical properties like piezoelectricity, which is used in sensors and transducers.

The structure of most ceramics varies from relatively simple to very complex. The microstructure can be entirely glassy (glasses only); entirely crystalline; or a combination of crystalline and glassy. In the latter case, the glassy phase usually surrounds small crystals, bonding them together. The main compositional classes of engineering ceramics are the oxides, nitrides and carbides.

Ceramic Glass

Ceramics with an entirely glassy structure have certain properties that are quite different from those of metals. Recall that when metal in the liquid state is cooled, a crystalline solid precipitates when the melting freezing point is reached. However, with a glassy material, as the liquid is cooled it becomes more and more viscous. There is no sharp melting or freezing point. It goes from liquid to a soft plastic solid and finally becomes hard and brittle. Because of this unique property, it can be blown into shapes, in addition to being cast, rolled, drawn and otherwise processed like a metal.

Glassy behaviour is related to the atomic structure of the material. If pure silica (SiO_2) is fused together, a glass called vitreous silica is formed on cooling. The basic unit structure of this glass is the silica tetrahedron, which is composed of a single silicon atom surrounded by four equidistant oxygen atoms. The silicon atoms occupy the openings (interstitials) between the oxygen atoms and share four valence electrons with the oxygen atoms through covalent bonding. The silica atom has four valence electrons and each of the oxygen atoms has two valence electrons so the silica tetrahedron has four extra valence electrons to share with adjacent tetrahedral. The silicate structures can link together by sharing the atoms in two corners of the SiO_2 tetrahedrons, forming chain or ring structures. A network of silica tetrahedral chains form, and at high temperatures these chains easily slide past each other. As the melt cools, thermal vibrational energy decreases and the chains can not move as easily so the structure becomes more rigid. Silica is the most important constituent of glass, but other oxides are added to change certain physical characteristics or to lower the melting point.

Ceramic Crystalline or Partially Crystalline Material

Most ceramics usually contain both metallic and nonmetallic elements with ionic or covalent bonds. Therefore, the structure the metallic atoms, the structure of the nonmetallic atoms, and the balance of charges produced by the valence electrons must be considered. As with metals, the unit cell is used in describing the atomic structure of ceramics. The cubic and the hexagonal cells are most common. Additionally, the difference in radii between the metallic and nonmetallic ions plays an important role in the arrangement of the unit cell.

In metals, the regular arrangement of atoms into densely packed planes led to the occurrence of slip under stress, which gives metal their characteristic ductility. In ceramics, brittle fracture rather than slip is common because both the arrangement of the atoms and the type of bonding is different. The fracture or cleavage planes of ceramics are the result of planes of regularly arranged atoms.

The building criteria for the crystal structure are:

- maintain neutrality
- charge balance dictates chemical formula
- achieve closest packing

A few of the different types of ceramic materials outside of the glass family are described below.

Silicate Ceramics

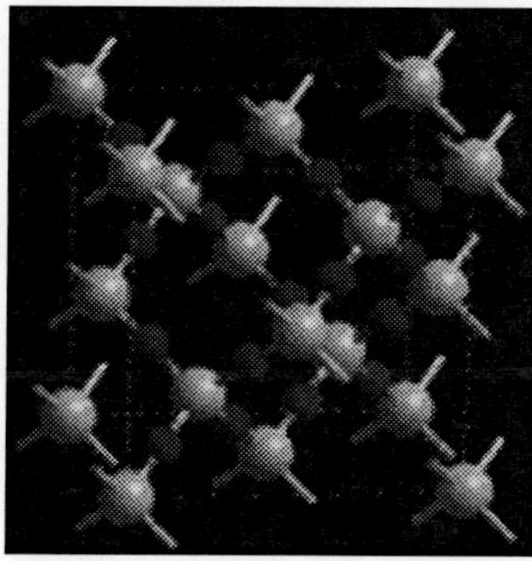

As mentioned previously, the silica structure is the basic structure for many ceramics, as well as glass. It has an internal arrangement consisting of pyramid (tetrahedral or four-sided) units. Four large oxygen (0) atoms surround each smaller silicon (Si) atom. When silica tetrahedrons share three corner atoms, they produce layered silicates (talc, kaolinite clay, mica). Clay is the basic raw material for many building products such as brick and tile. When silica tetrahedrons share four comer atoms, they produce framework silicates (quartz, tridymite). Quartz is formed when the tetrahedra in this material are arranged in a regular, orderly fashion. If silica in the molten state is cooled very slowly it crystallizes at the freezing point. But if molten silica is cooled more rapidly, the resulting solid is a disorderly arrangement which is glass.

Cement

Cement (Portland cement) is one of the main ingredients of concrete. There are a number of different grades of cement but a typical Portland cement will contain 19 to 25% SiO_2 , 5 to 9% Al_2O_3, 60 to 64% CaO and 2 to 4% FeO. Cements are prepared by grinding the clays and limestone in proper proportion, firing in a kiln, and regrinding. When water is added, the minerals either decompose or combine with water, and a new phase grows throughout the mass. The reaction is solution, recrystallization, and precipitation of a silicate structure. It is usually important to control the amount of water to prevent an excess that would not be part of the structure and would weaken it. The heat of hydration (heat of reaction in the adsorption of water) in setting of the cement can be large and can cause damage in large structures.

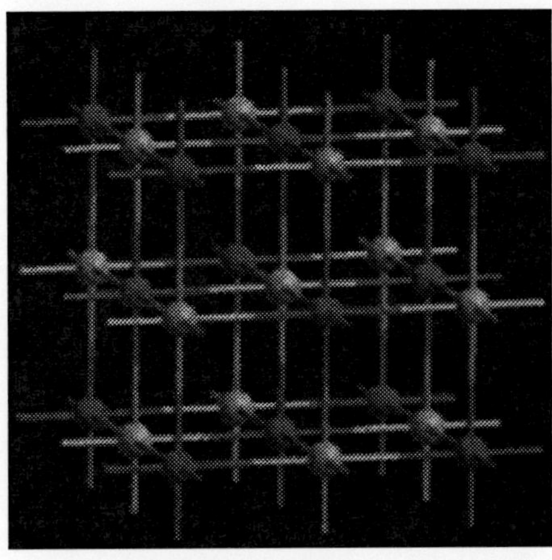

Nitride Ceramics

Nitrides combine the superior hardness of ceramics with high thermal and mechanical stability, making them suitable for applications as cutting tools, wear-resistant parts and structural components at high temperatures. TiN has a cubic structure which is perhaps the simplest and best known of structure types. Cations and anions both lie at the nodes of separate fcc lattices. The structure is unchanged if the Ti and N atoms (lattices) are interchanged.

Ferroelectric Ceramics

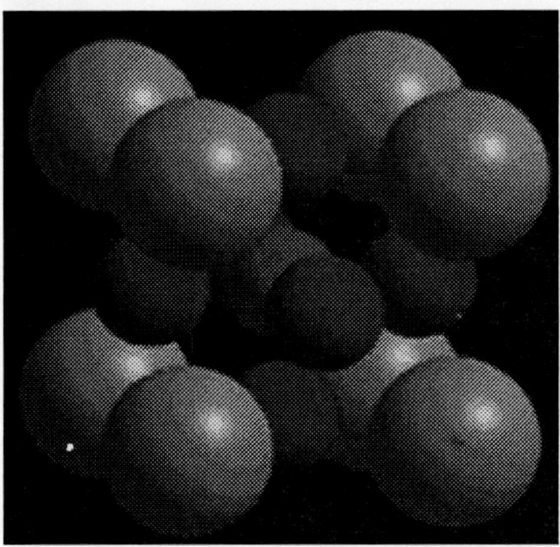

Depending on the crystal structure, in some crystal lattices, the centres of the positive and negative charges do not coincide even without the application of external electric field. In this case, it is said that there exists spontaneous polarization in the crystal. When the polarization of the dielectric can be altered by an electric field, it is called ferroelectric.

A typical ceramic ferroelectric is barium titanate, $BaTiO_3$. Ferroelectric materials, especially polycrystalline ceramics, are very promising for varieties of application fields such as piezoelectric/ electrostrictive transducers, and electrooptic.

Phase Diagram

The phase diagram is important in understanding the formation and control of the microstructure of the microstructure of polyphase ceramics, just as it is with polyphase metallic materials. Also, nonequilibrium structures are even more prevalent in ceramics because the more complex crystal structures are more difficult to nucleate and to grow from the melt.

Imperfections in Ceramics

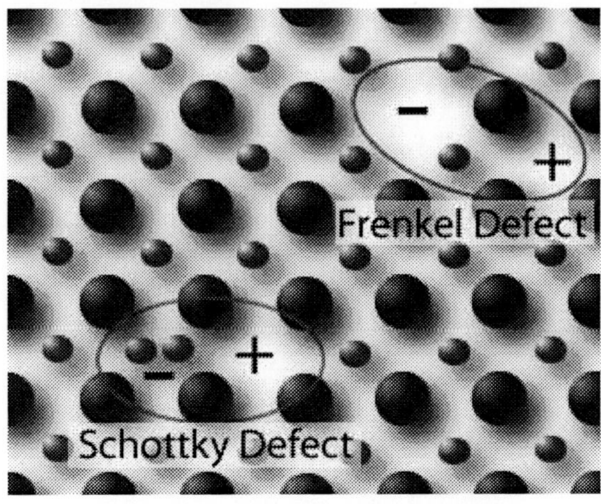

Imperfections in ceramic crystals include point defects and impurities like in metals. However, in ceramics defect formation is strongly affected by the condition of charge neutrality because the creation of areas of unbalanced charges requires an expenditure of a large amount of energy. In ionic crystals, charge neutrality often results in defects that come as pairs of ions with opposite charge or several nearby point defects in which the sum of all charges is zero. Charge neutral defects include the Frenkel and Schottky defects. A Frenkel-defect occurs when a host atom moves into a nearby interstitial position to create a vacancy-interstitial pair of cations. A Schottky-defect is a pair of nearby cation and anion vacancies. Schottky defect occurs when a host atom leaves its position and moves to the surface creating a vacancy-vacancy pair.

Sometimes, the composition may alter slightly to arrive at a more balanced atomic charge. Solids such as SiO_2, which have a well-defined chemical formula, are called stoichiometric compounds. When the composition of a solid deviates from the standard chemical formula, the resulting solid is said to be nonstoichiometric. Nonstoichiometry and the existence of point defects in a solid are often closely related. Anion vacancies are the source of the nonstoichiometry in SiO_{2-x},

Introduction of impurity atoms in the lattice is likely in conditions where the charge is maintained. This is the case of electronegative impurities that substitute a lattice anion or electropositive substitutional impurities. This is more likely for similar ionic radii since this minimizes the energy required for lattice distortion. Defects will appear if the charge of the impurities is not balanced.

Silicate Structures

The relative abundance of elements in the Earth's crust determines what minerals will form and what minerals will be common. Because Oxygen and Silicon are the most abundant elements, the silicate minerals are the most common.

Thus, we will spend some time here discussing the structure, chemistry, and occurrence of silicate minerals. Our systematic discussion of the common rock forming minerals will follow in the lectures throughout the remainder of the course.

Element	*Wt%*	*Atomic%*	*Volume%*
O	46.60	62.55	~94
Si	27.72	21.22	~6
Al	8.13	6.47	
Fe	5.00	1.92	
Ca	3.63	1.94	
Na	2.83	2.34	
K	2.59	1.42	
Mg	2.09	1.84	
Total	98.59	100.00	100

In order to discuss the silicates and their structures it is first necessary to remember that the way atoms are packed together or coordinated by larger anions, like oxygen depends on the radius ratio of the cation to the anion, Rx/Rz.

Rx/Rz	*C.N.*	*Type*
1.0	12	Hexagonal or Cubic
Closest Packing		
1.0 - 0.732	8	Cubic
0.732 - 0.414	6	Octahedral
0.414 - 0.225	4	Tetrahedral
0.225 - 0.155	3	Triangular
<0.155	2	Linear

Since oxygen is the most abundant element in the crust, oxygen will be the major anion that coordinates the other cations. Thus, for the major ions that occur in the crust, we can make the following table showing the coordination and coordination polyhedra that are expected for each of the common cations.

Ion	*C.N. (with Oxygen)*	*Coord. Polyhedron*	*Ionic Radius, Å*
K^+	8 - 12	cubic to closest	1.51 (8) - 1.64 (12)
Na^+	8 - 6	cubic to octahedral	1.18 (8) - 1.02 (6)
Ca^{+2}	8 - 6		1.12 (8) - 1.00 (6)
Mn^{+2}	6	Octahedral	0.83
Fe^{+2}	6		0.78
Mg^{+2}	6		0.72
Fe^{+3}	6		0.65
Ti^{+4}	6		0.61
Al^{+3}	6		0.54
Al^{+3}	4	Tetrahedral	0.39
Si^{+4}	4		0.26
C^{+4}	3	Triangular	0.08

The radius ratio of Si^{+4} to O^{-2} requires that Si^{+4} be coordinated by 4 O^{-2} ions in tetrahedral coordination.

In order to neutralize the +4 charge on the Si cation, one negative charge from each of the Oxygen ions will reach the Si cation. Thus, each Oxygen will be left with a net charge of -1, resulting in a SiO_4^{-4} tetrahedral group that can be bonded to other cations. It is this SiO_4^{-4} tetrahedron that forms the basis of the silicate minerals.

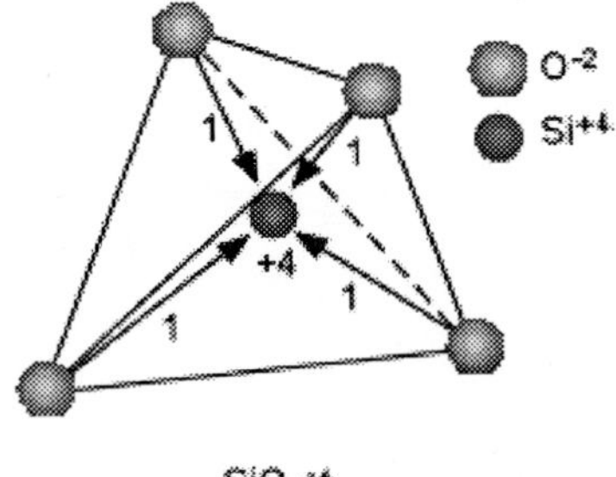

SiO_4^{-4}

Since Si^{+4} is a highly charged cation, Pauling's rules state that it should be separated a far as possible from other Si^{+4} ions. Thus, when these SiO_4^{-4} tetrahedrons are linked together, only corner oxygens will be shared with other SiO_4^{-4} groups. Several possibilities exist and give rise to the different silicate groups.

Nesosilicates (Island Silicates)

If the corner oxygens are not shared with other SiO_4^{-4} tetrahedrons, each tetrahedron will be isolated. Thus, this group is often referred to as the island silicate group. The basic structural unit is then SiO_4^{-4}. In this group the oxygens are shared with octahedral groups that contain other cations like Mg^{+2}, Fe^{+2}, or Ca^{+2}. Olivine is a good example: $(Mg,Fe)_2SiO_4$.

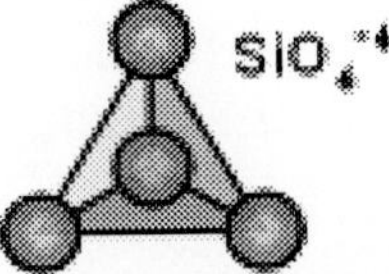

Sorosilicates (Double Island Silicates)

If one of the corner oxygens is shared with another tetrahedron, this gives rise to the sorosilicate group. It is often referred to as the double island group because there are two linked tetrahedrons isolated from all other tetrahedrons. In this case, the basic structural unit is $Si_2O_7^{-6}$. A good example of a sorosilicate is the mineral hemimorphite - $Zn_4Si_2O_7(OH){\cdot}H_2O$. Some sorosilicates are a combination of single and double islands, like in epidote - $Ca_2(Fe^{+3},Al)Al_2(SiO_4)(Si_2O_7)(OH)$.

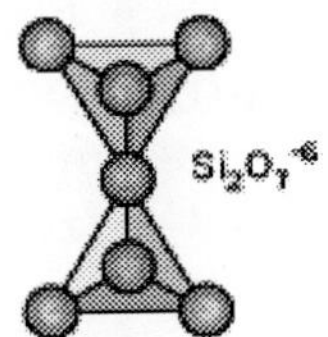

Cyclosilicates (Ring Silicates)

If two of the oxygens are shared and the structure is arranged in a ring, such as that shown here, we get the basic structural unit of the cyclosilcates or ring silicates. Shown here is a six membered ring forming the structural group $Si_6O_{18}^{-12}$. Three membered rings, $Si_3O_9^{-6}$, four membered rings, $Si_4O_{12}^{-8}$, and five membered rings $Si_5O_{15}^{-10}$ are also possible. A good example of a cyclosilicate is the mineral Beryl - $Be_3Al_2Si_6O_{18}$.

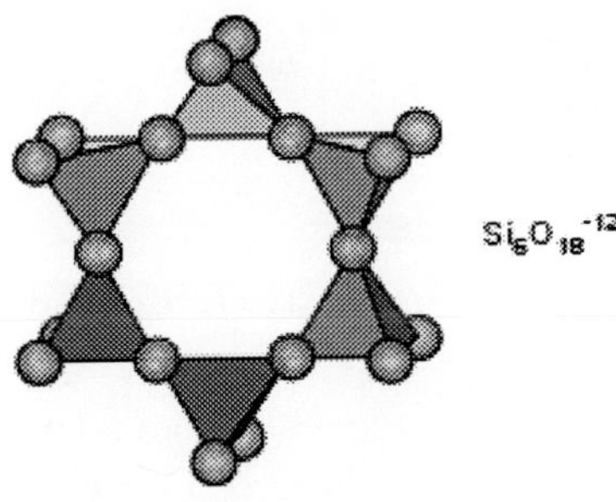

Inosilicates (Single Chain Silicates)

If two of the oxygens are shared in a way to make long single chains of linked SiO_4 tetrahedra, we get the single chain silicates or inosilicates. In this case the basic structural unit is $Si_2O_6^{-4}$ or SiO_3^{-2}. This group is the basis for the pyroxene group of minerals, like the orthopyroxenes $(Mg,Fe)SiO_3$ or the clinopyroxenes $Ca(Mg,Fe)Si_2O_6$.

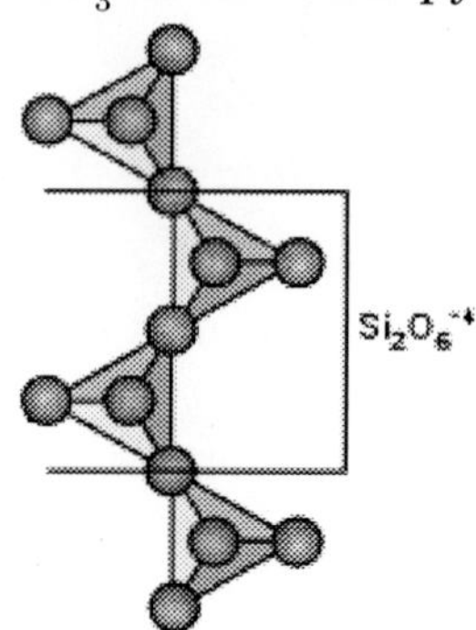

Inosilicates (Double Chain Silicates)

If two chains are linked together so that each tetrahedral group shares 3 of its oxygens, we can from double chains, with the basic structural group being $Si_4O_{11}^{-6}$. The amphibole group of minerals are double chain silicates, for example the tremolite - ferroactinolite series -$Ca_2(Mg,Fe)_5Si_8O_{22}(OH)_2$.

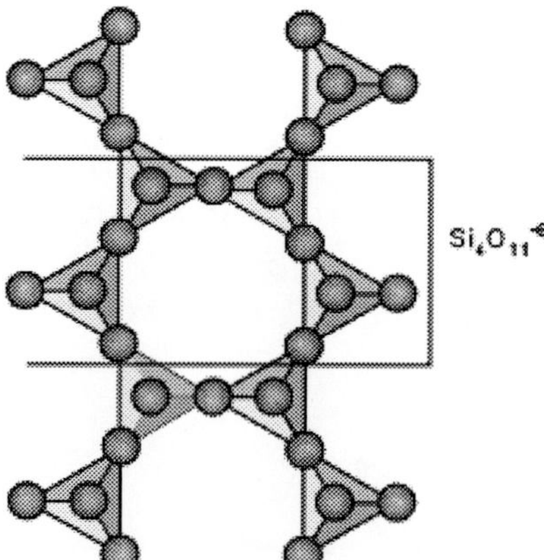

Phyllosilicates (Sheet Silicates)

If 3 of the oxygens from each tetrahedral group are shared such that an infinite sheet of SiO_4 tetrahedra are shared we get the basis for the phyllosilicates or sheet silicates. In this case the basic structural group is $Si_2O_5^{-2}$. The micas, clay minerals, chlorite, talc, and serpentine minerals are all based on this structure. A good example is biotite - $K(Mg,Fe)_3(AlSi_3)O_{10}(OH)_2$. Note that in this structure, Al is substituting for Si in one of the tetrahedral groups.

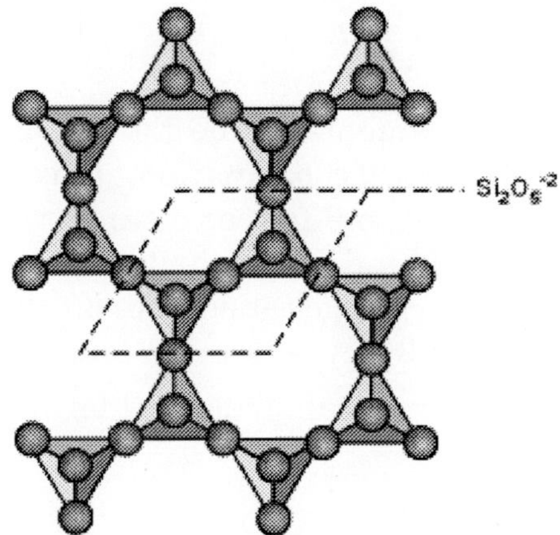

Tectosilicates (Framework Silicates)

If all of the corner oxygens are shared with another SiO_4 tetrahedron, then a framework structure develops. The basic structural group then becomes SiO_2. The minerals quartz, cristobalite, and tridymite all are based on this structure. If some of the Si^{+4} ions are replaced by Al^{+3} then this produces a charge imbalance and allows for other ions to be found coordinated in different arrangements within the framework structure. Thus, the feldspar and feldspathoid minerals are also based on the tectosilicate framework.

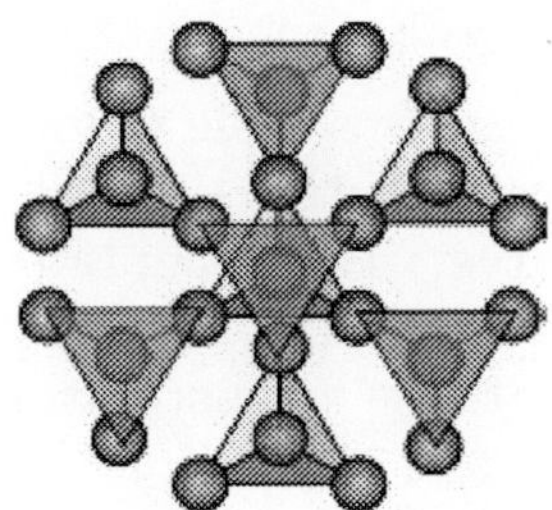

General Formula for Silicates

Based on these basic structural units, we can construct a general structural chemical formula for the silicates. But one substitution in particular tends to mess things up a bit. This is Al^{+3}, the third most abundant element in the Earth's crust. Al^{+3} has an ionic radius that varies between 0.54 and 0.39 depending on the coordination number.

Thus, it could either fit in 6-fold coordination with oxygen or 4-fold coordination with oxygen. Because Al^{+3} will go into 4-fold coordination with oxygen, it sometimes substitutes for Si^{+4}. If such a substitution takes place, it creates a charge imbalance that must be made up elsewhere in the silicate structure.

The other common elements in the Earth's crust that enter the silicates do so in other types of coordination. Ions like Al^{+3}, Mg^{+2}, Fe^{+2}, Fe^{+3}, Mn^{+2}, and Ti^{+4} enter into 6-fold or octahedral sites. Larger ions like Ca^{+2}, and Na^{+1}, are found in octahedral coordination or 8-fold, cubic coordination sites. Very large cations like K^{+1}, Ba^{+2}, and sometimes Na^{+1} are coordinated by 12 oxygens in 12-fold coordination sites.

We can thus write a general structural formula for the silicates as follows:

$$X_mY_n(Z_pO_q)W_r$$

where X represents an 8 to 12 fold coordination site for large cations like K^+, Rb^+, Ba^{+2}, Na^+, and Ca^{+2}.

Y represents a 6-fold (octahedral) site for intermediate sized cations like Al^{+3}, Mg^{+2}, Fe^{+2}, Fe^{+3}, Mn^{+2}, and Ti^{+4}. Z represents the tetrahedral site containing Si^{+4}, and Al^{+3}. the ratio p:q depends on the degree of polymerization of the silica (or alumina) tetrahedrons, or the silicate structural type as discussed above. O is oxygen, and W is a hyrdoxyl (OH^{-1}) site into which can substitute large anions like F^{-1} or Cl^{-1}.

The subscripts m, n, and r depend on the ratio of p to q and are chosen to maintain charge balance.

This is summarized in the table shown here. In this table note that there is very little substitution that takes place between ions that enter the X, Y, and Z sites. The exceptions are mainly substitution of Al^{+3} for Si^{+4}, which is noted in the Table, and whether the X site is large enough to accept the largest cations like K^{+1}, Ba^{+2}, or Rb^{+1}.

Processing of Ceramics

Ceramic engineering or processing is the science and technology of creating objects from inorganic, non-metallic materials. This is done either by the action of heat, or at lower temperatures using precipitation reactions from high purity chemical solutions. The term includes the purification of raw materials, the study and production of the chemical compounds concerned, their formation into components and the study of their structure, composition and properties.

Ceramic materials may have a crystalline or partly crystalline structure, with long-range order on atomic scale. Glass ceramics may have an amorphous or glassy structure, with limited or short-range atomic order. They are either formed from a molten mass that solidifies on cooling, formed and matured by the action of heat, or chemically synthesized at low temperatures using, for example, hydrothermal or sol-gel synthesis.

The special character of ceramic materials gives rise to many applications in materials engineering, electrical engineering, chemical engineering and mechanical engineering. As ceramics are heat resistant, they can be used for many tasks that materials like metal and polymers are unsuitable for. Ceramic materials are used in a wide range of industries, including mining, aerospace, medicine, refinery, food and chemical industries, packaging science, electronics, industrial and transmission electricity, and guided lightwave transmission.

History

The word "ceramic" is derived from the Greek word *keramikos* meaning pottery. It is related to the older Indo-European language root "to burn", "Ceramic" may be used as a noun in the singular to refer to a ceramic material or the product of ceramic manufacture, or as an adjective. The plural "ceramics" may be used to refer the making of things out of ceramic materials. Ceramic engineering, like many sciences, evolved from a different discipline by today's standards. Materials science engineering is grouped with ceramics engineering to this day.

Abraham Darby first used coke in 1709 in Shropshire, England, to improve the yield of a smelting process. Coke is now widely used to produce carbide ceramics. Potter Josiah Wedgwood opened the first modern ceramics factory in Stoke-on-Trent, England, in 1759. Austrian chemist Carl Josef Bayer, working for the textile industry in Russia, developed a process to separate alumina from bauxite ore in 1888. The Bayer process is still used to purify alumina for the ceramic and aluminum industries. Brothers Pierre and Jacques Curie discovered piezoelectricity in Rochelle salt circa 1880. Piezoelectricity is one of the key properties of electroceramics.

E.G. Acheson heated a mixture of coke and clay in 1893, and invented carborundum, or synthetic silicon carbide. Henri Moissan also synthesized SiC and tungsten carbide in his electric arc furnace in Paris about the same time as Acheson. Karl Schröter used liquid-phase sintering to bond or "cement" Moissan's tungsten carbide particles with cobalt in 1923 in Germany. Cemented (metal-bonded) carbide edges greatly increase the durability of hardened steel cutting tools. W.H. Nernst developed cubic-stabilized zirconia in the 1920s in Berlin. This material is used as an oxygen sensor in exhaust systems. The main limitation on the use of ceramics in engineering is brittleness.

Military

The military requirements of World War II encouraged developments, which created a need for high-performance materials and helped speed the development of ceramic science and engineering. Throughout the 1960s and 1970s, new types of ceramics were developed in response to advances in atomic energy, electronics, communications, and space travel. The discovery of ceramic superconductors in 1986 has spurred intense research to develop superconducting ceramic parts for electronic devices, electric motors, and transportation equipment.

There is an increasing need in the military sector for high-strength, robust materials which have the capability to transmit light around the visible (0.4–0.7 micrometers) and mid-infrared (1–5 micrometers) regions of the spectrum. These materials are needed for applications requiring transparent armor. Transparent armor is a material or system of materials designed to be optically transparent, yet protect from fragmentation or ballistic impacts.

The primary requirement for a transparent armor system is to not only defeat the designated threat but also provide a multi-hit capability with minimized distortion of surrounding areas. Transparent armor windows must also be compatible with night vision equipment. New materials that are thinner, lightweight, and offer better ballistic performance are being sought. Such solid-state components have found widespread use for various applications in the electro-optical field including: optical fibres for guided lightwave transmission, optical switches, laser amplifiers and lenses, hosts for solid-state lasers and optical window materials for gas lasers, and infrared (IR) heat seeking devices for missile guidance systems and IR night vision.

Education

India

- Indian Institute Of Technology, Banaras Hindu University, Varanasi

The founder of Banaras Hindu University, Pandit Madan Mohan Malviyaji instituted a course in Ceramic Technology as early as 1924 with the objective of advancing glass and ceramic technology in India. In 1956, Department of Glass Technology and Department of Ceramic Technology were merged to form the Department of Silicate Technology, offering a four-year degree course by injecting into its curriculum balanced engineering and scientific contents. In 1968 it was renamed Department of Ceramic Engineering. Presently this department offers B.Tech., M.Tech. and Ph.D. programmes in the areas of ceramic engineering and technology.

- College of Ceramic Technology, Calcutta

Government College of Ceramic Technology, GCT was established in the Year 1941, by Professor Sasadhar Ray, an eminent educationist of that time with the objective of imparting knowledge and training for developing skilled technical manpower for serving the Ceramic and Allied Industries. GCT is the oldest Ceramic College in India and one of the few colleges that has got the opportunity to be headed by people who have got international recognition. The college has reached its magnificence under the admirable guidance of Principals like Prof. Sasadhar Ray (founder of the college), a student of Prof. Satyen Bose (F.R.S.) and Prof. K. S. Krishnan (F.R.S.), and Prof. (Dr.) B. N. Sammadar, who did his research work at M. I. T. (USA) and worked together with Prof. (Dr.) W. D. Kingery (considered as Father of Modern Ceramics).

Czech Republic

- The Secondary Technical School Of Ceramics was founded in 1872 in Znojmo. In 1922 it moved to Karlovy Vary.
- The Ceramic Technical School at Bechyne was founded in 1884.

Japan – The Ceramic Society of Japan was founded in 1891 in Tokyo.

Germany

- The Ceramic Society Of Germany was founded in Berlin in 1919.
- Staatliche Fachschule fur Porzellan (Government Technical College for Porcelain) was founded in Selb in 1908. In 1973 it was transferred to Nuremberg Polytechnic, when it was incorporated into a professional training organisation for

ceramics which also includes the Staatliche Fachschule fur Keramtechnik and a college for block release courses in ceramic trades, testing and laboratory work.

France – the "Ecole Nationale Supérieure de Céramique Industrielle" (ENSCI) funded in Sèvres (Paris) in 1893 and transferred to Limoges in 1979 educate students in engineered and traditional ceramic, glass and cement fields. This school proposes exchange programmes with other European and International schools like Alfred University.

Poland – the Bunzlau Ceramic Technical College operated from 1887 to 1945.

Spain

- The 'Official Ceramic School' open in Madrid in 1911.
- The Ceramic School Of Manises – was founded in 1914.

United States – the first ceramic engineering course and department in the USA were established by Edward Orton, Jr., a professor of geology and mining engineering, at Ohio State University in 1894.

Orton and eight other refractory professionals founded the American Ceramic Society (ACerS) at the 1898 National Brick Manufacturers' Association convention in Pittsburgh. Orton was the first ACerS General Secretary, and his office at OSU served as the society headquarters in the beginning. Charles F. Binns established the New York State School of Clay-Working and Ceramics, now Alfred University, in 1900. Binns was the third ACerS president, and Orton the 32nd.

Modern Industry

Now a multi-billion dollar a year industry, ceramic engineering and research has established itself as an important field of science. Applications continue to expand as researchers develop new kinds of ceramics to serve different purposes.

- Zirconium dioxide ceramics are used in the manufacture of knives. The blade of the ceramic knife will stay sharp for much longer than that of a steel knife, although it is more brittle and can be snapped by dropping it on a hard surface.
- Ceramics such as alumina, boron carbide and silicon carbide have been used in bulletproof vests to repel large-caliber rifle fire. Such plates are known commonly as small-arms protective

inserts (SAPI). Similar material is used to protect cockpits of some military airplanes, because of the low weight of the material.

- Silicon nitride parts are used in ceramic ball bearings. Their higher hardness means that they are much less susceptible to wear and can offer more than triple lifetimes. They also deform less under load meaning they have less contact with the bearing retainer walls and can roll faster. In very high speed applications, heat from friction during rolling can cause problems for metal bearings; problems which are reduced by the use of ceramics. Ceramics are also more chemically resistant and can be used in wet environments where steel bearings would rust. The major drawback to using ceramics is a significantly higher cost. In many cases their electrically insulating properties may also be valuable in bearings.
- In the early 1980s, Toyota researched production of an adiabatic ceramic engine which can run at a temperature of over 6000 °F (3300 °C). Ceramic engines do not require a cooling system and hence allow a major weight reduction and therefore greater fuel efficiency. Fuel efficiency of the engine is also higher at high temperature, as shown by Carnot's theorem. In a conventional metallic engine, much of the energy released from the fuel must be dissipated as waste heat in order to prevent a meltdown of the metallic parts. Despite all of these desirable properties, such engines are not in production because the manufacturing of ceramic parts in the requisite precision and durability is difficult. Imperfection in the ceramic leads to cracks, which can lead to potentially dangerous equipment failure. Such engines are possible in laboratory settings, but mass-production is not feasible with current technology.
- Work is being done in developing ceramic parts for gas turbine engines. Currently, even blades made of advanced metal alloys used in the engines' hot section require cooling and careful limiting of operating temperatures. Turbine engines made with ceramics could operate more efficiently, giving aircraft greater range and payload for a set amount of fuel.
- Recently, there have been advances in ceramics which include bio-ceramics, such as dental implants and synthetic bones. Hydroxyapatite, the natural mineral component of bone, has been made synthetically from a number of biological and

chemical sources and can be formed into ceramic materials. Orthopedic implants made from these materials bond readily to bone and other tissues in the body without rejection or inflammatory reactions. Because of this, they are of great interest for gene delivery and tissue engineering scaffolds. Most hydroxyapatite ceramics are very porous and lack mechanical strength and are used to coat metal orthopedic devices to aid in forming a bond to bone or as bone fillers. They are also used as fillers for orthopedic plastic screws to aid in reducing the inflammation and increase absorption of these plastic materials. Work is being done to make strong, fully dense nano crystalline hydroxyapatite ceramic materials for orthopedic weight bearing devices, replacing foreign metal and plastic orthopedic materials with a synthetic, but naturally occurring, bone mineral. Ultimately these ceramic materials may be used as bone replacements or with the incorporation of protein collagens, synthetic bones.

- High-tech ceramic is used in watchmaking for producing watch cases. The material is valued by watchmakers for its light weight, scratch-resistance, durability and smooth touch. IWC is one of the brands that initiated the use of ceramic in watchmaking. The case of the IWC 2007 Top Gun edition of the Pilot's Watch Double chronograph is crafted in high-tech black ceramic.

Glass-ceramics

Figure: *A high strength glass-ceramic cooktop with negligible thermal expansion.*

Glass-ceramic materials share many properties with both glasses and ceramics. Glass-ceramics have an amorphous phase and one or more crystalline phases and are produced by a so-called "controlled crystallization", which is typically avoided in glass manufacturing. Glass-ceramics often contain a crystalline phase which constitutes anywhere from 30% [m/m] to 90% [m/m] of its composition by volume, yielding an array of materials with interesting thermomechanical properties.

In the processing of glass-ceramics, molten glass is cooled down gradually before reheating and annealing. In this heat treatment the glass partly crystallizes. In many cases, so-called 'nucleation agents' are added in order to regulate and control the crystallization process. Because there is usually no pressing and sintering, glass-ceramics do not contain the volume fraction of porosity typically present in sintered ceramics.

The term mainly refers to a mix of lithium and aluminosilicates which yields an array of materials with interesting thermomechanical properties. The most commercially important of these have the distinction of being impervious to thermal shock. Thus, glass-ceramics have become extremely useful for countertop cooking. The negative thermal expansion coefficient (TEC) of the crystalline ceramic phase can be balanced with the positive TEC of the glassy phase. At a certain point (~70% crystalline) the glass-ceramic has a net TEC near zero. This type of glass-ceramic exhibits excellent mechanical properties and can sustain repeated and quick temperature changes up to 1000 °C.

Processing Steps

The traditional ceramic process generally follows this sequence: Milling → Batching → Mixing → Forming → Drying → Firing → Assembly.

- Milling is the process by which materials are reduced from a large size to a smaller size. Milling may involve breaking up cemented material (in which case individual particles retain their shape) or pulverization (which involves grinding the particles themselves to a smaller size). Milling is generally done by mechanical means, including *attrition* (which is particle-to-particle collision that results in agglomerate break up or particle shearing), *compression* (which applies a forces that results in fracturing), and *impact* (which employs a milling medium or the particles themselves to cause fracturing).

Attrition milling equipment includes the wet scrubber (also called the planetary mill or wet attrition mill), which has paddles in water creating vortexes in which the material collides and break up. Compression mills include the jaw crusher, roller crusher and cone crusher. Impact mills include the ball mill, which has media that tumble and fracture the material. Shaft impactors cause particle-to particle attrition and compression.

- Batching is the process of weighing the oxides according to recipes, and preparing them for mixing and drying.
- Mixing occurs after batching and is performed with various machines, such as dry mixing ribbon mixers (a type of cement mixer), Mueller mixers, and pug mills. Wet mixing generally involves the same equipment.
- Forming is making the mixed material into shapes, ranging from toilet bowls to spark plug insulators. Forming can involve: (1) Extrusion, such as extruding "slugs" to make bricks, (2) Pressing to make shaped parts, (3) Slip casting, as in making toilet bowls, wash basins and ornamentals like ceramic statues. Forming produces a "green" part, ready for drying. Green parts are soft, pliable, and over time will lose shape. Handling the green product will change its shape. For example, a green brick can be "squeezed", and after squeezing it will stay that way.
- Drying is removing the water or binder from the formed material. Spray drying is widely used to prepare powder for pressing operations. Other dryers are tunnel dryers and periodic dryers. Controlled heat is applied in this two-stage process. First, heat removes water. This step needs careful control, as rapid heating causes cracks and surface defects. The dried part is smaller than the green part, and is brittle, necessitating careful handling, since a small impact will cause crumbling and breaking.
- Firing is where the dried parts pass through a controlled heating process, and the oxides are chemically changed to cause sintering and bonding. The fired part will be smaller than the dried part.

Forming Methods

Ceramic forming techniques include throwing, slipcasting, tape casting, injection molding, dry pressing, isostatic pressing, hot isostatic pressing (HIP) and others. Methods for forming ceramic powders into complex shapes are desirable in many areas of technology. Such methods are required for producing advanced, high-temperature structural parts such as heat engine components and turbines. Materials other than ceramics which are used in these processes may include: wood, metal, water, plaster and epoxy—most of which will be eliminated upon firing.

These forming techniques are well known for providing tools and other components with dimensional stability, surface quality, high (near theoretical) density and microstructural uniformity. The increasing use and diversity of speciality forms of ceramics adds to the diversity of process technologies to be used.

Thus, reinforcing fibres and filaments are mainly made by polymer, sol-gel, or CVD processes, but melt processing also has applicability. The most widely used speciality form is layered structures, with tape casting for electronic substrates and packages being preeminent. Photolithography is of increasing interest for precise patterning of conductors and other components for such packaging. Tape casting or forming processes are also of increasing interest for other applications, ranging from open structures such as fuel cells to ceramic composites.

The other major layer structure is coating, where melt spraying is very important, but chemical and physical vapor deposition and chemical (e.g., sol-gel and polymer pyrolysis) methods are all seeing increased use. Besides open structures from formed tape, extruded structures, such as honeycomb catalyst supports, and highly porous structures, including various foams, for example, reticulated foam, are of increasing use.

Densification of consolidated powder bodies continues to be achieved predominantly by (pressureless) sintering. However, the use of pressure sintering by hot pressing is increasing, especially for non-oxides and parts of simple shapes where higher quality (mainly microstructural homogeneity) is needed, and larger size or multiple parts per pressing can be an advantage.

The Sintering Process

The principles of sintering-based methods are simple ("sinter" has roots in the English "cinder"). The firing is done at a temperature below the melting point of the ceramic. Once a roughly-held-together object called a "green body" is made, it is baked in a kiln, where atomic and molecular diffusion processes give rise to significant changes in the primary microstructural features. This includes the gradual elimination of porosity, which is typically accompanied by a net shrinkage and overall densification of the component. Thus, the pores in the object may close up, resulting in a denser product of significantly greater strength and fracture toughness.

Another major change in the body during the firing or sintering process will be the establishment of the polycrystalline nature of the solid. This change will introduce some form of grain size distribution, which will have a significant impact on the ultimate physical properties of the material. The grain sizes will either be associated with the initial particle size, or possibly the sizes of aggregates or particle clusters which arise during the initial stages of processing.

The ultimate microstructure (and thus the physical properties) of the final product will be limited by and subject to the form of the structural template or precursor which is created in the initial stages of chemical synthesis and physical forming. Hence the importance of chemical powder and polymer processing as it pertains to the synthesis of industrial ceramics, glasses and glass-ceramics.

There are numerous possible refinements of the sintering process. Some of the most common involve pressing the green body to give the densification a head start and reduce the sintering time needed. Sometimes organic binders such as polyvinyl alcohol are added to hold the green body together; these burn out during the firing (at 200–350 °C). Sometimes organic lubricants are added during pressing to increase densification. It is common to combine these, and add binders and lubricants to a powder, then press. (The formulation of these organic chemical additives is an art in itself. This is particularly important in the manufacture of high performance ceramics such as those used by the billions for electronics, in capacitors, inductors, sensors, etc.)

A slurry can be used in place of a powder, and then cast into a desired shape, dried and then sintered. Indeed, traditional pottery is done with this type of method, using a plastic mixture worked with the hands. If a mixture of different materials is used together in a ceramic, the sintering temperature is sometimes above the melting point of one minor component – a *liquid phase* sintering. This results in shorter sintering times compared to solid state sintering.

Strength of Ceramics

A material's strength is dependent on its microstructure. The engineering processes to which a material is subjected can alter this microstructure. The variety of strengthening mechanisms that alter the strength of a material include the mechanism of grain boundary strengthening. Thus, although yield strength is maximized with decreasing grain size, ultimately, very small grain sizes make the material brittle. Considered in tandem with the fact that the yield strength is the parameter that predicts plastic deformation in the material, one can make informed decisions on how to increase the strength of a material depending on its microstructural properties and the desired end effect.

The relation between yield stress and grain size is described mathematically by the Hall-Petch equation which is

$$\sigma_y = \sigma_0 + \frac{k_y}{\sqrt{d}}$$

where k_y is the strengthening coefficient (a constant unique to each material), σ_o is a materials constant for the starting stress for dislocation movement (or the resistance of the lattice to dislocation motion), d is the grain diameter, and σ_y is the yield stress.

Theoretically, a material could be made infinitely strong if the grains are made infinitely small. This is, unfortunately, impossible because the lower limit of grain size is a single unit cell of the material. Even then, if the grains of a material are the size of a single unit cell, then the material is in fact amorphous, not crystalline, since there is no long range order, and dislocations can not be defined in an amorphous material. It has been observed experimentally that the microstructure with the highest yield strength is a grain size of about 10 nanometers, because grains smaller than this undergo another yielding mechanism, grain boundary sliding. Producing engineering materials with this ideal grain size is difficult because of the limitations of initial particle sizes inherent to nanomaterials and nanotechnology.

Theory of Chemical Processing

Microstructural Uniformity: In the processing of fine ceramics, the irregular particle sizes and shapes in a typical powder often lead to non-uniform packing morphologies that result in packing density variations in the powder compact. Uncontrolled agglomeration of powders due to attractive van der Waals forces can also give rise to in microstructural inhomogeneities.

Differential stresses that develop as a result of non-uniform drying shrinkage are directly related to the rate at which the solvent can be removed, and thus highly dependent upon the distribution of porosity. Such stresses have been associated with a plastic-to-brittle transition in consolidated bodies, and can yield to crack propagation in the unfired body if not relieved.

In addition, any fluctuations in packing density in the compact as it is prepared for the kiln are often amplified during the sintering process, yielding inhomogeneous densification. Some pores and other structural defects associated with density variations have been shown to play a detrimental role in the sintering process by growing and thus limiting end-point densities. Differential stresses arising from inhomogeneous densification have also been shown to result in the propagation of internal cracks, thus becoming the strength-controlling flaws.

It would therefore appear desirable to process a material in such a way that it is physically uniform with regard to the distribution of components and porosity, rather than using particle size distributions which will maximize the green density. The containment of a uniformly dispersed assembly of strongly interacting particles in suspension requires total control over particle-particle interactions. Monodisperse colloids provide this potential.

Monodisperse powders of colloidal silica, for example, may therefore be stabilized sufficiently to ensure a high degree of order in the colloidal crystal or polycrystalline colloidal solid which results from aggregation. The degree of order appears to be limited by the time and space allowed for longer-range correlations to be established.

Such defective polycrystalline colloidal structures would appear to be the basic elements of submicrometer colloidal materials science, and, therefore, provide the first step in developing a more rigorous understanding of the mechanisms involved in microstructural evolution in inorganic systems such as polycrystalline ceramics.

Self-assembly

Self-assembly is the most common term in use in the modern scientific community to describe the spontaneous aggregation of particles (atoms, molecules, colloids, micelles, etc.) without the influence of any external forces. Large groups of such particles are known to assemble themselves into thermodynamically stable, structurally well-defined arrays, quite reminiscent of one of the 7 crystal systems found in metallurgy and mineralogy (e.g. face-centred cubic, body-centred cubic, etc.). The fundamental difference in equilibrium structure is in the spatial scale of the unit cell (or lattice parameter) in each particular case. Thus, self-assembly is emerging as a new strategy in chemical synthesis and nanotechnology. Molecular self-assembly has been observed in various biological systems and underlies the formation of a wide variety of complex biological structures.

Molecular crystals, liquid crystals, colloids, micelles, emulsions, phase-separated polymers, thin films and self-assembled monolayers all represent examples of the types of highly ordered structures which are obtained using these techniques. The distinguishing feature of these methods is self-organization in the absence of any external forces.

In addition, the principal mechanical characteristics and structures of biological ceramics, polymer composites, elastomers, and cellular materials are being re-evaluated, with an emphasis on bioinspired materials and structures. Traditional approaches focus on design methods of biological materials using conventional synthetic materials. This includes an emerging class of mechanically superior biomaterials based on microstructural features and designs found in nature. The new horizons have been identified in the synthesis of bioinspired materials through processes that are characteristic of biological systems in nature. This includes the nanoscale self-assembly of the components and the development of hierarchical structures.

Ceramic Composites

Substantial interest has arisen in recent years in fabricating ceramic composites. While there is considerable interest in composites with one or more non-ceramic constituents, the greatest attention is on composites in which all constituents are ceramic. These typically comprise two ceramic constituents: a continuous matrix, and a dispersed phase of ceramic particles, whiskers, or short (chopped) or continuous ceramic fibres. The challenge, as in wet chemical processing, is to obtain a uniform or homogeneous distribution of the dispersed particle or fibre phase.

Consider first the processing of particulate composites. The particulate phase of greatest interest is tetragonal zirconia because of the toughening that can be achieved from the phase transformation from the metastable tetragonal to the monoclinic crystalline phase, aka transformation toughening. There is also substantial interest in dispersion of hard, non-oxide phases such as SiC, TiB, TiC, boron, carbon and especially oxide matrices like alumina and mullite. There is also interest too incorporating other ceramic particulates, especially those of highly anisotropic thermal expansion. Examples include Al_2O_3, TiO_2, graphite, and boron nitride.

Figure: *Silicon carbide single crystal*

In processing particulate composites, the issue is not only homogeneity of the size and spatial distribution of the dispersed and matrix phases, but also control of the matrix grain size. However, there is some built-in self-control due to inhibition of matrix grain growth by the dispersed phase.

Particulate composites, though generally offer increased resistance to damage, failure, or both, are still quite sensitive to inhomogeneities of composition as well as other processing defects such as pores. Thus they need good processing to be effective.

Particulate composites have been made on a commercial basis by simply mixing powders of the two constituents. Although this approach is inherently limited in the homogeneity that can be achieved, it is the most readily adaptable for existing ceramic production technology. However, other approaches are of interest.

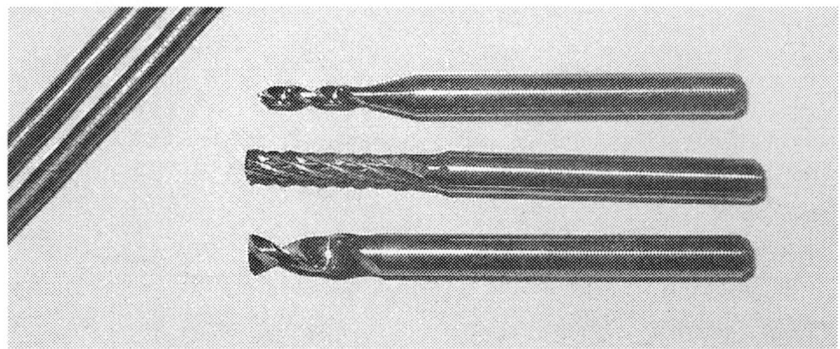

Figure: *Tungsten carbide milling bits*

From the technological standpoint, a particularly desirable approach to fabricating particulate composites is to coat the matrix or its precursor onto fine particles of the dispersed phase with good control of the starting dispersed particle size and the resultant matrix coating thickness.

One should in principle be able to achieve the ultimate in homogeneity of distribution and thereby optimize composite performance. This can also have other ramifications, such as allowing more useful composite performance to be achieved in a body having porosity, which might be desired for other factors, such as limiting thermal conductivity.

There are also some opportunities to utilize melt processing for fabrication of ceramic, particulate, whisker and short-fibre, and continuous-fibre composites. Clearly, both particulate and whisker composites are conceivable by solid-state precipitation after solidification of the melt. This can also be obtained in some cases by sintering, as for precipitation-toughened, partially stabilized zirconia. Similarly, it is known that one can directionally solidify ceramic eutectic mixtures and hence obtain uniaxially aligned fibre composites. Such composite processing has typically been limited to very simple shapes and thus suffers from serious economic problems due to high machining costs.

Clearly, there are possibilities of using melt casting for many of these approaches. Potentially even more desirable is using melt-derived particles. In this method, quenching is done in a solid solution or in a fine eutectic structure, in which the particles are then processed by more typical ceramic powder processing methods into a useful body. There have also been preliminary attempts to use melt spraying as a means of forming composites by introducing the dispersed particulate, whisker, or fibre phase in conjunction with the melt spraying process.

Other methods besides melt infiltration to manufacture ceramic composites with long fibre reinforcement are chemical vapor infiltration and the infiltration of fibre preforms with organic precursor, which after pyrolysis yield an amorphous ceramic matrix, initially with a low density. With repeated cycles of infiltration and pyrolysis one of those types of ceramic matrix composites is produced. Chemical vapor infiltration is used to manufacture carbon/carbon and silcon carbide reinforced with carbon or silicon carbide fibres.

Besides many process improvements, the first of two major needs for fibre composites is lower fibre costs. The second major need is fibre compositions or coatings, or composite processing, to reduce degradation that results from high-temperature composite exposure under oxidizing conditions.

Applications

The products of technical ceramics include tiles used in the Space Shuttle programme, gas burner nozzles, ballistic protection, nuclear fuel uranium oxide pellets, bio-medical implants, jet engine turbine blades, and missile nose cones. Its products are often made from materials other than clay, chosen for their particular physical properties. These may be classified as follows:

- Oxides: silica, alumina, zirconia
- Non-oxides: carbides, borides, nitrides, silicides
- Composites: particulate or whisker reinforced matrices, combinations of oxides and non-oxides (e.g. polymers).

Ceramics can be used in many technological industries. One application is the ceramic tiles on NASA's Space Shuttle, used to protect it and the future supersonic space planes from the searing heat of reentry into the Earth's atmosphere. They are also used widely in electronics and optics. In addition to the applications listed here, ceramics are also used as a coating in various engineering cases. An example would be a ceramic bearing coating over a titanium frame used for an airplane. Recently the field has come to include the studies of single crystals or glass fibres, in addition to traditional polycrystalline materials, and the applications of these have been overlapping and changing rapidly.

Aerospace

- Engines; Shielding a hot running airplane engine from damaging other components.
- Airframes; Used as a high-stress, high-temp and lightweight bearing and structural component.
- Missile nose-cones; Shielding the missile internals from heat.
- Space Shuttle tiles
- Space-debris ballistic shields – ceramic fibre woven shields offer better protection to hypervelocity (~7 km/s) particles than aluminum shields of equal weight.

- Rocket nozzles, withstands and focuses the exhaust of the rocket booster.
- Unmanned Air Vehicles; Implications of ceramic engine utilization in aeronautical applications (such as Unmanned Air Vehicles) may result in enhanced performance characteristics and less operational costs.

Biomedical

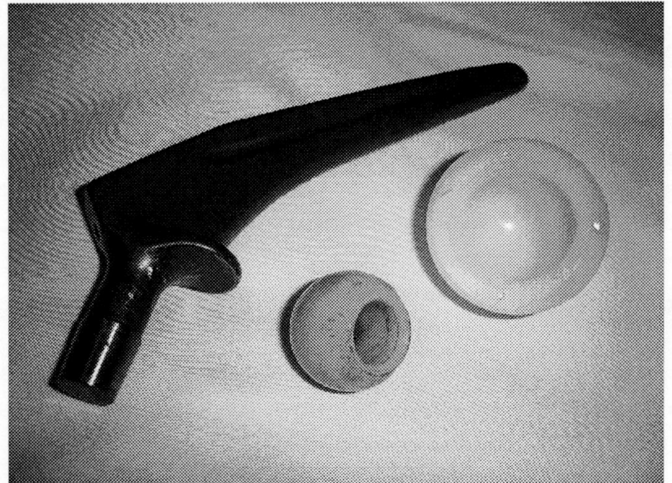

Figure: *A titanium hip prosthesis, with a ceramic head and polyethylene acetabular cup.*

- Artificial bone; Dentistry applications, teeth.
- Biodegradable splints; Reinforcing bones recovering from osteoporosis
- Implant material

Electronics

- Capacitors
- Integrated circuit packages
- Transducers
- Insulators

Optical

- Optical fibres, guided lightwave transmission
- Switches
- Laser amplifiers

- Lenses
- Infrared heat-seeking devices

Biomaterials

Silicification is quite common in the biological world and occurs in bacteria, single-celled organisms, plants, and animals (invertebrates and vertebrates). Crystalline minerals formed in such environment often show exceptional physical properties (e.g. strength, hardness, fracture toughness) and tend to form hierarchical structures that exhibit microstructural order over a range of length or spatial scales. The minerals are crystallized from an environment that is undersaturated with respect to silicon, and under conditions of neutral pH and low temperature (0–40 °C). Formation of the mineral may occur either within or outside of the cell wall of an organism, and specific biochemical reactions for mineral deposition exist that include lipids, proteins and carbohydrates.

Most natural (or biological) materials are complex composites whose mechanical properties are often outstanding, considering the weak constituents from which they are assembled. These complex structures, which have risen from hundreds of million years of evolution, are inspiring the design of novel materials with exceptional physical properties for high performance in adverse conditions. Their defining characteristics such as hierarchy, multifunctionality, and the capacity for self-healing, are currently being investigated.

The basic building blocks begin with the 20 amino acids and proceed to polypeptides, polysaccharides, and polypeptides–saccharides. These, in turn, compose the basic proteins, which are the primary constituents of the 'soft tissues' common to most biominerals. With well over 1000 proteins possible, current research emphasizes the use of collagen, chitin, keratin, and elastin. The 'hard' phases are often strengthened by crystalline minerals, which nucleate and grow in a biomediated environment that determines the size, shape and distribution of individual crystals. The most important mineral phases have been identified as hydroxyapatite, silica, and aragonite. Using the classification of Wegst and Ashby, the principal mechanical characteristics and structures of biological ceramics, polymer composites, elastomers, and cellular materials have been presented. Selected systems in each class are being investigated with emphasis on the relationship between their microstructure over a range of length scales and their mechanical response.

Thus, the crystallization of inorganic materials in nature generally occurs at ambient temperature and pressure. Yet the vital organisms through which these minerals form are capable of consistently producing extremely precise and complex structures. Understanding the processes in which living organisms control the growth of crystalline minerals such as silica could lead to significant advances in the field of materials science, and open the door to novel synthesis techniques for nanoscale composite materials, or nanocomposites.

High-resolution SEM observations were performed of the microstructure of the mother-of-pearl (or nacre) portion of the abalone shell. Those shells exhibit the highest mechanical strength and fracture toughness of any non-metallic substance known. The nacre from the shell of the abalone has become one of the more intensively studied biological structures in materials science. Clearly visible in these images are the neatly stacked (or ordered) mineral tiles separated by thin organic sheets along with a macrostructure of larger periodic growth bands which collectively form what scientists are currently referring to as a hierarchical composite structure. (The term hierarchy simply implies that there are a range of structural features which exist over a wide range of length scales).

Future developments reside in the synthesis of bio-inspired materials through processing methods and strategies that are characteristic of biological systems. These involve nanoscale self-assembly of the components and the development of hierarchical structures.

GLASSES

Glass-ceramics are polycrystalline materials produced through controlled crystallization of base glass. Glass-ceramic materials share many properties with both glasses and ceramics. Glass-ceramics have an amorphous phase and one or more crystalline phases and are produced by a so-called "controlled crystallization" in contrast to a spontaneous crystallization, which is usually not wanted in glass manufacturing. Glass-ceramics have the fabrication advantage of glass as well as special properties of ceramics. Glass-ceramics usually have between 30% [m/m] to 90% [m/m] crystallinity and yield an array of materials with interesting properties like zero porosity, high strength, toughness, translucency or opacity, pigmentation, opalescence, low or even negative thermal expansion, high temperature stability, fluorescence, machinability, ferromagnetism, resorbability or high chemical durability, biocompatibility, bio-activity, ion conductivity, superconductivity, isolation capabilities, low dielectric constant and loss, high resistivity and break down voltage. These properties can be tailored by controlling the base glass composition and by controlled heat treatment/crystallization of base glass.

Glass-ceramics are mostly produced in two steps: First, a glass is formed by a glass manufacturing process. The glass is cooled down and is then reheated in a second step. In this heat treatment the glass partly crystallizes. In most cases nucleation agents are added to the base composition of the glass-ceramic. These nucleation agents aid and control the crystallization process. Because there is usually no pressing and sintering, glass-ceramics have, unlike sintered ceramics, no pores.

A wide variety of glass-ceramic systems exists, e.g., the Li_2O x Al_2O_3 x $nSiO_2$-System (LAS-System), the MgO x Al_2O_3 x $nSiO_2$-System (MAS-System), the ZnO x Al_2O_3 x $nSiO_2$-System (ZAS-System).

LAS System

The commercially most important system is the Li_2O x Al_2O_3 x $nSiO_2$-System (LAS-System). The LAS-system mainly refers to a mix of lithium-, silicon-, and aluminum-oxides with additional components e.g., glass-phase forming agents such as Na_2O, K_2O and CaO and refining agents. As nucleation agents most commonly zirconium(IV)-oxide in combination with titanium(IV)-oxide is used. This important system was studied first and intensively by Hummel, and Smoke.

After crystallization the dominant crystal-phase in this type of glass-ceramic is a high-quartz solid solution (HQ s.s.). If the glass-ceramic is subjected to a more intense heat treatment, this HQ s.s. transforms into a keatite-solid solution (K s.s., sometimes wrongly named as beta-spodumene). This transition is non-reversible and reconstructive, which means bonds in the crystal-lattice are broken and new arranged. However, these two crystal phases show a very similar structure as Li could show.

The most interesting properties of these glass-ceramics are their thermomechanical properties. Glass-ceramic from the LAS-System is a mechanically strong material and can sustain repeated and quick temperature changes up to 800–1000 °C. The dominant crystalline phase of the LAS-glass-ceramics, HQ s.s., has a strong *negative* coefficient of thermal expansion (CTE), keatite-solid solution as still a negative CTE but much higher than HQ s.s.. These negative CTE's of the crystal-phase contrasts with the positive CTE of the residual glass. Adjusting the proportion of these phases offers a wide range of possible CTE's in the finished composite. Mostly for today's applications a low or even zero CTE is desired. Also a negative CTE is possible, which means, in contrast to most materials when heated up, such a glass-ceramic contracts. At a certain point, generally between 60% [m/m] and 80% [m/m] crystallinity, the two coefficients balance such that the glass-ceramic as a whole has a thermal expansion coefficient that is very close to zero. Also, when an interface between material will be subject to thermal fatigue, glass-ceramics can be adjusted to match the coefficient of the material they will be bonded to.

Originally developed for use in the mirrors and mirror mounts of astronomical telescopes, LAS-glass-ceramics have become known and entered the domestic market through its use in glass-ceramic cooktops, as well as cookware and bakeware or as high performance reflectors for digital projectors.

Cooktops

Glass-ceramic from the LAS-System is a mechanically strong material and can sustain repeated and quick temperature changes. However, it is not totally unbreakable. Because it is still a brittle material as glass and ceramics are, it can be broken. There have been instances where users reported damage to their cooktops when the surface was struck with a hard or blunt object (such as a can falling from above or other heavy items).

At the same time, it has a very low heat conduction coefficient and can be made nearly transparent (15–20% loss in a typical cooktop) for radiation in the infrared wavelengths.

In the visible range glass-ceramics can be transparent, translucent or opaque and even coloured by colouring agents.

Today, there are two major types of electrical stoves with cooktops made of glass-ceramic:

- A glass-ceramic stove uses radiant heating coils or infrared halogen lamps as the heating elements. The surface of the glass-ceramic cooktop above the burner heats up, but the adjacent surface remains cool because of the low heat conduction coefficient of the material.
- An induction stove heats a metal pot's bottom directly through electromagnetic induction.

It is interesting to note that this technology is not entirely new, as glass-ceramic ranges were first introduced in the 1970s using Corningware tops instead of the more durable material used today. These first generation smoothtops were problematic and could only be used with flat-bottomed cookware as the heating was primarily conductive rather than radiative.

Compared to conventional kitchen stoves, glass-ceramic cooktops are relatively simple to clean, due to their flat surface. However, glass-ceramic cooktops can be scratched very easily, so care must be taken not to slide the cooking pans over the surface. Food with a high sugar content (such as jam) should never be allowed to dry on the surface if it spills, otherwise damage will occur. Cleaning is best carried out by using a soft cloth along with a special glass-ceramic cleaner that applies a thin protective film on the glass.

For best results and maximum heat transfer, all cookware should be flat-bottomed and matched to the same size as the burner zone.

Brands and Manufacturers

Some well-known brands of glass-ceramics are *Ceran* (cooktops), *Eurokera* (cooktop, stoves and fireplaces), *Zerodur* (telescope mirrors), and *Macor*. German manufacturer Schott introduced Zerodur in 1968, Ceran followed in 1971. Nippon Electric Glass of Japan is another worldwide manufacturer of glass ceramics, whose related products in this area include Firelite and Neoceram fire-rated glass. Glass-ceramics manufactured in the Soviet Union/Russia are known under the name *Sitall*.

The same class of material was also used, until the late 1990s, in Corningware dishes, which could be taken from the freezer directly to the oven with no risk of thermal shock.

Composite Materials

Composite materials (also called composition materials or shortened to composites) are materials made from two or more constituent materials with significantly different physical or chemical properties, that when combined, produce a material with characteristics different from the individual components. The individual components remain separate and distinct within the finished structure. The new material may be preferred for many reasons: common examples include materials which are stronger, lighter or less expensive when compared to traditional materials.

Typical engineered composite materials include:

- Composite building materials such as cements, concrete
- Reinforced plastics such as fibre-reinforced polymer
- Metal Composites
- Ceramic Composites (composite ceramic and metal matrices)

Composite materials are generally used for buildings, bridges and structures such as boat hulls, swimming pool panels, race car bodies, shower stalls, bathtubs, and storage tanks, imitation granite and cultured marble sinks and countertops. The most advanced examples perform routinely on spacecraft in demanding environments.

History

The earliest man-made composite materials were straw and mud combined to form bricks for building construction. This ancient brick-making process was documented by Egyptian tomb paintings. Wattle and daub is one of the oldest man-made composite materials, at over 6000 years old.. Concrete is also a composite material, and is used more than any other man-made material in the world. As of 2006, about 7.5 billion cubic metres of concrete are made each year—more than one cubic metre for every person on Earth.

- Woody plants, both true wood from trees and such plants as palms and bamboo, yield natural composites that were used prehistorically by mankind and are still used widely in construction and scaffolding.
- Plywood 3400 B.C. by the Ancient Mesopotamians; gluing wood at different angles gives better properties than natural wood
- Cartonnage layers of linen or papyrus soaked in plaster dates to the First Intermediate Period of Egypt c. 2181-2055 BC and was used for death masks
- Concrete was described by Vitruvius, writing around 25 BC in his *Ten Books on Architecture*, distinguished types of aggregate appropriate for the preparation of lime mortars. For structural mortars, he recommended *pozzolana*, which were volcanic sands from the sandlike beds of Puteoli brownish-yellow-gray in colour near Naples and reddish-brown at Rome. Vitruvius specifies a ratio of 1 part lime to 3 parts pozzolana for cements used in buildings and a 1:2 ratio of lime to pulvis Puteolanus for underwater work, essentially the same ratio

mixed today for concrete used at sea. Natural cement-stones, after burning, produced cements used in concretes from post-Roman times into the 20th century, with some properties superior to manufactured Portland cement.

- Papier-mβché, a composite of paper and glue, has been used for hundreds of years
- The first artificial fibre reinforced plastic was bakelite which dates to 1907, although natural polymers such as shellac predate it

Examples

Materials: Concrete is the most common artificial composite material of all and typically consists of loose stones (aggregate) held with a matrix of cement. Concrete is a very robust material, much more robust than cement, and will not compress or shatter even under quite a large compressive force. However, concrete cannot survive tensile loading (i.e. if stretched it will quickly break apart). Therefore to give concrete the ability to resist being stretched, steel bars, which can resist high stretching forces, are often added to concrete to form reinforced concrete.

Fibre-reinforced polymers or FRPs include carbon-fibre reinforced plastic or CFRP, and glass-reinforced plastic or GRP. If classified by matrix then there are thermoplastic composites, short fibre thermoplastics, long fibre thermoplastics or long fibre-reinforced thermoplastics. There are numerous thermoset composites, but advanced systems usually incorporate aramid fibre and carbon fibre in an epoxy resin matrix.

Shape memory polymer composites are high-performance composites, formulated using fibre or fabric reinforcement and shape memory polymer resin as the matrix. Since a shape memory polymer resin is used as the matrix, these composites have the ability to be easily manipulated into various configurations when they are heated above their activation temperatures and will exhibit high strength and stiffness at lower temperatures. They can also be reheated and reshaped repeatedly without losing their material properties. These composites are ideal for applications such as lightweight, rigid, deployable structures; rapid manufacturing; and dynamic reinforcement. Composites can also use metal fibres reinforcing other metals, as in metal matrix composites (MMC) or ceramic matrix composites (CMC), which includes bone (hydroxyapatite reinforced with collagen fibres), cermet (ceramic and metal) and concrete. Ceramic matrix composites are built primarily for fracture toughness, not for strength. Organic matrix/ceramic aggregate composites include asphalt concrete, mastic asphalt, mastic roller hybrid, dental composite, syntactic foam and mother of pearl. Chobham armour is a special type of composite armour used in military applications.

Additionally, thermoplastic composite materials can be formulated with specific metal powders resulting in materials with a density range from 2 g/cm^3 to 11 g/cm^3 (same density as lead). The most common name for this type of material is "high gravity compound" (HGC), although "lead replacement" is also used. These materials can be used in place of traditional materials such as aluminium, stainless steel, brass, bronze, copper, lead, and even tungsten in weighting, balancing (for example, modifying the centre of gravity of a tennis racquet), vibration damping, and radiation shielding applications. High density composites are an economically viable option when certain materials are deemed hazardous and are banned (such as lead) or when secondary operations costs (such as machining, finishing, or coating) are a factor.

A sandwich-structured composite is a special class of composite material that is fabricated by attaching two thin but stiff skins to a lightweight but thick core. The core material is normally low strength material, but its higher thickness provides the sandwich composite with high bending stiffness with overall low density.

Wood is a naturally occurring composite comprising cellulose fibres in a lignin and hemicellulose matrix. Engineered wood includes a wide variety of different products such as wood fibre board, plywood, oriented strand board, wood plastic composite (recycled wood fibre in polyethylene matrix), Pykrete (sawdust in ice matrix), Plastic-impregnated or laminated paper or textiles, Arborite, Formica (plastic) and Micarta. Other engineered laminate composites, such as Mallite, use a central core of end grain balsa wood, bonded to surface skins of light alloy or GRP. These generate low-weight, high rigidity materials.

Products

Fibre-reinforced composite materials have gained popularity (despite their generally high cost) in high-performance products that need to be lightweight, yet strong enough to take harsh loading conditions such as aerospace components (tails, wings, fuselages, propellers), boat and scull hulls, bicycle frames, swimming pool panels and racing car bodies. Other uses include fishing rods, storage tanks, and baseball bats. The new Boeing 787 structure including the wings and fuselage is composed largely of composites. Composite materials are also becoming more common in the realm of orthopedic surgery.

Carbon composite is a key material in today's launch vehicles and heat shields for the re-entry phase of spacecraft. It is widely used in solar panel substrates, antenna reflectors and yokes of spacecraft. It is also used in payload adapters, inter-stage structures and heat shields of launch vehicles. Furthermore disk brake systems of airplanes and racing cars are using carbon/carbon material, and the composite material with carbon fibres and silicon carbide matrix has been introduced in luxury vehicles and sports cars.

In 2006, a fibre-reinforced composite pool panel was introduced for in-ground swimming pools, residential as well as commercial, as a non-corrosive alternative to galvanized steel.

In 2007, an all-composite military Humvee was introduced by TPI Composites Inc and Armor Holdings Inc, the first all-composite military vehicle. By using composites the vehicle is lighter, allowing higher payloads. In 2008, carbon fibre and DuPont Kevlar (five times stronger than steel) were combined with enhanced thermoset resins to make military transit cases by ECS Composites creating 30-percent lighter cases with high strength.

Pipes and fittings for various purpose like transportation of potable water, fire-fighting, irrigation, seawater, desalinated water, chemical and industrial waste, and sewage are now manufactured in glass reinforced plastics.

Overview

Composites are made up of individual materials referred to as constituent materials. There are two main categories of constituent materials: matrix and reinforcement. At least one portion of each type is required.

The matrix material surrounds and supports the reinforcement materials by maintaining their relative positions. The reinforcements impart their special mechanical and physical properties to enhance the matrix properties. A synergism produces material properties unavailable from the individual constituent materials, while the wide variety of matrix and strengthening materials allows the designer of the product or structure to choose an optimum combination.

Engineered composite materials must be formed to shape. The matrix material can be introduced to the reinforcement before or after the reinforcement material is placed into the mould cavity or onto the mould surface. The matrix material experiences a melding event, after which the part shape is essentially set. Depending upon the nature of the matrix material, this melding event can occur in various ways such as chemical polymerization or solidification from the melted state.

A variety of moulding methods can be used according to the end-item design requirements. The principal factors impacting the methodology are the natures of the chosen matrix and reinforcement materials. Another important factor is the gross quantity of material to be produced. Large quantities can be used to justify high capital expenditures for rapid and automated manufacturing technology. Small production quantities are accommodated with lower capital expenditures but higher labour and tooling costs at a correspondingly slower rate.

Many commercially produced composites use a polymer matrix material often called a resin solution. There are many different polymers available depending upon the starting raw ingredients. There are several broad categories, each with numerous variations. The most common are known as polyester, vinyl ester, epoxy, phenolic, polyimide, polyamide, polypropylene, PEEK, and others. The reinforcement materials are often fibres but also commonly ground minerals. The various methods described below have been developed to reduce the resin content of the final product, or the fibre content is increased. As a rule of thumb, lay up results in a product containing 60% resin and 40% fibre, whereas vacuum infusion gives a final product with 40% resin and 60% fibre content. The strength of the product is greatly dependent on this ratio.

Martin Hubbe and Lucian A Lucia consider Wood to be a natural composite of cellulose fibres in a matrix of lignin.

Constituents

Matrices

Resins: Typically, most common polymer-based composite materials, including fibreglass, carbon fibre, and Kevlar, include at least two parts, the substrate and the resin. Polyester resin tends to have yellowish tint, and is suitable for most backyard projects. Its weaknesses are that it is UV sensitive and can tend to degrade over time, and thus generally is also coated to help preserve it. It is often used in the making of surfboards and for marine applications. Its hardener is a peroxide, often MEKP (methyl ethyl ketone peroxide). When the peroxide is mixed with the resin, it decomposes to generate free radicals, which initiate the curing reaction. Hardeners in these systems are commonly called catalysts, but since they do not re-appear unchanged at the end of the reaction, they do not fit the strictest chemical definition of a catalyst.

Vinylester resin tends to have a purplish to bluish to greenish tint. This resin has lower viscosity than polyester resin, and is more transparent. This resin is often billed as being fuel resistant, but will melt in contact with gasoline. This resin tends to be more resistant over time to degradation than polyester resin, and is more flexible. It uses the same hardeners as polyester resin (at a similar mix ratio) and the cost is approximately the same.

Epoxy resin is almost totally transparent when cured. In the aerospace industry, epoxy is used as a structural matrix material or as a structural glue.

Shape memory polymer (SMP) resins have varying visual characteristics depending on their formulation. These resins may be epoxy-based, which can be used for auto body and outdoor equipment repairs; cyanate-ester-based, which are used in space applications; and acrylate-based, which can be used in very cold temperature applications, such as for sensors that indicate whether perishable goods have warmed above a certain maximum temperature.

These resins are unique in that their shape can be repeatedly changed by heating above their glass transition temperature (T_g). When heated, they become flexible and elastic, allowing for easy configuration.

Once they are cooled, they will maintain their new shape. The resins will return to their original shapes when they are reheated above their T_g. The advantage of shape memory polymer resins is that they can be shaped and reshaped repeatedly without losing their material properties. These resins can be used in fabricating shape memory composites.

Other Matrices

Common matrices include mud (wattle and daub), cement (concrete), polymers (fibre reinforced plastics), metals and ceramics.

Road surfaces are often made from asphalt concrete which uses bitumen as a matrix. Unusual matrices such as ice are sometime proposed as in pykecrete.

Reinforcement

Fibre:

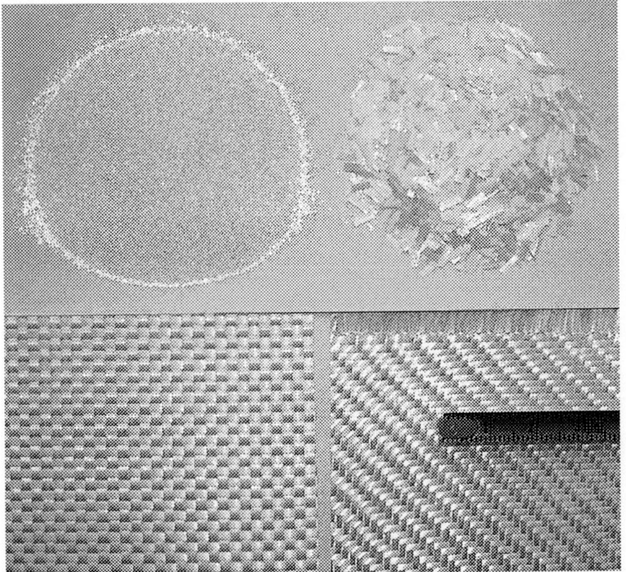

Figure: *Differences in the way the fibres are laid out give different strengths and ease of manufacture*

Reinforcement usually adds rigidity and greatly impedes crack propagation. Thin fibres can have very high strength, and provided they are mechanically well attached to the matrix they can greatly improve the composite's overall properties.

Fibre-reinforced composite materials can be divided into two main categories normally referred to as short fibre-reinforced materials and continuous fibre-reinforced materials. Continuous reinforced materials will often constitute a layered or laminated structure. The woven and continuous fibre styles are typically available in a variety of forms, being pre-impregnated with the given matrix (resin), dry, uni-directional tapes of various widths, plain weave, harness satins, braided, and stitched. The short and long fibres are typically employed in compression moulding and sheet moulding operations. These come in the form of flakes, chips, and random mate (which can also be made from a continuous fibre laid in random fashion until the desired thickness of the ply / laminate is achieved).

Common fibres used for reinforcement include glass fibres, carbon fibres, cellulose (wood/paper fibre and straw) and high strength polymers for example aramid.

Other Reinforcement

Concrete uses aggregate, and reinforced concrete additionally uses steel bars (rebar) to tension the concrete. Steel mesh or wires are also used in some glass and plastic products.

Cores

Many composite layup designs also include a co-curing or post-curing of the prepreg with various other mediums, such as honeycomb or foam. This is commonly called a sandwich structure. This is a more common layup process for the manufacture of radomes, doors, cowlings, or non-structural parts.

Open- and closed-cell-structured foams like polyvinylchloride, polyurethane, polyethylene or polystyrene foams, balsa wood, syntactic foams, and honeycombs are commonly used core materials. Open- and closed-cell metal foam can also be used as core materials.

Fabrication Methods

Fabrication usually involves wetting or mixing or saturating the reinforcement with the matrix, and then causing the matrix to bind together (with heat or a chemical reaction) into a rigid structure. The operation is usually done in an open or closed forming mold, but the order and ways of introducing the ingredients varies considerably.

Mold Overview

Within a mold, the reinforcing and matrix materials are combined, compacted, and cured (processed) to undergo a melding event. After the melding event, the part shape is essentially set, although it can deform under certain process conditions. For a thermoset polymeric matrix material, the melding event is a curing reaction that is initiated by the application of additional heat or chemical reactivity such as an organic peroxide. For a thermoplastic polymeric matrix material, the melding event is a solidification from the melted state. For a metal matrix material such as titanium foil, the melding event is a fusing at high pressure and a temperature near the melting point.

For many moulding methods, it is convenient to refer to one mould piece as a "lower" mould and another mould piece as an "upper" mould. Lower and upper refer to the different faces of the moulded panel, not the mould's configuration in space. In this convention, there is always a lower mould, and sometimes an upper mould. Part construction begins by applying materials to the lower mould. Lower mould and upper mould are more generalized descriptors than more common and specific terms such as male side, female side, a-side, b-side, tool side, bowl, hat, mandrel, etc. Continuous manufacturing processes use a different nomenclature.

The moulded product is often referred to as a panel. For certain geometries and material combinations, it can be referred to as a casting. For certain continuous processes, it can be referred to as a profile.

Vacuum Bag Moulding

Vacuum bag moulding uses a flexible film to enclose the part and seal it from outside air. A vacuum is then drawn on the vacuum bag and atmospheric pressure compresses the part during the cure process. Vacuum bag material is available in a tube shape or a sheet of material. When a tube shaped bag is used, the entire part can be enclosed within the bag. When using sheet bagging materials, the edges of the vacuum bag are sealed against the edges of the mould surface to enclose the part against an air-tight mould. When bagged in this way, the lower mold is a rigid structure and the upper surface of the part is formed by the flexible membrane vacuum bag. The flexible membrane can be a reusable silicone material or an extruded polymer film. After sealing the part inside the vacuum bag, a vacuum is drawn on the part (and held) during cure. This process can be performed at either ambient or elevated temperature with ambient atmospheric pressure acting upon the vacuum bag. A vacuum pump is typically used to draw a vacuum. An economical method of drawing a vacuum is with a venturi vacuum and air compressor.

A vacuum bag is a bag made of strong rubber-coated fabric or a polymer film used to compress the part during a cure or hardening process. In some applications the bag encloses the entire material, or in other applications a mold is used to form one face of the laminate with the bag being a single layer to seal to the outer edge of the mold face. When using a tube shaped bag, the ends of the bag are sealed and the air is drawn out of the bag through a nipple using a vacuum pump. As a result, uniform pressure approaching one atmosphere is applied to the surfaces of the object inside the bag, holding parts together while the adhesive cures. The entire bag may be placed in a temperature-controlled oven, oil bath or water bath and gently heated to accelerate curing. Vacuum bagging is widely used in the composites industry as well. Carbon fibre fabric and fibreglass, along with resins and epoxies are common materials laminated together with a vacuum bag operation.

Woodworking Applications

In commercial woodworking facilities, vacuum bags are used to laminate curved and irregular shaped workpieces.

Typically, polyurethane or vinyl materials are used to make the bag. A tube shaped bag is open at both ends. The piece, or pieces to be glued are placed into the bag and the ends sealed. One method of sealing the open ends of the bag is by placing a clamp on each end of the bag. A plastic rod is laid across the end of the bag, the bag is then folded over the rod. A plastic sleeve with an opening in it, is then snapped over the rod. This procedure forms a seal at both ends of the bag, when the vacuum is ready to be drawn. A "platen" is sometimes used inside the bag for the piece being glued to lay on. The platen has a series of small slots cut into it, to allow the air under it to be evacuated. The platen must have rounded edges and corners to prevent the vacuum from tearing the bag. When a curved part is to be glued in a vacuum bag, it is important that the pieces being glued be placed over a solidly built form, or have an air bladder placed under the form. This air bladder has access to "free air" outside the bag. It is used to create an equal pressure under the form, preventing it from being crushed.

Pressure Bag Moulding

This process is related to vacuum bag moulding in exactly the same way as it sounds. A solid female mould is used along with a flexible male mould. The reinforcement is placed inside the female mould with just enough resin to allow the fabric to stick in place (wet lay up). A measured amount of resin is then liberally brushed indiscriminately into the mould and the mould is then clamped to a machine that contains the male flexible mould. The flexible male membrane is then inflated with heated compressed air or possibly steam. The female mould can also be heated. Excess resin is forced out along with trapped air. This process is extensively used in the production of composite helmets due to the lower cost of unskilled labour. Cycle times for a helmet bag moulding machine vary from 20 to 45 minutes, but the finished shells require no further curing if the moulds are heated.

Autoclave Moulding

A process using a two-sided mould set that forms both surfaces of the panel. On the lower side is a rigid mould and on the upper side is a flexible membrane made from silicone or an extruded polymer film such as nylon. Reinforcement materials can be placed manually or robotically. They include continuous fibre forms fashioned into textile constructions. Most often, they are pre-impregnated with the resin in the form of prepreg fabrics or unidirectional tapes. In some instances, a resin film is placed upon the lower mould and dry reinforcement is placed above.

The upper mould is installed and vacuum is applied to the mould cavity. The assembly is placed into an autoclave. This process is generally performed at both elevated pressure and elevated temperature. The use of elevated pressure facilitates a high fibre volume fraction and low void content for maximum structural efficiency.

Resin Transfer Moulding (RTM)

RTM is a process using a rigid two-sided mould set that forms both surfaces of the panel. The mould is typically constructed from aluminum or steel, but composite molds are sometimes used. The two sides fit together to produce a mould cavity. The distinguishing feature of resin transfer moulding is that the reinforcement materials are placed into this cavity and the mould set is closed prior to the introduction of matrix material.

Resin transfer moulding includes numerous varieties which differ in the mechanics of how the resin is introduced to the reinforcement in the mould cavity. These variations include everything from the RTM methods used in out of autoclave composite manufacturing for high-tech aerospace components to vacuum infusion to vacuum assisted resin transfer moulding (VARTM). This process can be performed at either ambient or elevated temperature.

Other Fabrication Methods

Other types of fabrication include press moulding, transfer moulding, pultrusion moulding, filament winding, casting, centrifugal casting, continuous casting and slip forming. There are also forming capabilities including CNC filament winding, vacuum infusion, wet lay-up, compression moulding, and thermoplastic moulding, to name a few. The use of curing ovens and paint booths is also needed for some projects.

Finishing Methods

The finishing of the composite parts is also critical in the final design. Many of these finishes will include rain-erosion coatings or polyurethane coatings.

Tooling

The mold and mold inserts are referred to as "tooling." The mold/tooling can be constructed from a variety of materials. Tooling materials include invar, steel, aluminium, reinforced silicone rubber, nickel, and carbon fibre. Selection of the tooling material is typically based on, but not limited to, the coefficient of thermal expansion, expected number of cycles, end item tolerance, desired or required surface condition, method of cure, glass transition temperature of the material being moulded, moulding method, matrix, cost and a variety of other considerations.

Physical Properties

The physical properties of composite materials are generally not isotropic (independent of direction of applied force) in nature, but rather are typically anisotropic (different depending on the direction of the applied force or load). For instance, the stiffness of a composite panel will often depend upon the orientation of the applied forces and/or moments.

Panel stiffness is also dependent on the design of the panel. For instance, the fibre reinforcement and matrix used, the method of panel build, thermoset versus thermoplastic, type of weave, and orientation of fibre axis to the primary force.

In contrast, isotropic materials (for example, aluminium or steel), in standard wrought forms, typically have the same stiffness regardless of the directional orientation of the applied forces and/or moments.

The relationship between forces/moments and strains/curvatures for an isotropic material can be described with the following material properties: Young's Modulus, the shear Modulus and the Poisson's ratio, in relatively simple mathematical relationships. For the anisotropic material, it requires the mathematics of a second order tensor and up to 21 material property constants. For the special case of orthogonal isotropy, there are three different material property constants for each of Young's Modulus, Shear Modulus and Poisson's ratio—a total of 9 constants to describe the relationship between forces/moments and strains/curvatures.

Techniques that take advantage of the anisotropic properties of the materials include mortise and tenon joints (in natural composites such as wood) and Pi Joints in synthetic composites.

Failure

Shock, impact, or repeated cyclic stresses can cause the laminate to separate at the interface between two layers, a condition known as delamination. Individual fibres can separate from the matrix e.g. fibre pull-out. Composites can fail on the microscopic or macroscopic scale. Compression failures can occur at both the macro scale or at each individual reinforcing fibre in compression buckling. Tension failures can be net section failures of the part or degradation of the composite at a microscopic scale where one or more of the layers in the composite fail in tension of the matrix or failure of the bond between the matrix and fibres.

Some composites are brittle and have little reserve strength beyond the initial onset of failure while others may have large deformations and have reserve energy absorbing capacity past the onset of damage. The variations in fibres and matrices that are available and the mixtures that can be made with blends leave a very broad range of properties that can be designed into a composite structure. The best known failure of a brittle ceramic matrix composite occurred when the carbon-carbon composite tile on the leading edge of the wing of the Space Shuttle Columbia fractured when impacted during take-off. It led to catastrophic break-up of the vehicle when it re-entered the Earth's atmosphere on 1 February 2003. Compared to metals, composites have relatively poor bearing strength.

Testing

To aid in predicting and preventing failures, composites are tested before and after construction. Pre-construction testing may use finite element analysis (FEA) for ply-by-ply analysis of curved surfaces and predicting wrinkling, crimping and dimpling of composites. Materials may be tested after construction through several nondestructive methods including ultrasonics, thermography, shearography and X-ray radiography

Concrete Material

Concrete is a composite material composed of coarse granular material (the aggregate or filler) embedded in a hard matrix of material (the cement or binder) that fills the space among the aggregate particles and glues them together. Concrete is widely used for making architectural structures, foundations, brick/block walls, pavements, bridges/overpasses, highways, runways, parking structures, dams, pools/reservoirs, pipes, footings for gates, fences and poles and even boats.

Famous concrete structures include the Burj Khalifa (world's tallest building), Hoover Dam, the Panama Canal and the Roman Pantheon.

Concrete technology was known by the Ancient Romans and was widely used within the Roman Empire—the Colosseum is largely built of concrete and the concrete dome of the Pantheon is the world's largest. After the Empire passed, use of concrete became scarce until the technology was re-pioneered in the mid-18th century.

History

The word concrete comes from the Latin word "*concretus*" (meaning compact or condensed), the perfect passive participle of "*concrescere*", from "*con-*" (together) and "*crescere*" (to grow).

Perhaps the earliest known occurrence of cement was twelve million years ago, when a natural deposit formed after an occurrence of oil shale naturally combusted while adjacent to a bed of limestone. These ancient deposits were investigated in the 1960s and 1970s. On a human time-scale, lime mortars were used in Greece, Crete, and Cyprus in 800 BC, and the Assyrian Jerwan Aqueduct (688 BC) made use of fully waterproof concrete. The Romans used concrete extensively from 300 BC to 476 AD, a span of more than seven hundred years.

Concrete was used for Construction in Many Ancient Structures.

During the Roman Empire, Roman concrete (or *opus caementicium*) was made from quicklime, pozzolana and an aggregate of pumice. Its widespread use in many Roman structures, a key event in the history of architecture termed the Roman Architectural Revolution, freed Roman construction from the restrictions of stone and brick material and allowed for revolutionary new designs in terms of both structural complexity and dimension.

Concrete, as the Romans knew it, was a new and revolutionary material. Laid in the shape of arches, vaults and domes, it quickly hardened into a rigid mass, free from many of the internal thrusts and strains that troubled the builders of similar structures in stone or brick. Modern tests show that *opus caementicium* had as much compressive strength as modern Portland-cement concrete (ca. 200 kg/cm^2). However, due to the absence of reinforcement, its tensile strength was far lower than modern reinforced concrete, and its mode of application was also different:

Modern structural concrete differs from Roman concrete in two important details. First, its mix consistency is fluid and homogeneous, allowing it to be poured into forms rather than requiring hand-layering together with the placement of aggregate, which, in Roman practice, often consisted of rubble. Second, integral reinforcing steel gives modern concrete assemblies great strength in tension, whereas Roman concrete could depend only upon the strength of the concrete bonding to resist tension.

The widespread use of concrete in many Roman structures has ensured that many survive to the present day. The Baths of Caracalla in Rome are just one example. Many Roman aqueducts and bridges have masonry cladding on a concrete core, as does the dome of the Pantheon.

After the Roman Empire, the use of burning lime and pozzolana was greatly reduced until the technique was all but forgotten between 500 AD and the 1300s. Between the 1300s until the mid-1700s, the use of cement gradually returned. The *Canal du Midi* was built using concrete in 1670, and there are concrete structures in Finland that date from the 16th century.

Perhaps the greatest driver behind the modern usage of concrete was the third Eddystone Lighthouse in Devon, England. To create this structure, between 1756 and 1793, British engineer John Smeaton pioneered the use of hydraulic lime in concrete, using pebbles and powdered brick as aggregate.

A method for producing Portland cement was patented by Joseph Aspdin on 1824. In 1889 the first concrete reinforced bridge was built, and the first large concrete dams were built in 1936, Hoover Dam and Grand Coulee Dam. Reinforced concrete was invented in 1849 by Joseph Monier.

Ancient Additives

Concrete additives have been used since Roman and Egyptian times, when it was discovered that adding volcanic ash to the mix allowed it to set underwater. Similarly, the Romans knew that adding horse hair made concrete less liable to crack while it hardened, and adding blood made it more frost-resistant.

Modern Additives

In modern times, researchers have experimented with the addition of other materials to create concrete with improved properties, such as higher strength or electrical conductivity.

Impact of Modern Concrete Use

The usage of concrete, world wide, is twice as much as steel, wood, plastics, and aluminum combined. Concrete's use in the modern world is only exceeded by the usage of naturally occurring water.

Concrete is also the basis of a large commercial industry, with all the positives and negatives that entails. In the United States alone, concrete production is a $30 billion per year industry, considering only the value of the ready-mixed concrete sold each year. Given the size of the concrete industry, and the fundamental way concrete is used to shape the infrastructure of the modern world, it is difficult to overstate the role this material plays today.

Composition of Concrete

There are many types of concrete available, created by varying the proportions of the main ingredients below. In this way or by substitution for the cementitious and aggregate phases, the finished product can be tailored to its application with varying strength, density, or chemical and thermal resistance properties.

"Aggregate" consists of large chunks of material in a concrete mix, generally a coarse gravel or crushed rocks such as limestone, or granite, along with finer materials such as sand.

"Cement", commonly Portland cement, and other cementitious materials such as fly ash and slag cement, serve as a binder for the aggregate.

Water is then mixed with this dry composite, which produces a semi-liquid that workers can shape (typically by pouring it into a form). The concrete solidifies and hardens to rock-hard strength through a chemical process called hydration. The water reacts with the cement, which bonds the other components together, creating a robust stone-like material.

"Chemical admixtures" are added to achieve varied properties. These ingredients may speed or slow down the rate at which the concrete hardens, and impart many other useful properties. "Reinforcements" are often added to concrete. Concrete can be formulated with high compressive strength, but always has lower tensile strength. For this reason it is usually reinforced with materials that are strong in tension (often steel). Concrete can be damaged by many processes, such as the freezing of trapped water.

"Mineral admixtures" are becoming more popular in recent decades. The use of recycled materials as concrete ingredients has been gaining popularity because of increasingly stringent environmental legislation, and the discovery that such materials often have complimentary and valuable properties. The most conspicuous of these are fly ash, a by-product of coal-fired power plants, and silica fume, a byproduct of industrial electric arc furnaces. The use of these materials in concrete reduces the amount of resources required as the ash and fume acts as a cement replacement. This displaces some cement production, an energetically expensive and environmentally problematic process, while reducing the amount of industrial waste that must be disposed of.

The *mix design* depends on the type of structure being built, how the concrete is mixed and delivered, and how it is placed to form the structure.

Cement

Portland cement is the most common type of cement in general usage. It is a basic ingredient of concrete, mortar and plaster. English masonry worker Joseph Aspdin patented Portland cement in 1824. It was named because of the similarity of its colour to Portland limestone, quarried from the English Isle of Portland and used extensively in London architecture. It consists of a mixture of oxides of calcium, silicon and aluminium. Portland cement and similar materials are made by heating limestone (a source of calcium) with clay and grinding this product (called *clinker*) with a source of sulphate (most commonly gypsum).

In modern cement kilns many advanced features are used to lower the fuel consumption per ton of clinker produced. Cement kilns are extremely large, complex, and inherently dirty industrial installations, and have many undesirable emissions. Of the various ingredients used in concrete the cement is the most energetically expensive. Even complex and efficient kilns require 3.3 to 3.6 gigajoules of energy to produce a ton of clinker and then grind it into cement. Many kilns can be fuelled with difficult to dispose of wastes, the most common being used tires. The extremely high temperatures and long periods of time at those temperatures allows cement kilns to efficiently and completely burn even difficult to use fuels. In recent years, alternatives have been developed to help replace cement. Products such as PLC (Portland Limestone Cement), which incorporate limestone into the mix, are being tested. This is due to cement production being one of the largest producers of global greenhouse gas emissions (about 5 to 10%).

Water

Combining water with a cementitious material forms a cement paste by the process of hydration. The cement paste glues the aggregate together, fills voids within it, and makes it flow more freely.

A lower water to concrete ratio yields a stronger, more durable concrete, while more water gives a freer-flowing concrete with a higher slump. Impure water used to make concrete can cause problems when setting or in causing premature failure of the structure. Hydration involves many different reactions, often occurring at the same time. As the reactions proceed, the products of the cement hydration process gradually bond together the individual sand and gravel particles and other components of the concrete, to form a solid mass.

Reaction

Cement chemist notation: $C_3S + H \rightarrow C\text{-}S\text{-}H + CH$

Standard notation: $Ca_3SiO_5 + H_2O \rightarrow (CaO)\cdot(SiO_2)\cdot(H_2O)(gel) + Ca(OH)_2$

Balanced: $2Ca_3SiO_5 + 7H_2O \rightarrow 3(CaO)\cdot 2(SiO_2)\cdot 4(H_2O)(gel) + 3Ca(OH)_2$

Aggregates

Fine and coarse aggregates make up the bulk of a concrete mixture. Sand, natural gravel and crushed stone are used mainly for this purpose. Recycled aggregates (from construction, demolition and excavation waste) are increasingly used as partial replacements of natural aggregates, while a number of manufactured aggregates, including air-cooled blast furnace slag and bottom ash are also permitted.

The presence of aggregate greatly increases the durability of concrete above that of cement, which is a brittle material in its pure state. Thus concrete is a true composite material. Redistribution of aggregates after compaction often creates inhomogeneity due to the influence of vibration. This can lead to strength gradients. Decorative stones such as quartzite, small river stones or crushed glass are sometimes added to the surface of concrete for a decorative "exposed aggregate" finish, popular among landscape designers. In addition to being decorative, exposed aggregate adds robustness to a concrete driveway.

Reinforcement

Concrete is strong in compression, as the aggregate efficiently carries the compression load. However, it is weak in tension as the cement holding the aggregate in place can crack, allowing the structure to fail. Reinforced concrete adds either steel reinforcing bars, steel fibres, glass fibre, or plastic fibre to carry tensile loads.

Mineral Admixtures and Blended Cements

There are inorganic materials that also have pozzolanic or latent hydraulic properties. These very fine-grained materials are added to the concrete mix to improve the properties of concrete (mineral admixtures), or as a replacement for Portland cement (blended cements).

- Fly ash: A by-product of coal-fired electric generating plants, it is used to partially replace Portland cement (by up to 60% by mass). The properties of fly ash depend on the type of coal burnt. In general, siliceous fly ash is pozzolanic, while calcareous fly ash has latent hydraulic properties.
- Ground granulated blast furnace slag (GGBFS or GGBS): A by-product of steel production is used to partially replace Portland cement (by up to 80% by mass). It has latent hydraulic properties.
- Silica fume: A by-product of the production of silicon and ferrosilicon alloys. Silica fume is similar to fly ash, but has a particle size 100 times smaller. This results in a higher surface to volume ratio and a much faster pozzolanic reaction. Silica fume is used to increase strength and durability of concrete, but generally requires the use of superplasticizers for workability.
- High reactivity Metakaolin (HRM): Metakaolin produces concrete with strength and durability similar to concrete made with silica fume. While silica fume is usually dark gray or black in colour, high-reactivity metakaolin is usually bright white in colour, making it the preferred choice for architectural concrete where appearance is important.

Concrete Production

Concrete production is the process of mixing together the various ingredients—water, aggregate, cement, and any additives—to produce concrete. Concrete production is time-sensitive. Once the ingredients are mixed, workers must put the concrete in place before it hardens.

There is a wide variety of equipment for processing concrete—from hand tools to heavy industrial machinery. Whichever equipment builders use however, the objective is to produce the desired building material—ingredients must be properly mixed, placed, shaped, and retained within time constraints. Once the mix is where it should be, the curing process must be controlled to ensure the concrete attains desired attributes. During concrete preparation, various technical details may affect the quality and nature of the product.

When initially mixed, Portland cement and water rapidly form a gel of tangled chains of interlocking crystals, and components of the gel continue to react over time. Initially the gel is fluid, which improves workability and aids in placement of the material, but as the concrete sets, the chains of crystals join into a rigid structure, counteracting the fluidity of the gel and fixing the particles of aggregate in place. During curing, the cement continues to react with the residual water in a process of hydration. In properly formulated concrete, once this curing process has terminated the product has the desired physical and chemical properties. Among the qualities typically desired, are mechanical strength, low moisture permeability, and chemical and volumetric stability.

Mixing Concrete

Thorough mixing is essential for the production of uniform, high quality concrete. For this reason equipment and methods should be capable of effectively mixing concrete materials containing the largest specified aggregate to produce *uniform mixtures* of the lowest slump practical for the work.

Separate paste mixing has shown that the mixing of cement and water into a paste before combining these materials with aggregates can increase the compressive strength of the resulting concrete. The paste is generally mixed in a *high-speed*, shear-type mixer at a w/cm (water to cement ratio) of 0.30 to 0.45 by mass. The cement paste premix may include admixtures such as accelerators or retarders, superplasticizers, pigments, or silica fume. The premixed paste is then blended with aggregates and any remaining batch water and final mixing is completed in conventional concrete mixing equipment. High-energy mixed (HEM) concrete is produced by means of high-speed mixing of cement, water and sand with net specific energy consumption of at least 5 kilojoules per kilogram of the mix. A plasticizer or a superplasticizer is then added to the activated mixture, which can later be mixed with aggregates in a conventional concrete mixer. In this process, sand provides dissipation of energy and creates high-shear conditions on the surface of cement particles. This results in the full volume of water interacting with cement. The liquid activated mixture can be used by itself or foamed (expanded) for lightweight concrete. HEM concrete hardens in low and subzero temperature conditions and possesses an increased volume of gel, which drastically reduces capillarity in solid and porous materials.

Workability

Workability is the ability of a fresh (plastic) concrete mix to fill the form/mold properly with the desired work (vibration) and without reducing the concrete's quality. Workability depends on water content, aggregate (shape and size distribution), cementitious content and age (level of hydration) and can be modified by adding chemical admixtures, like superplasticizer. Raising the water content or adding chemical admixtures increases concrete workability. Excessive water leads to increased bleeding (surface water) and/or segregation of aggregates (when the cement and aggregates start to separate), with the resulting concrete having reduced quality. The use of an aggregate with an undesirable gradation can result in a very harsh mix design with a very low slump, which cannot be readily made more workable by addition of reasonable amounts of water.

Workability can be measured by the concrete slump test, a simplistic measure of the plasticity of a fresh batch of concrete following the ASTM C 143 or EN 12350-2 test standards. Slump is normally measured by filling an "Abrams cone" with a sample from a fresh batch of concrete. The cone is placed with the wide end down onto a level, non-absorptive surface. It is then filled in three layers of equal volume, with each layer being tamped with a steel rod to consolidate the layer. When the cone is carefully lifted off, the enclosed material slumps a certain amount due to gravity. A relatively dry sample slumps very little, having a slump value of one or two inches (25 or 50 mm) out of one foot (305 mm). A relatively wet concrete sample may slump as much as eight inches. Workability can also be measured by using the flow table test.

Slump can be increased by addition of chemical admixtures such as plasticizer or superplasticizer without changing the water-cement ratio. Some other admixtures, especially air-entraining admixture, can increase the slump of a mix.

High-flow concrete, like self-consolidating concrete, is tested by other flow-measuring methods. One of these methods includes placing the cone on the narrow end and observing how the mix flows through the cone while it is gradually lifted.

After mixing, concrete is a fluid and can be pumped to the location where needed.

Curing

In all but the least critical applications, care must be taken to properly *cure* concrete, to achieve best strength and hardness. This happens after the concrete has been placed. Cement requires a moist, controlled environment to gain strength and harden fully. The cement paste hardens over time, initially setting and becoming rigid though very weak and gaining in strength in the weeks following. In around 4 weeks, typically over 90% of the final strength is reached, though strengthening may continue for decades. The conversion of calcium hydroxide in the concrete into calcium carbonate from absorption of CO_2 over several decades further strengthen the concrete and make it more resilient to damage. However, this reaction, called carbonation, lowers the pH of the cement pore solution and can cause the reinforcement bars to corrode.

Hydration and hardening of concrete during the first three days is critical. Abnormally fast drying and shrinkage due to factors such as evaporation from wind during placement may lead to increased tensile stresses at a time when it has not yet gained sufficient strength, resulting in greater shrinkage cracking. The early strength of the concrete can be increased if it is kept damp during the curing process. Minimizing stress prior to curing minimizes cracking. High-early-strength concrete is designed to hydrate faster, often by increased use of cement that increases shrinkage and cracking. The strength of concrete changes (increases) for up to three years. It depends on cross-section dimension of elements and conditions of structure exploitation.

During this period concrete must be kept under controlled temperature and humid atmosphere. In practice, this is achieved by spraying or ponding the concrete surface with water, thereby protecting the concrete mass from ill effects of ambient conditions. The pictures to the right show two of many ways to achieve this, ponding – submerging setting concrete in water and wrapping in plastic to contain the water in the mix. Additional common curing methods include wet burlap and/or plastic sheeting covering the fresh concrete, or by spraying on a water-impermeable temporary curing membrane.

Properly curing concrete leads to increased strength and lower permeability and avoids cracking where the surface dries out prematurely. Care must also be taken to avoid freezing or overheating due to the exothermic setting of cement. Improper curing can cause scaling, reduced strength, poor abrasion resistance and cracking.

Chapter 4

Properties of Concrete

Concrete has relatively high compressive strength, but significantly lower tensile strength, and as such is usually reinforced with materials that are strong in tension (often steel). The elasticity of concrete is relatively constant at low stress levels but starts decreasing at higher stress levels as matrix cracking develops. Concrete has a very low coefficient of thermal expansion, and as it matures concrete shrinks. All concrete structures will crack to some extent, due to shrinkage and tension. Concrete which is subjected to long-duration forces is prone to creep. Tests can be made to ensure the properties of concrete correspond to specifications for the application.

The density of concrete varies, but is around 2,400 kg/m^3 (150 pounds per cubic foot or 4,050 lb/yd^3).

As a result, without compensating, concrete would almost always fail from tensile stresses – even when loaded in compression. The practical implication of this is that concrete elements subjected to tensile stresses must be reinforced with materials that are strong in tension. Reinforced concrete is the most common form of concrete. The reinforcement is often steel, rebar (mesh, spiral, bars and other forms). Structural fibres of various materials are available.

Concrete can also be prestressed (reducing tensile stress) using internal steel cables (tendons), allowing for beams or slabs with a longer span than is practical with reinforced concrete alone. Inspection of concrete structures can be non-destructive if carried out with equipment such as a Schmidt hammer, which is used to estimate concrete strength.

The ultimate strength of concrete is influenced by the water-cementitious ratio *(w/cm)*, the design constituents, and the mixing, placement and curing methods employed. All things being equal, concrete with a lower water-cement (cementitious) ratio makes a stronger concrete than that with a higher ratio. The total quantity of cementitious materials (portland cement, slag cement, pozzolans) can affect strength, water demand, shrinkage, abrasion resistance and density. All concrete will crack independent of whether or not it has sufficient compressive strength.

In fact, high Portland cement content mixtures can actually crack more readily due to increased hydration rate. As concrete transforms from its plastic state, hydrating to a solid, the material undergoes shrinkage. Plastic shrinkage cracks can occur soon after placement but if the evaporation rate is high they often can actually occur during finishing operations, for example in hot weather or a breezy day. In very high-strength concrete mixtures (greater than 70 MPa) the crushing strength of the aggregate can be a limiting factor to the ultimate compressive strength. In lean concretes (with a high water-cement ratio) the crushing strength of the aggregates is not so significant.

The internal forces in common shapes of structure, such as arches, vaults, columns and walls are predominantly compressive forces, with floors and pavements subjected to tensile forces. Compressive strength is widely used for specification requirement and quality control of concrete. The engineer knows his target tensile (flexural) requirements and will express these in terms of compressive strength.

Wired.com reported on April 13, 2007 that a team from the University of Tehran, competing in a contest sponsored by the American Concrete Institute, demonstrated several blocks of concretes with abnormally high compressive strengths between 340 and 410 MPa (49,000 and 59,000 psi) at 28 days. The blocks appeared to use an aggregate of steel fibres and quartz – a mineral with a compressive strength of 1100 MPa, much higher than typical high-strength aggregates such as granite (100–140 MPa or 15,000–20,000 psi).

Reactive Powder Concrete, also known as Ultra-High Performance Concrete, can be even stronger, with strengths of up to 800 MPa (116,000 PSI). These are made by eliminating large aggregate completely, carefully controlling the size of the fine aggregates to ensure the best possible packing, and incorporating steel fibres (sometimes produced by grinding steel wool) into the matrix. Reactive

Powder Concretes may also make use of silica fume as a fine aggregate. Commercial Reactive Powder Concretes are available in the 170–210 MPa (25,000–30,000 psi) strength range.

Elasticity

The modulus of elasticity of concrete is a function of the modulus of elasticity of the aggregates and the cement matrix and their relative proportions. The modulus of elasticity of concrete is relatively constant at low stress levels but starts decreasing at higher stress levels as matrix cracking develops. The elastic modulus of the hardened paste may be in the order of 10-30 GPa and aggregates about 45 to 85 GPa. The concrete composite is then in the range of 30 to 50 GPa.

The American Concrete Institute allows the modulus of elasticity to be calculated using the following equation:

$$E_c = 33 w_c^{1.5} \sqrt{f_c'} \text{ (psi)}$$

where

w_c = weight of concrete (pounds per cubic foot) and where

$$90 \frac{\text{lb}}{\text{ft}^3} \leq w_c \leq 155 \frac{\text{lb}}{\text{ft}^3}$$

f_c' = compressive strength of concrete at 28 days (psi)

This equation is completely empirical and is not based on theory. Note that the value of E_c found is in units of psi. For normalweight concrete (defined as concrete with a w_c of 150 lb/ft^3 and subtracting 5 lb/ft^3 for steel) E_c is permitted to be taken as $57000\sqrt{f_c'}$.

Expansion and Shrinkage

Concrete has a very low coefficient of thermal expansion. However, if no provision is made for expansion, very large forces can be created, causing cracks in parts of the structure not capable of withstanding the force or the repeated cycles of expansion and contraction. The coefficient of thermal expansion of Portland cement concrete is 0.000008 to 0.000012 (per degree Celsius) (8 to 12 microstrains/°C)(8-12 1/MK).

As concrete matures it continues to shrink, due to the ongoing reaction taking place in the material, although the rate of shrinkage falls relatively quickly and keeps reducing over time (for all practical purposes concrete is usually considered to not shrink due to hydration any further after 30 years). The relative shrinkage and expansion of concrete and brickwork require careful accommodation when the two forms of construction interface. Because concrete is continuously shrinking for years after it is initially placed, it is generally accepted that under thermal loading it will never expand to its originally placed volume.

Due to its low thermal conductivity, a layer of concrete is frequently used for fireproofing of steel structures.

Cracking

All concrete structures will crack to some extent. One of the early designers of reinforced concrete, Robert Maillart, employed reinforced concrete in a number of arched bridges. His first bridge was simple, using a large volume of concrete. He then realized that much of the concrete was very cracked, and could not be a part of the structure under compressive loads, yet the structure clearly worked. His later designs simply removed the cracked areas, leaving slender, beautiful concrete arches. The Salginatobel Bridge is an example of this.

Concrete cracks due to tensile stress induced by shrinkage or stresses occurring during setting or use. Various means are used to overcome this. Fibre reinforced concrete uses fine fibres distributed throughout the mix or larger metal or other reinforcement elements to limit the size and extent of cracks.

In many large structures joints or concealed saw-cuts are placed in the concrete as it sets to make the inevitable cracks occur where they can be managed and out of sight. Water tanks and highways are examples of structures requiring crack control.

Shrinkage Cracking

Shrinkage cracks occur when concrete members undergo restrained volumetric changes (shrinkage) as a result of either drying, autogenous shrinkage or thermal effects. Restraint is provided either externally (i.e. supports, walls, and other boundary conditions) or internally (differential drying shrinkage, reinforcement). Once the tensile strength of the concrete is exceeded, a crack will develop.

The number and width of shrinkage cracks that develop are influenced by the amount of shrinkage that occurs, the amount of restraint present and the amount and spacing of reinforcement provided.These are minor indications and have no real structural impact on the concrete member. Plastic-shrinkage cracks are immediately apparent, visible within 0 to 2 days of placement, while drying-shrinkage cracks develop over time. Autogenous shrinkage also occurs when the concrete is quite young and results from the volume reduction resulting from the chemical reaction of the Portland cement.

Tension Cracking

Concrete members may be put into tension by applied loads. This is most common in concrete beams where a transversely applied load will put one surface into compression and the opposite surface into tension due to induced bending. The portion of the beam that is in tension may crack. The size and length of cracks is dependent on the magnitude of the bending moment and the design of the reinforcing in the beam at the point under consideration. Reinforced concrete beams are designed to crack in tension rather than in compression. This is achieved by providing reinforcing steel which yields before failure of the concrete in compression occurs and allowing remediation, repair, or if necessary, evacuation of an unsafe area.

Creep

Creep is the permanent movement or deformation of a material in order to relieve stresses within the material. Concrete that is subjected to long-duration forces is prone to creep. Short-duration forces (such as wind or earthquakes) do not cause creep. Creep can sometimes reduce the amount of cracking that occurs in a concrete structure or element, but it also must be controlled. The amount of primary and secondary reinforcing in concrete structures contributes to a reduction in the amount of shrinkage, creep and cracking.

Concrete Testing

Engineers usually specify the required compressive strength of concrete, which is normally given as the 28 day compressive strength in megapascals (MPa) or pounds per square inch (psi). Twenty eight days is a long wait to determine if desired strengths are going to be obtained, so three-day and seven-day strengths can be useful to predict the ultimate 28-day compressive strength of the concrete. A 25% strength gain between 7 and 28 days is often observed with 100% OPC (ordinary Portland cement) mixtures, and up to 40% strength gain can be realized with the inclusion of pozzolans and supplementary cementitious materials (SCMs) such as fly ash and/or slag cement. Strength gain depends on the type of mixture, its constituents, the use of standard curing, proper testing and care of cylinders in transport, etc. It is imperative to accurately test the fundamental properties of concrete in its fresh, plastic state. Concrete is typically sampled while being placed, with testing protocols requiring that test samples be cured under laboratory conditions (standard cured). Additional samples may be field cured (non-standard) for the purpose of early 'stripping' strengths, that is, form removal, evaluation of curing, etc. but the standard cured cylinders comprise acceptance criteria. Concrete tests can measure the "plastic" (unhydrated) properties of concrete prior to, and during placement. As these properties affect the hardened compressive strength and durability of concrete (resistance to freeze-thaw), the properties of workability (slump/flow), temperature, density and age are monitored to ensure the production and placement of 'quality' concrete. Tests are performed per ASTM International, European Committee for Standardization or Canadian Standards Association. As measurement of quality must represent the potential of concrete material delivered and placed, it is imperative that concrete technicians performing concrete tests are certified to do so according to these standards. Structural design, material design and properties are often specified in accordance with national/regional design codes such as American Concrete Institute.

Compressive strength tests are conducted using an instrumented hydraulic ram to compress a standard cylindrical or cubic sample to failure. Tensile strength tests are conducted either by three-point bending of a prismatic beam specimen or by compression along the sides of a standard cylindrical specimen. These are not to be equated with nondestructive testing using a rebound hammer or probe systems which are hand-held indicators, for relative strength of the top few millimeters, of comparative concretes in the field.

Testing Fraud in New York City

American Standard Testing and Consulting Laboratories, company president Alan Fortich, and five other executives admitted filing false documents on "thousands" of New York construction projects including Yankee Stadium, the Second Avenue subway, and Memorial Sloan-Kettering Cancer Centre over 10 years. ASTCL had replaced Testwell Inc., another firm indicted for faking concrete tests, in 2008.

Concrete Degradation

Concrete can be damaged by many processes, such as the expansion of corrosion products of the steel reinforcement bars, freezing of trapped water, fire or radiant heat, aggregate expansion, sea water effects, bacterial corrosion, leaching, erosion by fast-flowing water, physical damage and chemical damage (from carbonatation, chlorides, sulphates and distillate water).

Environmental and Health

The manufacture and use of concrete produce a wide range of environmental and social consequences. Some are harmful, some welcome, and some both, depending on circumstances. A major component of concrete is cement, which similarly exerts environmental and social effects.

The cement industry is one of the three primary producers of carbon dioxide, a major greenhouse gas. The other two are the energy production and transportation industries.

Concrete is used to create hard surfaces that contribute to surface runoff, which can cause heavy soil erosion, water pollution, and flooding, but conversely can be used to divert, dam, and control flooding. Concrete is a primary contributor to the urban heat island effect, though less so than asphalt.

Workers who cut, grind or polish concrete are at risk of inhaling airborne silica, which can lead to silicosis. Concrete dust released by building demolition and natural disasters can be a major source of dangerous air pollution.

The presence of some substances in concrete, including useful and unwanted additives, can cause health concerns due to toxicity and radioactivity. Wet concrete is highly alkaline and must be handled with proper protective equipment.

Concrete Recycling

Concrete recycling is an increasingly common method of disposing of concrete structures. Concrete debris was once routinely shipped to landfills for disposal, but recycling is increasing due to improved environmental awareness, governmental laws and economic benefits.

Concrete, which must be free of trash, wood, paper and other such materials, is collected from demolition sites and put through a crushing machine, often along with asphalt, bricks and rocks.

Reinforced concrete contains rebar and other metallic reinforcements, which are removed with magnets and recycled elsewhere. The remaining aggregate chunks are sorted by size. Larger chunks may go through the crusher again. Smaller pieces of concrete are used as gravel for new construction projects. Aggregate base gravel is laid down as the lowest layer in a road, with fresh concrete or asphalt placed over it. Crushed recycled concrete can sometimes be used as the dry aggregate for brand new concrete if it is free of contaminants, though the use of recycled concrete limits strength and is not allowed in many jurisdictions. On 3 March 1983, a government funded research team (the VIRL research.codep) approximated that almost 17% of worldwide landfill was by-products of concrete based waste.

Use of Concrete in Infrastructure

Mass Concrete Structures: These large structures typically include gravity dams, such as the Hoover Dam, the Itaipu Dam and the Three Gorges Dam, arch dams, navigation locks and large breakwaters. Such large structures, even though individually placed in formed horizontal blocks, generate excessive heat and associated expansion; to mitigate these effects post-cooling is commonly provided in the design. An early example at Hoover Dam, installed a network of pipes between vertical concrete placements to circulate cooling water during the curing process to avoid damaging overheating. Similar systems are still used; depending on volume of the pour, the concrete mix used, and ambient air temperature, the cooling process may last for many months after the concrete is placed. Various methods also are used to pre-cool the concrete mix in mass concrete structures.

Another approach to mass concrete structures that is becoming more widespread is the use of roller-compacted concrete, which uses much lower levels of cement and water than traditional concrete mixtures, and is generally not poured into place. Instead it is placed in thick layers as a semi-dry material and compacted into a dense, strong mass with rolling compactors. Because it uses less cementitious material, Roller Compacted Concrete has a much lower cooling requirement than traditional concrete.

Concrete that is poured all at once in one form (so that there are no weak points where the concrete is "welded" together) is used for tornado shelters.

Prestressed Concrete Structures

Prestressed concrete is a form of reinforced concrete that builds in compressive stresses during construction to oppose those found when in use. This can greatly reduce the weight of beams or slabs, by better distributing the stresses in the structure to make optimal use of the reinforcement. For example a horizontal beam tends to sag. Prestressed reinforcement along the bottom of the beam counteracts this. In pre-tensioned concrete, the prestressing is achieved by using steel or polymer tendons or bars that are subjected to a tensile force prior to casting, or for post-tensioned concrete, after casting.

Concrete Textures

When one thinks of concrete, the image of a dull, gray concrete wall often comes to mind. With the use of form liner, concrete can be cast and molded into different textures and used for decorative concrete applications. Sound/retaining walls, bridges, office buildings and more serve as the optimal canvases for concrete art. For example, the Pima Freeway/Loop 101 retaining and sound walls in Scottsdale, Arizona, feature desert flora and fauna, a 67-foot (20 m) lizard and 40-foot (12 m) cacti along the 8-mile (13 km) stretch. The project, titled "The Path Most Travelled," is one example of how concrete can be shaped using elastomeric form liner.

Building with Concrete

Concrete is one of the most durable building materials. It provides superior fire resistance, compared with wooden construction and can gain strength over time. Structures made of concrete can have a long service life. Concrete is the most widely used construction material in the world with annual consumption estimated at between 21 and 31 billion tonnes.

Concrete is used more than any other man-made material in the world. As of 2006, about 7.5 billion cubic metres of concrete are made each year—more than one cubic metre for every person on Earth.

Concrete powers a US$35 billion industry, employing more than two million workers in the United States alone. More than 55,000 miles (89,000 km) of highways in the United States are paved with this material. Reinforced concrete, prestressed concrete and precast concrete are the most widely used types of concrete functional extensions in modern days.

Energy Efficiency

Energy requirements for transportation of concrete are low because it is produced locally from local resources, typically manufactured within 100 kilometers of the job site. Similarly, relatively little energy is used in producing and combining the raw materials (although large amounts of CO_2 are produced by the chemical reactions in cement manufacture). The overall embodied energy of concrete is therefore lower than for most structural materials other than wood.

Once in place, concrete offers significant energy efficiency over the lifetime of a building. Concrete walls leak air far less than those made of wood-frames. Air leakage accounts for a large percentage of energy loss from a home. The thermal mass properties of concrete increase the efficiency of both residential and commercial buildings. By storing and releasing the energy needed for heating or cooling, concrete's thermal mass delivers year-round benefits by reducing temperature swings inside and minimizing heating and cooling costs. While insulation reduces energy loss through the building envelope, thermal mass uses walls to store and release energy. Modern concrete wall systems use both external insulation and thermal mass to create an energy-efficient building. Insulating Concrete Forms (ICFs) are hollow blocks or panels made of either insulating foam or rastra that are stacked to form the shape of the walls of a building and then filled with reinforced concrete to create the structure.

Pervious Concrete

Pervious concrete is a mix of specially graded coarse aggregate, cement, water and little-to-no fine aggregates. This concrete is also known as "no-fines" or porous concrete. Mixing the ingredients in a carefully controlled process creates a paste that coats and bonds the aggregate particles. The hardened concrete contains interconnected air voids totalling approximately 15 to 25 percent. Water runs through the voids in the pavement to the soil underneath. Air entrainment admixtures are often used in freeze-thaw climates to minimize the possibility of frost damage.

Fire Safety

Concrete buildings are more resistant to fire than those constructed using steel frames, since concrete has lower heat conductivity than steel and therefore could last longer under the same fire conditions. Concrete is sometimes used as a fire protection for steel frames, for the same effect as above. Concrete as a fire shield, for example Fondu fyre, can also be used in extreme environments like a missile launch pad. Options for non-combustible construction include floors, ceilings and roofs made of cast-in-place and hollow-core precast concrete. For walls, concrete masonry technology and Insulating Concrete Forms (ICFs) are additional options. ICFs are hollow blocks or panels made of fire-proof insulating foam that are stacked to form the shape of the walls of a building and then filled with reinforced concrete to create the structure.

Concrete also provides a good resistance against external forces applied such as high winds, hurricanes, tornadoes due to its lateral stiffness that results in minimal horizontal movement. However this stiffness can work against certain types of structure being constructed in concrete, particularly where a relatively higher flexing structure is require to resist more extreme forces.

Earthquake Safety

As discussed above, concrete is very strong in compression, but weak in tension. Larger earthquakes can generate very large shear loads on structures. These shear loads subject the structure to both tensile and compressional loads. Concrete structures without reinforcing, like other unreinforced masonry structures, can fail during severe earthquake shaking. Unreinforced masonry structures constitute one of the largest earthquake risks globally. These risks can be reduced through seismic retrofitting of at-risk buildings, (e.g. School buildings in Istanbul, Turkey).

Useful Life

Concrete can be viewed as a form of artificial sedimentary rock. As a type of mineral, the compounds of which it is composed are extremely stable. Many concrete structures are built with an expected lifetime of approximately 100 years, but researchers have suggested that adding silica fume could extend the useful life of bridges and other concrete uses to as long as 16,000 years. Coatings are also available to protect concrete from damage, and extend the useful life. Epoxy coatings may only be applied to interior surfaces, though, as they would otherwise trap moisture in the concrete.

Self healing concrete has been developed, which can also last longer than traditional concrete.

Large dams, such as the Hoover Dam, and the Three Gorges Dam are intended to last “forever”, a period that is not quantified.

World Records

The world record for the largest concrete pour in a single project is the Three Gorges Dam in Hubei Province, China by the Three Gorges Corporation. The amount of concrete used in the construction of the dam is estimated at 16 million cubic metres over 17 years. The previous record was 12.3 million cubic metres held by Itaipu hydropower station in Brazil. The world record for concrete pumping was set on 7 August 2009 during the construction of the Parbati Hydroelectric Project, near the village of Suind, Himachal Pradesh, India, when the concrete mix was pumped through a vertical height of 715 m (2,346 ft).

The world record for the largest continuously poured concrete raft was achieved in August 2007 in Abu Dhabi by contracting firm Al Habtoor-CCC Joint Venture and the concrete supplier is Unibeton Ready Mix. The pour (a part of the foundation for the Abu Dhabi's Landmark Tower) was 16,000 cubic metres of concrete poured within a two-day period. The previous record, 13,200 cubic metres poured in 54 hours despite a severe tropical storm requiring the site to be covered with tarpaulins to allow work to continue, was achieved in 1992 by joint Japanese and South Korean consortiums Hazama Corporation and the Samsung C&T Corporation for the construction of the Petronas Towers in Kuala Lumpur, Malaysia.

The world record for largest continuously poured concrete floor was completed 8 November 1997, in Louisville, Kentucky by design-build firm EXXCEL Project Management. The monolithic placement consisted of 225,000 square feet (20,900 m^2) of concrete placed within a 30-hour period, finished to a flatness tolerance of F_F 54.60 and a levelness tolerance of F_L 43.83. This surpassed the previous record by 50% in total volume and 7.5% in total area.

The record for the largest continuously placed underwater concrete pour was completed 18 October 2010, in New Orleans, Louisiana by contractor C. J. Mahan Construction Company, LLC of Grove City, Ohio. The placement consisted of 10,251 cubic yards of concrete placed in a 58.5 hour period using two concrete pumps and two dedicated concrete batch plants. Upon curing, this placement allows the 50,180-square-foot (4,662 m^2) cofferdam to be dewatered approximately 26 feet (7.9 m) below sea level to allow the construction of the IHNC GIWW Sill & Monolith Project to be completed in the dry.

Metal Matrix Composite

A metal matrix composite (MMC) is composite material with at least two constituent parts, one being a metal. The other material may be a different metal or another material, such as a ceramic or organic compound. When at least three materials are present, it is called a hybrid composite. An MMC is complementary to a cermet.

Composition

MMCs are made by dispersing a reinforcing material into a metal matrix. The reinforcement surface can be coated to prevent a chemical reaction with the matrix. For example, carbon fibres are commonly used in aluminum matrix to synthesize composites showing low density and high strength.

However, carbon reacts with aluminum to generate a brittle and water-soluble compound Al_4C_3 on the surface of the fibre. To prevent this reaction, the carbon fibres are coated with nickel or titanium boride.

Matrix

The matrix is the monolithic material into which the reinforcement is embedded, and is completely continuous. This means that there is a path through the matrix to any point in the material, unlike two materials sandwiched together. In structural applications, the matrix is usually a lighter metal such as aluminum, magnesium, or titanium, and provides a compliant support for the reinforcement. In high temperature applications, cobalt and cobalt-nickel alloy matrices are common.

Reinforcement

The reinforcement material is embedded into the matrix. The reinforcement does not always serve a purely structural task (reinforcing the compound), but is also used to change physical properties such as wear resistance, friction coefficient, or thermal conductivity. The reinforcement can be either continuous, or discontinuous. Discontinuous MMCs can be isotropic, and can be worked with standard metalworking techniques, such as extrusion, forging or rolling. In addition, they may be machined using conventional techniques, but commonly would need the use of polycrystaline diamond tooling (PCD).

Continuous reinforcement uses monofilament wires or fibres such as carbon fibre or silicon carbide. Because the fibres are embedded into the matrix in a certain direction, the result is an anisotropic structure in which the alignment of the material affects its strength. One of the first MMCs used boron filament as reinforcement. Discontinuous reinforcement uses “whiskers”, short fibres, or particles. The most common reinforcing materials in this category are alumina and silicon carbide.

Manufacturing and Forming Methods

MMC manufacturing can be broken into three types: solid, liquid, and vapor.

Solid state methods

- Powder blending and consolidation (powder metallurgy): Powdered metal and discontinuous reinforcement are mixed and then bonded through a process of compaction, degassing, and thermo-mechanical treatment (possibly via hot isostatic pressing (HIP) or extrusion).
- *Foil diffusion bonding:* Layers of metal foil are sandwiched with long fibres, and then pressed through to form a matrix.

Liquid state methods

- *Electroplating / Electroforming:* A solution containing metal ions loaded with reinforcing particles is co-deposited forming a composite material.
- *Stir casting:* Discontinuous reinforcement is stirred into molten metal, which is allowed to solidify.
- *Squeeze casting:* Molten metal is injected into a form with fibres preplaced inside it.
- *Spray deposition:* Molten metal is sprayed onto a continuous fibre substrate.
- *Reactive processing:* A chemical reaction occurs, with one of the reactants forming the matrix and the other the reinforcement.

Vapor deposition

- *Physical vapor deposition:* The fibre is passed through a thick cloud of vaporized metal, coating it.

In situ fabrication technique

- Controlled unidirectional solidification of a eutectic alloy can result in a two-phase microstructure with one of the phases, present in lamellar or fibre form, distributed in the matrix.

Applications

- Carbide drills are often made from a tough cobalt matrix with hard tungsten carbide particles inside.
- Some tank armors may be made from metal matrix composites, probably steel reinforced with boron nitride. Boron nitride is a good reinforcement for steel because it is very stiff and it does not dissolve in molten steel.

- Some automotive disc brakes use MMCs. Early Lotus Elise models used aluminum MMC rotors, but they have less than optimal heat properties and Lotus has since switched back to cast-iron. Modern high-performance sport cars, such as those built by Porsche, use rotors made of carbon fibre within a silicon carbide matrix because of its high specific heat and thermal conductivity. 3M sells a preformed aluminum matrix insert for strengthening cast aluminum disc brake calipers, allowing them to weigh as much as 50% less while increasing stiffness. 3M has also used alumina preforms for AMC pushrods.
- Ford offers a Metal Matrix Composite (MMC) driveshaft upgrade. The MMC driveshaft is made of an aluminum matrix reinforced with boron carbide, allowing the critical speed of the driveshaft to be raised by reducing inertia. The MMC driveshaft has become a common modification for racers, allowing the top speed to be increased far beyond the safe operating speeds of a standard aluminum driveshaft.
- Honda has used aluminum metal matrix composite cylinder liners in some of their engines, including the B21A1, H22A and H23A, F20C and F22C, and the C32B used in the NSX.
- Toyota has since used metal matrix composites in the Yamaha-designed 2ZZ-GE engine which is used in the later Lotus Lotus Elise S2 versions as well as Toyota car models, including the eponymous Toyota Matrix. Porsche also uses MMCs to reinforce the engine's cylinder sleeves in the Boxster and 911.
- The F-16 Fighting Falcon uses monofilament silicon carbide fibres in a titanium matrix for a structural component of the jet's landing gear.
- Specialized Bicycles has used aluminum MMC compounds for its top of the range bicycle frames for several years. Griffen Bicycles also makes boron carbide-aluminum MMC bike frames, and Univega briefly did so as well.
- Some equipment in particle accelerators such as Radio Frequency Quadrupoles (RFQs) or electron targets use copper MMC compounds such as Glidcop to retain the material properties of copper at high temperatures and radiation levels.
- Copper-silver alloy matrix containing 55 vol.% diamond particles, known as Dymalloy, is used as a substrate for high-power and high density multi-chip modules in electronics for its very high thermal conductivity.

MMCs are nearly always more expensive than the more conventional materials they are replacing. As a result, they are found where improved properties and performance can justify the added cost. Today these applications are found most often in aircraft components, space systems and high-end or "boutique" sports equipment. The scope of applications will certainly increase as manufacturing costs are reduced.

In comparison with conventional polymer matrix composites, MMCs are resistant to fire, can operate in wider range of temperatures, do not absorb moisture, have better electrical and thermal conductivity, are resistant to radiation damage, and do not display outgassing. On the other hand, MMCs tend to be more expensive, the fibre-reinforced materials may be difficult to fabricate, and the available experience in use is limited.

Ceramic Matrix Composite

Ceramic matrix composites (CMCs) are a subgroup of composite materials as well as a subgroup of technical ceramics. They consist of ceramic fibres embedded in a ceramic matrix, thus forming a ceramic fibre reinforced ceramic (CFRC) material. The matrix and fibres can consist of any ceramic material, whereby carbon and carbon fibres can also be considered a ceramic material.

The motivation to develop CMCs was to overcome the problems associated with the conventional technical ceramics like alumina, silicon carbide, aluminium nitride, silicon nitride or zirconia – they fracture easily under mechanical or thermo-mechanical loads because of cracks initiated by small defects or scratches. The crack resistance is – like in glass – very low. To increase the crack resistance or fracture toughness, particles (so-called monocrystalline *whiskers* or *platelets*) were embedded into the matrix. However, the improvement was limited, and the products have found application only in some ceramic cutting tools. So far only the integration of long multi-strand fibres has drastically increased the crack resistance, elongation and thermal shock resistance, and resulted in several new applications.

Carbon (C), special silicon carbide (SiC), alumina (Al_2O_3) and mullite (Al_2O_3–SiO_2) fibres are most commonly used for CMCs. The matrix materials are usually the same, that is C, SiC, alumina and mullite. Generally, CMC names include a combination of *type of fibre/type of matrix*. For example, *C/C* stands for carbon-fibre-reinforced carbon (carbon/carbon), or *C/SiC* for carbon-fibre-reinforced silicon carbide. Sometimes the manufacturing process is included, and a C/SiC composite manufactured with the liquid polymer infiltration (LPI) process is abbreviated as *LPI-C/SiC*.

The important commercially available CMCs are C/C, C/SiC, SiC/SiC and Al_2O_3/Al_2O_3. They differ from conventional ceramics in the following properties, presented in more detail below:

- Elongation to rupture up to 1%
- Strongly increased fracture toughness
- Extreme thermal shock resistance
- Improved dynamical load capability
- Anisotropic properties following the orientation of fibres

Manufacture

The manufacturing processes usually consist of the following three steps:

1. Lay-up and fixation of the fibres, shaped as the desired component
2. Infiltration of the matrix material
3. Final machining and, if required, further treatments like coating or impregnation of the intrinsic porosity.

The first and the last step are almost the same for all CMCs: In step one, the fibres, often named rovings, are arranged and fixed using techniques used in fibre-reinforced plastic materials, such as lay-up of fabrics, filament winding, braiding and knotting. The result of this procedure is called *fibre-preform* or simply *preform*.

For the second step, five different procedures are used to fill the ceramic matrix in between the fibres of the preform:

1. Deposition out of a gas mixture
2. Pyrolysis of a pre-ceramic polymer
3. Chemical reaction of elements
4. Sintering at a relatively low temperature in the range 1000–1200 °C
5. Electrophoretic deposition of a ceramic powder

Procedures one, two and three find applications with non-oxide CMCs, whereas the fourth one is used for oxide CMCs; combinations of these procedures are also practiced. The fifth procedure is not yet established in industrial processes. All procedures have sub-variations, which differ in technical details. All procedures yield a porous material. The third and final step of machining – grinding, drilling, lapping or milling – has to be done with diamond tools. CMCs can also be processed with a water jet or a laser.

Ceramic Fibres

Ceramic fibres in CMCs can have a polycrystalline structure, as in conventional ceramics. They can also be amorphous or have inhomogeneous chemical composition, which develops upon pyrolysis of organic precursors. The high process temperatures required for making CMCs preclude the use of organic, metallic or glass fibres. Only fibres stable at temperatures above 1000 °C can be used, such as fibres of alumina, mullite, SiC, zirconia or carbon. Amorphous SiC fibres have an elongation capability above 2% – much larger than in conventional ceramic materials (0.05 to 0.10%). The reason for this property of SiC fibres is that most of them contain additional elements like oxygen, titanium and/or aluminium yielding a tensile strength above 3 GPa. These enhanced elastic properties are required for various three-dimensional fibre arrangements in textile fabrication, where a small bending radius is essential.

Manufacturing Procedures

Matrix Deposition from a gas Phase: Chemical vapor deposition (CVD) is well suited for this purpose. In the presence of a fibre preform, CVD takes place in between the fibres and their individual filaments and therefore is called chemical vapor infiltration (CVI). One example is the manufacture of C/C composites: a C-fibre preform is exposed to a mixture of argon and a hydrocarbon gas (methane, propane, etc.) at a pressure of around or below 100 kPa and a temperature above 1000 °C. The gas decomposes depositing carbon on and between the fibres. Another example is the deposition of silicon carbide, which is usually conducted from a mixture of hydrogen and methyl-trichlorosilane (MTS, CH_3SiCl_3; it is also common in silicone production). Under defined condition this gas mixture deposits fine and crystalline silicon carbide on the hot surface within the preform. This CVI procedure leaves a body with a porosity of about 10–15%, as access of reactants to the interior of the preform is increasingly blocked by deposition on the exterior.

Matrix Forming via Pyrolysis of C- and Si-containing Polymers

Hydrocarbon polymers shrink during pyrolysis, and upon outgassing form carbon with an amorphous, glass-like structure, which by additional heat treatment can be changed to a more graphite-like structure. Other special polymers, where some carbon atoms are replaced by silicon atoms, the so-called polycarbosilanes, yield amorphous silicon carbide of more or less stoichiometric composition. A large variety of such SiC-, SiNC-, or SiBNC-producing precursors already exist and more are being developed. To manufacture a CMC material, the fibre preform is infiltrated with the chosen polymer. Subsequent curing and pyrolysis yield a highly porous matrix, which is undesirable for most applications. Further cycles of polymer infiltration and pyrolysis are performed until the final and desired quality is achieved. Usually five to eight cycles are necessary. The process is called *liquid polymer infiltration* (LPI), or *polymer infiltration and pyrolysis* (PIP). Here also a porosity of about 15% is common due to the shrinking of the polymer. The porosity is reduced after every cycle.

Matrix Forming via Chemical Reaction

With this method, one material located between the fibres reacts with a second material to form the ceramic matrix. Some conventional ceramics are also manufactured by chemical reactions. For example, reaction-bonded silicon nitride (RBSN) is produced through the reaction of silicon powder with nitrogen, and porous carbon reacts with silicon to form reaction bonded silicon carbide, a silicon carbide which contains inclusions of a silicon phase. An example of CMC manufacture, which was introduced for the production of ceramic brake discs, is the reaction of silicon with a porous preform of C/C. The process temperature is above 1414 °C, that is above the melting point of silicon, and the process conditions are controlled such that the carbon fibres of the C/C-preform almost completely retain their mechanical properties. This process is called *liquid silicon infiltration* (LSI). Sometimes, and because of its starting point with C/C, the material is abbreviated as *C/C-SiC*. The material produced in this process has a very low porosity of about 3%.

Matrix Forming via Sintering

This process is used to manufacture oxide fibre/oxide matrix CMC materials. Since most ceramic fibres can not withstand the normal sintering temperatures of above 1600 °C, special precursor liquids are used to infiltrate the preform of oxide fibres. These precursors allow sintering, that is ceramic-forming processes, at temperatures of 1000–1200 °C. They are, for example, based on mixtures of alumina powder with the liquids tetra-ethyl-orthosilicate (as Si donor) and aluminium-butylate (as Al donor), which yield a mullite matrix. Other techniques, such as sol-gel chemistry, are also used. CMCs obtained with this process usually have a high porosity of about 20%.

Matrix Formed via Electrophoresis

In the electrophoretic process, electrically charged particles dispersed in a special liquid are transported through an electric field into the preform, which has the opposite electrical charge polarity. This process is under development, and is not yet used industrially. Some remaining porosity must be expected here, too.

Basic Mechanism of Mechanical Properties

The high fracture toughness or crack resistance mentioned above is a result of the following mechanism: under load the ceramic matrix cracks, like any ceramic material, at an elongation of about 0.05%. In CMCs the embedded fibres bridge these cracks. This mechanism works only when the matrix can slide along the fibres, which means that there must be a weak bond between the fibres and matrix. A strong bond would require a very high elongation capability of the fibre bridging the crack, and would result in a brittle fracture, as with conventional ceramics. The production of CMC material with high crack resistance requires a step to weaken this bond between the fibres and matrix. This is achieved by depositing a thin layer of pyrolytic carbon or boron nitride on the fibres, which weakens the bond at the fibre/matrix interface (sometimes "interface"), leading to the fibre pull-out at crack surfaces, as shown in the SEM picture at the top of this article. In oxide-CMCs, the high porosity of the matrix is sufficient to establish the weak bond.

Properties Under Tensile and Bending Loads, Crack Resistance

The influence and quality of the fibre interface can be evaluated through mechanical properties. Measurements of the crack resistance were performed with notched specimens in so-called single-edge-notch-bend (SENB) tests. In fracture mechanics, the measured data (force, geometry and crack surface) are normalized to yield the so-called stress intensity factor (SIF), K_{Ic}. Because of the complex crack surface the real crack surface area can not be determined for CMC materials. The measurements therefore use the initial notch as the crack surface, yielding the *formal SIF*. This requires identical geometry for comparing different samples. The area under these curves thus gives a relative indication of the energy required to drive the crack tip through the sample (force times path length gives energy). The maxima indicate the load level necessary to propagate the crack through the sample. Compared to the sample of conventional SiSiC ceramic, two observations can be made:

- All tested CMC materials need up to several orders of magnitude more energy to propagate the crack through the material.
- The force required for crack propagation varies between different types of CMCs.

Type of material	***Al_2O_3/ Al_2O_3***	***Al_2O_3***	***CVI-C/ SiC***	***LPI-C/ SiC***	***LSI-C/ SiC***	***SiSiC***
Porosity (%)	35	<1	12	12	3	<1
Density (g/cm³)	2.1	3.9	2.1	1.9	1.9	3.1
Tensile strength (MPa)	65	250	310	250	190	200
Elongation (%)	0.12	0.1	0.75	0.5	0.35	0.05
Young's modulus (GPa)	50	400	95	65	60	395
Flexural strength (MPa)	80	450	475	500	300	400

In the table, CVI, LPI, and LSI denote the manufacturing process of the C/SiC-material. Data of the oxide CMC and SiSiC are taken from manufacturer data sheets. Tensile strength of SiSiC and Al_2O_3 were calculated from measurements of elongation to fracture and Young's modulus, since generally only bending strength data are available for those ceramics. Averaged values are given in the table, and significant differences, even within one manufacturing route, are possible.

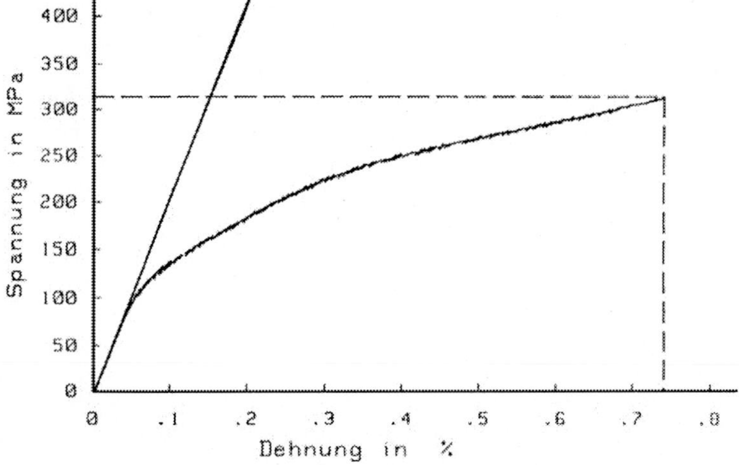

Figure: *Stress-strain curve of a tensile test for CVI-SiC/SiC.*

Tensile tests of CMCs usually show nonlinear stress-strain curves, which look as if the material deforms plastically. It is called *quasi-plastic*, because the effect is caused by the microcracks, which are formed and bridged with increasing load. Since the Young's modulus of the load-carrying fibres is generally lower than that of the matrix, the slope of the curve decreases with increasing load. Curves from bending tests look similar to those of the crack resistance measurements shown above. The following features are essential in evaluating bending and tensile data of CMCs:

- CMC materials with a low matrix content (down to zero) have a high tensile strength (close to the tensile strength of the fibre), but low bending strength.
- CMC materials with a low fibre content (down to zero) have a high bending strength (close to the strength of the monolithic ceramic), but no elongation beyond 0.05% under tensile load.

The primary quality criterion for CMCs is the crack resistance behaviour or fracture toughness.

Other Mechanical Properties

In many CMC components the fibres are arranged as 2-dimensional (2D) stacked plain or satin weave fabrics. Thus the resulting material is anisotropic or, more specifically, orthotropic. A crack between the layers is not bridged by fibres. Therefore, the interlaminar shear strength (ILS) and the strength perpendicular to the 2D fibre orientation are low for these materials. Delamination can occur easily under certain mechanical loads. Three-dimensional fibre structures can improve this situation.

Material	*CVI-C/SiC*	*LPI-C/SiC*	*LSI-C/SiC*	*CVI-SiC/SiC*
Interlaminar shear strength (MPa)	45	30	33	50
Tensile strength vertical to fabric plane (MPa)	6	4	–	7
Compressive strength vertical to fabric plane (MPa)	500	450	–	500

The compressive strengths shown in the table are lower than those of conventional ceramics, where values above 2000 MPa are common; this is a result of porosity.

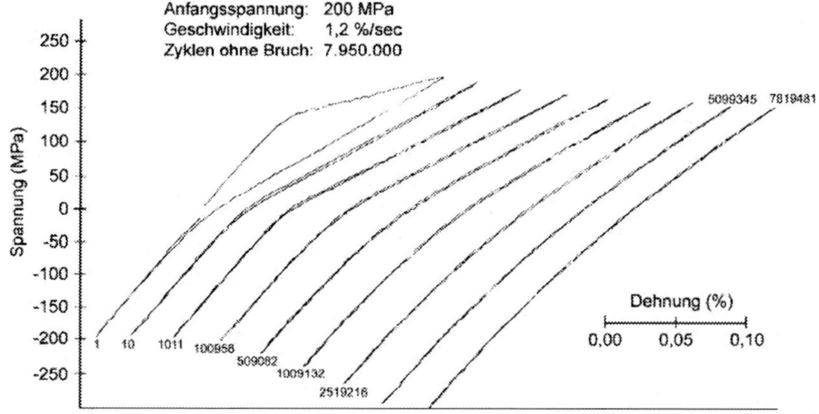

Figure: *Strain controlled LCF-test for a CVI-SiC/SiC-specimen.*

The composite structure allows high dynamical loads. In the so-called low-cycle-fatigue (LCF) or high-cycle-fatigue (HCF) tests the material experiences cyclic loads under tensile and compressive (LCF) or only tensile (HCF) load.

The higher the initial stress the shorter the lifetime and the smaller the number of cycles to rupture. With an initial load of 80% of the strength, a SiC/SiC sample survived about 8 million cycles.

The Poisson's ratio shows an anomaly when measured perpendicular to the plane of the fabric, because interlaminar cracks increase the sample thickness.

Thermal and Electrical Properties

The thermal and electrical properties of the composite are a result of its constituents, namely fibres, matrix and pores as well as their composition.

The orientation of the fibres yields anisotropic data. Oxide CMCs are very good electrical insulators, and because of their high porosity their thermal insulation is much better than that of conventional oxide ceramics.

The use of carbon fibres increases the electrical conductivity, provided the fibres contact each other and the voltage source. Silicon carbide matrix is a good thermal conductor. Electrically, it is a semiconductor, and its resistance therefore decreases with increasing temperature. Compared to (poly) crystalline SiC, the amorphous SiC fibres are relatively poor conductors of heat and electricity.

Material	*CVI-C/SiC*	*LPI-C/SiC*	*LSI-C/SiC*	*CVI-Si C/SiC*	*SiSiC*
Thermal conductivity (p) [W/(m·K)]	15	11	21	18	>100
Thermal conductivity (v) [W/(m·K)]	7	5	15	10	>100
Linear expansion (p) [10^{-6}·1/K]	1.3	1.2	0	2.3	4
Linear expansion (v) [10^{-6}·1/K]	3	4	3	3	4
Electrical resistivity (p) [Ω·cm]	–	–	–	–	50
Electrical resistivity (v) [Ω·cm]	0.4	–	–	5	50

Comments for the Table: (p) and (v) refer to data parallel and vertical to fibre orientation of the 2D-fibre structure, respectively. LSI material has the highest thermal conductivity because of its low porosity – an advantage when using it for brake discs. These data are subject to scatter depending on details of the manufacturing processes.

Conventional ceramics are very sensitive to thermal stress because of their high Young's modulus and low elongation capability. Temperature differences and low thermal conductivity create locally different elongations, which together with the high Young's modulus generate high stress. This results in cracks, rupture and brittle failure. In CMCs, the fibres bridge the cracks, and the components show no macroscopic damage, even if the matrix has cracked locally. The application of CMCs in brake disks demonstrates the effectiveness of ceramic composite materials under extreme thermal shock conditions.

Corrosion Properties

Data on the corrosion behaviour of CMCs are scarce except for oxidation at temperatures above 1000 °C. These properties are determined by the constituents, namely the fibres and matrix. Ceramic materials in general are very stable to corrosion. The broad spectrum of manufacturing techniques with different sintering additives, mixtures, glass phases and porosities are crucial for the results of corrosion tests. Less impurities and exact stoichiometry lead to less corrosion. Amorphous structures and non-ceramic chemicals frequently used as sintering aids are starting points of corrosive attack.

Alumina

Pure alumina shows excellent corrosion resistivity against most chemicals. Amorphous glass and silica phases at the grain boundaries determine the speed of corrosion in concentrated acids and bases and result in creep at high temperatures. These characteristics limit the use of alumina. For molten metals, alumina is used only with gold and platinum.

Alumina Fibres

These fibres demonstrate corrosion properties similar to alumina, but commercially available fibres are not very pure and therefore less resistant. Because of creep at temperatures above 1000 °C, there are only few applications for oxide CMCs.

Carbon

The most significant corrosion of carbon occurs in presence of oxygen above about 500 °C. It burns to form carbon dioxide and/or carbon monoxide. It also oxidizes in strong oxidizing agents like concentrated nitric acid. In molten metals it dissolves and forms metal carbides. Carbon fibres do not differ from carbon in their corrosion behaviour.

Silicon Carbide

Pure silicon carbide is one of the most corrosion-resistant materials. Only strong bases, oxygen above about 800 °C, and molten metals react with it to form carbides and silicides. The reaction with oxygen forms SiO_2 and CO_2, whereby a surface layer of SiO_2 slows down subsequent oxidation (*passive oxidation*). Temperatures above about 1600 °C and a low partial pressure of oxygen result in so-called *active oxidation*, in which CO, CO_2 and gaseous SiO are formed causing rapid loss of SiC. If the SiC matrix is produced other than by CVI, corrosion-resistance is not as good. This is a consequence of porosity in the amorphous LPI, and residual silicon in the LSI-matrix.

Silicon Carbide Fibres

Silicon carbide fibres are produced via pyrolysis of organic polymers, and therefore their corrosion properties are similar to those of the silicon carbide found in LPI-matrices. These fibres are thus more sensitive to bases and oxidizing media than pure silicon carbide.

Applications

CMC materials overcome the major disadvantages of conventional technical ceramics, namely brittle failure and low fracture toughness, and limited thermal shock resistance. Therefore, their applications are in fields requiring reliability at high-temperatures (beyond the capability of metals) and resistance to corrosion and wear. These include:

- Heat shield systems for space vehicles, which are needed during the re-entry phase, where high temperatures, thermal shock conditions and heavy vibration loads take place.
- Components for high-temperature gas turbines such as combustion chambers, stator vanes and turbine blades.
- Components for burners, flame holders, and hot gas ducts, where the use of oxide CMCs has found its way.
- Brake disks and brake system components, which experience extreme thermal shock (greater than throwing a glowing part of any material into water).
- Components for slide bearings under heavy loads requiring high corrosion and wear resistance.

In addition to the foregoing, CMCs can be used in applications, which employ conventional ceramics or in which metal components have limited lifetimes due to corrosion or high temperatures.

Developments for Applications in Space

During the re-entry phase of space vehicles, the heat shield system is exposed to temperatures above 1500 °C for a few minutes. Only ceramic materials are able to survive such conditions without significant damage, and among ceramics only CMCs can adequately handle thermal shocks. The development of CMC-based heat shield systems promises the following advantages:

- Reduced weight
- Higher load carrying capacity of the system
- Reusability for several re-entries
- Better steering during the re-entry phase with CMC flap systems

Figure: *NASA-space vehicle X-38 during a test flight*

In these applications the high temperatures preclude the use of oxide fibre CMCs, because under the expected loads the creep would be too high. Amorphous silicon carbide fibres lose their strength due to re-crystallization at temperatures above 1250 °C. Therefore carbon fibres in a silicon carbide matrix (C/SiC) are used in development programmes for these applications. The European programme HERMES of ESA, started in the 1980s and for financial reasons abandoned in 1992, has produced first results. Several follow-up programmes focused on the development, manufacture, and qualification of nose cap, leading edges and steering flaps for the NASA space vehicle X-38.

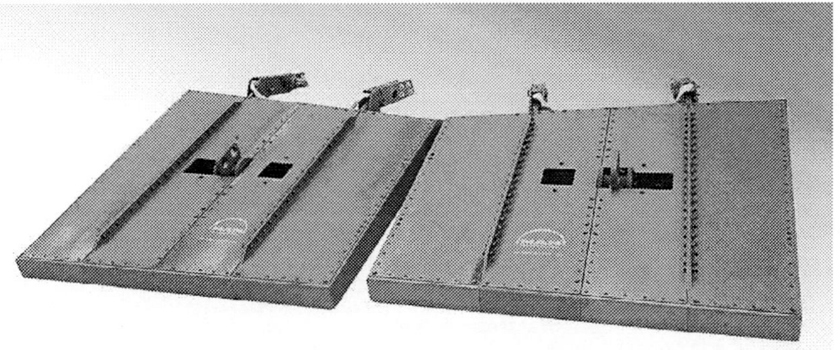

Figure: *Pair of steering flaps for the NASA-space vehicle X-38. Size: 1.5×1.5×0.15 m, mass: 68 kg each, various components are mounted using more than 400 CVI-C/SiC screws and nuts.*

This development programme has qualified the use of C/SiC bolts and nuts, and the bearing system of the flaps. The latter were ground-tested at the DLR in Stuttgart, Germany, under expected conditions of the re-entry phase: 1600 °C, 4 tonnes load, oxygen partial pressure similar to re-entry conditions, and simultaneous bearing movements of four cycles per second. A total of five re-entry phases was simulated. Furthermore, oxidation protection systems were developed and qualified to prevent burnout of the carbon fibres. After mounting of the flaps, mechanical ground tests were performed successfully by NASA in Houston, Texas, US. The next test – a real re-entry of the unmanned vehicle X-38 – was cancelled for financial reasons. One of the space shuttles would have brought the vehicle into orbit, from where it would have returned to the Earth.

These qualifications were promising for only this application. The high-temperature load lasts only around 20 minutes per re-entry, and for reusability, only about 30 cycles would be sufficient. For industrial applications in hot gas environment, though, several hundred cycles of thermal loads and up to many thousands hours of lifetime are required.

Developments for gas Turbine Components

The use of CMCs in gas turbines would permit higher turbine inlet temperatures, which would improve turbine efficiency. Because of the complex shape of stator vanes and turbine blades, the development was first focused on the combustion chamber.

In the US, a combustor made of SiC/SiC with a special SiC fibre of enhanced high-temperature stability was successfully tested for 15,000 hours. SiC oxidation was substantially reduced by the use of an oxidation protection coating consisting of several layers of oxides.

The engine collaboration between General Electric and Rolls-Royce is studying the use of CMC stator vanes in the hot section of the F136 turbofan engine, an engine which failed to beat the Pratt and Whitney F-135 for use in the Joint Strike Fighter.

The engine joint venture, CFM International is also considering the use of CMC parts to reduce weight in its Leap-X demonstrator engine programme, which is aimed at providing next-generation turbine engines for narrow-body airliners.

CMC parts are also being studied for stationary applications in both the cold and hot sections of the engines, since stresses imposed on rotating parts would require further development effort.

Generally, a successful application in turbines still needs a lot of technical and cost reduction work for all high-temperature components to justify the efficiency gain. Furthermore, cost reduction for fibres, manufacturing processes and protective coatings is essential.

Application of oxide CMC in Burner and Hot gas Ducts

Oxygen-containing gas at temperatures above 1000 °C is rather corrosive for metal and silicon carbide components. Such components, which are not exposed to high mechanical stress, can be made of oxide CMCs, which can withstand temperatures up to 1200 °C.

The gallery below shows the flame holder of a crisp bread bakery as tested after for 15,000 hours, which subsequently operated for a total of more than 20,000 hours.

Figure: *Oxide CMC flame holder*

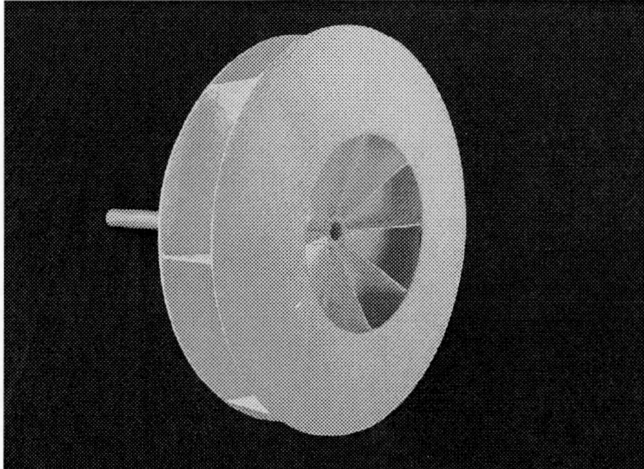

Figure: *Ventilator for hot gases*

Flaps and ventilators circulating hot, oxygen-containing gases can be fabricated in the same shape as their metal equivalents. The lifetime for these oxide CMC components is several times longer than for metals, which often deform. A further example is an oxide CMC lifting gate for a sintering furnace, which has survived more than 260,000 opening cycles.

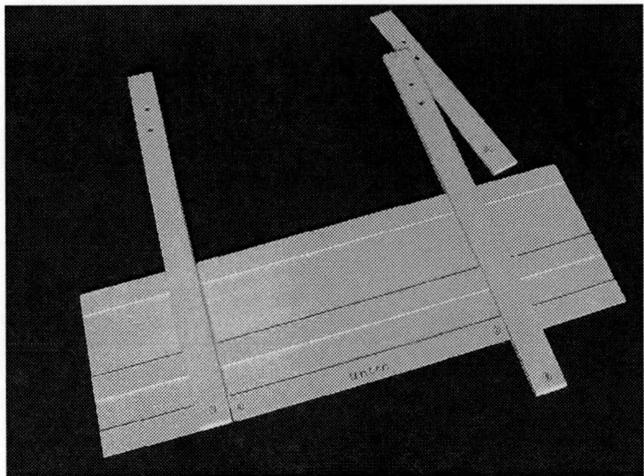

Figure: *Lifting gate, oxide CMC*

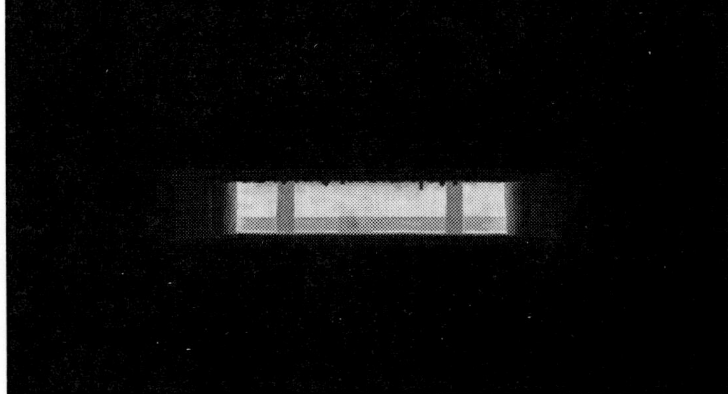

Figure: *Lifting gate in the field*

Application in Brake Disk

Carbon/carbon (C/C) materials have found their way into the disk brakes of racing cars and airplanes, and C/SiC brake disks manufactured by the LSI process were qualified and are commercially available for luxury vehicles. The advantages of these C/SiC disks are:

- Very little wear, resulting in lifetime use for a car with a normal driving load of 300,000 km, is forecast by manufacturers.
- No fading is experienced, even under high load.
- No surface humidity effect on the friction coefficient shows up, as in C/C brake disks.
- The corrosion resistance, for example to the road salt, is much better than for metal disks.

- The disk mass is only 40% of a metal disk. This translates into less unsprung and rotating mass.

The weight reduction improves shock absorber response, road-holding comfort, agility, fuel economy, and thus driving comfort.

The SiC-matrix of LSI has a very low porosity, which protects the carbon fibres quite well. Brake disks do not experience temperatures above 500 °C for more than a few hours in their lifetime. Oxidation is therefore not a problem in this application. The reduction of manufacturing costs will decide the success of this application for middle-class cars.

Application in Slide Bearings

Figure: *Components for a ceramic slide bearing; the picture shows a sintered SiC-bearing for a hydrostatic slide bearing and a CVI-SiC/SiC-shaft sleeve shrink-fitted on metal, a system tested with liquid oxygen as lubricant.*

Conventional SiC, or sometimes the less expensive SiSiC, have been used successfully for more than 25 years in slide or journal bearings of pumps. The pumped liquid itself provides the lubricant for the bearing. Very good corrosion resistance against practically all kinds of media, and very low wear and low friction coefficients are the basis of this success. These bearings consist of a static bearing, shrink-fitted in its metallic environment, and a rotating shaft sleeve, mounted on the shaft. Under compressive stress the ceramic static bearing has a low risk of failure, but a SiC shaft sleeve does not have this situation and must therefore have a large wall thickness and/or be specially designed. In large pumps with shafts 100–350 mm in diameter, the risk of failure is higher due to the changing requirements on the pump performance – for example, load changes during operation. The introduction of SiC/SiC as a shaft sleeve material has proven to be very successful. Test rig experiments showed an almost triple specific load capability of the bearing system with a shaft sleeve made of SiC/SiC, sintered SiC as static bearing, and water at 80 °C as lubricant. The specific load capacity of a bearing is usually given in W/mm^2 and calculated as a product of the load (MPa), surface speed of the bearing (m/s) and friction coefficient; it is equal to the power loss of the bearing system due to friction.

In boiler feedwater pumps of power stations, which pump several thousand cubic metres of hot water to a level of 2000 m, and in tubular casing pumps for water works or sea water desalination plants (pumping up to 40,000 m^3 to a level of around 20 m) this slide bearing concept, namely SiC/SiC shaft sleeve and SiC bearing, has been used since 1994. A picture of such shaft sleeves is shown at the top of this article. This bearing system has been tested in pumps for liquid oxygen, for example in oxygen turbopumps for thrust engines of space rockets, with the following results. SiC and SiC/SiC are compatible with liquid oxygen. In an auto-ignition test according to the French standard NF 28-763, no auto-ignition was observed with powdered SiC/SiC in 20 bar pure oxygen at temperatures up to 525 °C. Tests have shown that the friction coefficient is half, and wear one fiftieth of standard metals used in this environment. A hydrostatic bearing system has survived several hours at a speed up to 10,000 revolutions per minute, various loads, and 50 cycles of start/stop transients without any significant traces of wear.

Other Applications and Developments

- Thrust control flaps for military jet engines
- Components for fusion and fission reactors
- Friction systems for various applications

Electrical and Electronic Properties of Materials

Electrical Properties of Materials

Electricity is concerned with the movement of charged particles in electrical circuits. Therefore in order to gain a basic understanding of electricity we must first understand what a charged particle is. We also need to know about the electrical properties of the materials that are used to make electrical circuits and this requires a general understanding of the structure of these materials. In this section we will look at the structure and electrical properties of atoms and molecules. We will also look at how atoms and molecules combine to make different materials. Later we will see how this determines the electrical properties of the materials.

Atoms and Molecules

All materials are made up of molecules and molecules in turn are made up of atoms. In some substances such as Argon, the molecules consist of a single atom. Therefore in these substances molecules and atoms are the same thing. However, in most materials two or more atoms bond together to form molecules.

- Monatomic molecules contain one atom
- Diatomic molecules contain two atoms
- Triatomic molecules contain three atoms
- Polyatomic describes any molecules that contain more than one atom.

Examples

- Argon and Helium molecules contain one atom (Ar, He)
- Hydrogen and Oxygen molecules contain two atoms (H_2, O_2).
- One type of sulphur molecule contains eight atoms (S_8)

Electrical Conductivity

Electrical resistivity (also known as resistivity, specific electrical resistance, or volume resistivity) quantifies how strongly a given material opposes the flow of electric current. A low resistivity indicates a material that readily allows the movement of electric charge. Resistivity is commonly represented by the Greek letter ρ (rho). The SI unit of electrical resistivity is the ohm·metre (Ω·m) although other units like ohm·centimetre (Ω·cm) are also in use. As an example, if a 1m×1m×1m solid cube of material has sheet contacts on two opposite faces, and the resistance between these contacts is 1Ω, then the resistivity of the material is 1Ω·m.

Electrical conductivity or specific conductance is the reciprocal of electrical resistivity, and measures a material's ability to conduct an electric current. It is commonly represented by the Greek letter σ (sigma), but κ (kappa) (especially in electrical engineering) or γ (gamma) are also occasionally used. Its SI unit is siemens per metre ($S \cdot m^{-1}$) and CGSE unit is reciprocal second (s^{-1}).

Definition

Resistors or conductors with uniform cross-section:

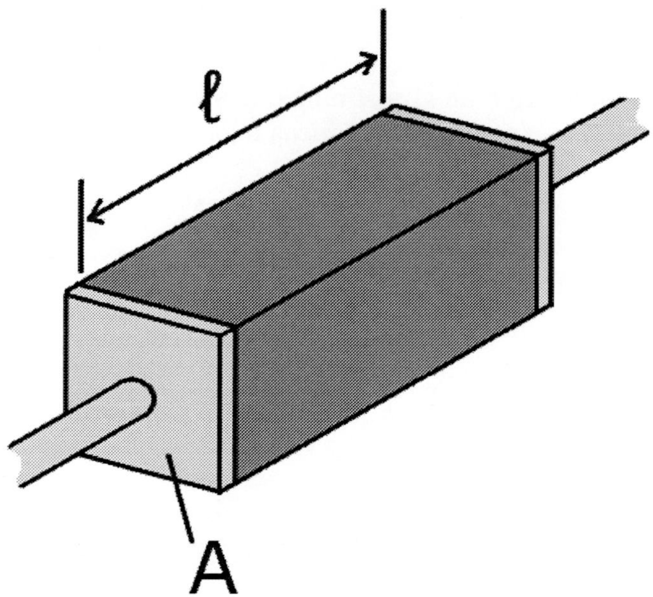

Figure: *A piece of resistive material with electrical contacts on both ends.*

Many resistors and conductors have a uniform cross section with a uniform flow of electric current and are made of one material. In this case, the electrical resistivity ρ (Greek: rho) is defined as:

$$\rho = R\frac{A}{\ell},$$

where

R is the electrical resistance of a uniform specimen of the material (measured in ohms, Ω)

ℓ is the length of the piece of material (measured in metres, m)

A is the cross-sectional area of the specimen (measured in square metres, m^2).

The reason resistivity is defined this way is that it makes resistivity a *material property*, unlike resistance. All copper wires, irrespective of their shape and size, have approximately the same *resistivity*, but a long, thin copper wire has a much larger *resistance* than a thick, short copper wire. Every material has its own characteristic resistivity – for example, rubber's resistivity is far larger than copper's.

In a hydraulic analogy, passing current through a high-resistivity material is like pushing water through a pipe full of sand, while passing current through a low-resistivity material is like pushing water through an empty pipe. If the pipes are the same size and shape, the pipe full of sand has higher resistance to flow. But resistance is not *solely* determined by the presence or absence of sand; it also depends on how wide the pipe is (it is harder to push water through a skinny pipe than a wide one) and how long it is (it is harder to push water through a long pipe than a short one.)

The above equation can be transposed to get Pouillet's law:

$$R = \rho\frac{\ell}{A}.$$

The resistance of a given material will increase with the length, but decrease with increasing cross-sectional area. From the above equations, resistivity has SI units of ohm·metre. Other units like ohm·cm or ohm·inch are also sometimes used.

The formula $R = \rho\ell / A$ can be used to intuitively understand the meaning of a resistivity value. For example, if $A = 1\text{m}^2$ and $\ell = 1\text{m}$ (forming a cube with perfectly-conductive contacts on opposite faces), then the resistance of this element in ohms is numerically equal to the resistivity of the material it is made of in ohm-meters. Likewise, a 1 ohmÅ"cm material would have a resistance of 1 ohm if contacted on opposite faces of a 1 cm×1 cm×1 cm cube.

Conductivity σ (Greek: sigma) is defined as the inverse of resistivity:

$$\sigma = \frac{1}{\rho}.$$

Conductivity has SI units of siemens per metre (S/m).

General Definition

The above definition was specific to resistors or conductors with a uniform cross-section, where current flows uniformly through them. A more basic and general definition starts from the fact that if there is electric field inside a material, it will cause electric current to flow. The electrical resistivity ρ is defined as the ratio of the electric field to the density of the current it creates:

$$\rho = \frac{E}{J},$$

where

ρ is the resistivity of the conductor material (measured in ohm·metres, $\Omega \cdot m$),

E is the magnitude of the electric field (in volts per metre, $V \cdot m^{-1}$),

J is the magnitude of the current density (in amperes per square metre, $A \cdot m^{-2}$), in which E and J are inside the conductor.

Conductivity is the inverse:

$$\sigma = \frac{1}{\rho} = \frac{J}{E}.$$

For example, rubber is a material with large ρ and small σ, because even a very large electric field in rubber will cause almost no current to flow through it. On the other hand, copper is a material with small ρ and large σ, because even a small electric field pulls a lot of current through it.

Causes of Conductivity

Band Theory Simplified: Quantum mechanics states that electrons in an atom cannot take on any arbitrary energy value. Rather, there are fixed energy levels which the electrons can occupy, and values in between these levels are impossible. When a large number of such allowed energy levels are spaced close together (in energy-space) i.e. have similar (minutely differing energies) then we can talk about these energy levels together as an "energy band". There can be many such energy bands in a material, depending on the atomic number (number of electrons) and their distribution (besides external factors like environment modifying the energy bands). Two such bands important in the discussion of conductivity of materials are: the valence band and the conduction band (the latter is generally above the former). Electrons in the conduction band may move freely throughout the material in the presence of an electrical field.

In insulators and semiconductors, the atoms in the substance influence each other so that between the valence band and the conduction band there exists a forbidden band of energy levels, which the electrons cannot occupy. In order for a current to flow, a relatively large amount of energy must be furnished to an electron for it to leap across this forbidden gap and into the conduction band. Thus, even large voltages can yield relatively small currents.

In metals

A metal consists of a lattice of atoms, each with an outer shell of electrons which freely dissociate from their parent atoms and travel through the lattice. This is also known as a positive ionic lattice. This 'sea' of dissociable electrons allows the metal to conduct electric current. When an electrical potential difference (a voltage) is applied across the metal, the resulting electric field causes electrons to move from one end of the conductor to the other.

Near room temperatures, metals have resistance. The primary cause of this resistance is the thermal motion of ions. This acts to scatter electrons (due to destructive interference of free electron waves on non-correlating potentials of ions). Also contributing to resistance in metals with impurities are the resulting imperfections in the lattice. In pure metals this source is negligible.

The larger the cross-sectional area of the conductor, the more electrons per unit length are available to carry the current. As a result, the resistance is lower in larger cross-section conductors. The number of scattering events encountered by an electron passing through a material is proportional to the length of the conductor. The longer the conductor, therefore, the higher the resistance. Different materials also affect the resistance.

In Semiconductors and Insulators

In metals, the Fermi level lies in the conduction band giving rise to free conduction electrons. However, in semiconductors the position of the Fermi level is within the band gap, approximately half-way between the conduction band minimum and valence band maximum for intrinsic (undoped) semiconductors.

This means that at 0 kelvins, there are no free conduction electrons and the resistance is infinite. However, the resistance will continue to decrease as the charge carrier density in the conduction band increases. In extrinsic (doped) semiconductors, dopant atoms increase the majority charge carrier concentration by donating electrons to the conduction band or accepting holes in the valence band. For both types of donor or acceptor atoms, increasing the dopant density leads to a reduction in the resistance, hence highly doped semiconductors behave metallically. At very high temperatures, the contribution of thermally generated carriers will dominate over the contribution from dopant atoms and the resistance will decrease exponentially with temperature.

In Ionic Liquids/electrolytes

In electrolytes, electrical conduction happens not by band electrons or holes, but by full atomic species (ions) travelling, each carrying an electrical charge. The resistivity of ionic liquids varies tremendously by the concentration – while distilled water is almost an insulator, salt water is a very efficient electrical conductor. In biological membranes, currents are carried by ionic salts. Small holes in the membranes, called ion channels, are selective to specific ions and determine the membrane resistance.

Superconductivity

The electrical resistivity of a metallic conductor decreases gradually as temperature is lowered. In ordinary conductors, such as copper or silver, this decrease is limited by impurities and other defects. Even near absolute zero, a real sample of a normal conductor shows some resistance. In a superconductor, the resistance drops abruptly to zero when the material is cooled below its critical temperature. An electric current flowing in a loop of superconducting wire can persist indefinitely with no power source.

In 1986, it was discovered that some cuprate-perovskite ceramic materials have a critical temperature above 90 K (–183 °C). Such a high transition temperature is theoretically impossible for a conventional superconductor, leading the materials to be termed high-temperature superconductors. Liquid nitrogen boils at 77 K, facilitating many experiments and applications that are less practical at lower temperatures. In conventional superconductors, electrons are held together in pairs by an attraction mediated by lattice phonons. The best available model of high-temperature superconductivity is still somewhat crude. There is a hypothesis that electron pairing in high-temperature superconductors is mediated by short-range spin waves known as paramagnons.

Temperature Dependence

Linear Approximation: The electrical resistivity of most materials changes with temperature. If the temperature T does not vary too much, a linear approximation is typically used:

$$\rho(T) = \rho_0[1 + \alpha(T - T_0)]$$

where α is called the *temperature coefficient of resistivity*, T_0 is a fixed reference temperature (usually room temperature), and ρ_0 is the resistivity at temperature T_0. The parameter is an empirical parameter fitted from measurement data. Because the linear approximation is only an approximation, α is different for different reference temperatures. For this reason it is usual to specify the temperature that α was measured at with a suffix, such as α_{15}, and the relationship only holds in a range of temperatures around the reference. When the temperature varies over a large temperature range, the linear approximation is inadequate and a more detailed analysis and understanding should be used.

Metals

In general, electrical resistivity of metals increases with temperature. Electron–phonon interactions can play a key role. At high temperatures, the resistance of a metal increases linearly with temperature. As the temperature of a metal is reduced, the temperature dependence of resistivity follows a power law function of temperature. Mathematically the temperature dependence of the resistivity ρ of a metal is given by the Bloch–Grüneisen formula:

$$\rho(T) = \rho(0) + A\left(\frac{T}{\Theta_R}\right)^n \int_0^{\frac{\Theta_R}{T}} \frac{x^n}{(e^x - 1)(1 - e^{-x})} dx$$

where $\rho(0)$ is the residual resistivity due to defect scattering, A is a constant that depends on the velocity of electrons at the Fermi surface, the Debye radius and the number density of electrons in the metal. Θ_R is the Debye temperature as obtained from resistivity measurements and matches very closely with the values of Debye temperature obtained from specific heat measurements. n is an integer that depends upon the nature of interaction:

1. n=5 implies that the resistance is due to scattering of electrons by phonons (as it is for simple metals)
2. n=3 implies that the resistance is due to s-d electron scattering (as is the case for transition metals)
3. n=2 implies that the resistance is due to electron–electron interaction.

If more than one source of scattering is simultaneously present, Matthiessen's Rule (first formulated by Augustus Matthiessen in the 1860s) says that the total resistance can be approximated by adding up several different terms, each with the appropriate value of *n*.

As the temperature of the metal is sufficiently reduced (so as to 'freeze' all the phonons), the resistivity usually reaches a constant value, known as the residual resistivity. This value depends not only on the type of metal, but on its purity and thermal history. The value of the residual resistivity of a metal is decided by its impurity concentration. Some materials lose all electrical resistivity at sufficiently low temperatures, due to an effect known as superconductivity. An investigation of the low-temperature resistivity of metals was the motivation to Heike Kamerlingh Onnes's experiments that led in 1911 to discovery of superconductivity.

Semiconductors

In general, resistivity of intrinsic semiconductors decreases with increasing temperature. The electrons are bumped to the conduction energy band by thermal energy, where they flow freely and in doing so leave behind holes in the valence band which also flow freely. The electric resistance of a typical intrinsic (non doped) semiconductor decreases exponentially with the temperature:

$$\rho = \rho_0 e^{-aT}$$

An even better approximation of the temperature dependence of the resistivity of a semiconductor is given by the Steinhart–Hart equation:

$$1/T = A + B\ln(\rho) + C(\ln(\rho))^3$$

where A, B and C are the so-called Steinhart–Hart coefficients.

This equation is used to calibrate thermistors.

Extrinsic (doped) semiconductors have a far more complicated temperature profile. As temperature increases starting from absolute zero they first decrease steeply in resistance as the carriers leave the donors or acceptors. After most of the donors or acceptors have lost their carriers the resistance starts to increase again slightly due to the reducing mobility of carriers (much as in a metal). At higher temperatures it will behave like intrinsic semiconductors as the carriers from the donors/acceptors become insignificant compared to the thermally generated carriers. In non-crystalline semiconductors, conduction can occur by charges quantum tunnelling from one localised site to another. This is known as variable range hopping and has the characteristic form of

$$\rho = A\exp(T^{-1/n}),$$

where n = 2, 3, 4, depending on the dimensionality of the system.

Complex Resistivity and Conductivity

When analyzing the response of materials to alternating electric fields, in applications such as electrical impedance tomography, it is necessary to replace resistivity with a complex quantity called impeditivity (in analogy to electrical impedance). Impeditivity is the sum of a real component, the resistivity, and an imaginary component, the reactivity (in analogy to reactance). The magnitude of Impeditivity is the square root of sum of squares of magnitudes of resistivity and reactivity. Conversely, in such cases the conductivity must be expressed as a complex number (or even as a matrix of complex numbers, in the case of anisotropic materials) called the *admittivity*. Admittivity is the sum of a real component called the conductivity and an imaginary component called the susceptivity. An alternative description of the response to alternating currents uses a real (but frequency-dependent) conductivity, along with a real permittivity. The larger the conductivity is, the more quickly the alternating-current signal is absorbed by the material (i.e., the more opaque the material is).

Tensor Equations for Anisotropic Materials

Some materials are anisotropic, meaning they have different properties in different directions. For example, a crystal of graphite consists microscopically of a stack of sheets, and current flows very easily through each sheet, but moves much less easily from one sheet to the next. For an anisotropic material, it is not generally valid to use the scalar equations

$$J = \sigma E \rightleftharpoons E = \rho J.$$

For example, the current may not flow in exactly the same direction as the electric field. Instead, the equations are generalized to the 3D tensor form

$$\mathrm{J} = \sigma \mathrm{E} \rightleftharpoons \mathrm{E} = \rho \mathrm{J}$$

where the conductivity σ and resistivity ρ are rank-2 tensors (in other words, 3×3 matrices). The equations are compactly illustrated in component form (using index notation and the summation convention):

$$J_i = \sigma_{ij} E_j \rightleftharpoons E_i = \rho_{ij} J_j.$$

The σ and ρ tensors are inverses (in the sense of a matrix inverse). The individual components are not necessarily inverses; for example, σ_{xx} may not be equal to $1/\rho_{xx}$.

Resistance Versus Resistivity in Complicated Geometries

If the material's resistivity is known, calculating the resistance of something made from it may, in some cases, be much more complicated than the formula $R = \rho \ell / A$ above. One example is

Spreading Resistance Profiling, where the material is inhomogeneous (different resistivity in different places), and the exact paths of current flow are not obvious.

In cases like this, the formulas

$$J = \sigma E \rightleftharpoons E = \rho J$$

need to be replaced with

$$J(r) = \sigma(r)E(r) \rightleftharpoons E(r) = \rho(r)J(r),$$

where E and J are now vector fields. This equation, along with the continuity equation for J and the Poisson equation for E, form a set of partial differential equations. In special cases, an exact or approximate solution to these equations can be worked out by hand, but for very accurate answers in complex cases, computer methods like finite element analysis may be required.

Resistivity Density Products

In some applications where the weight of an item is very important resistivity density products are more important than absolute low resistivity – it is often possible to make the conductor thicker to make up for a higher resistivity; and then a low resistivity density product material (or equivalently a high conductance to density ratio) is desirable. For example, for long distance overhead power lines, aluminium is frequently used rather than copper because it is lighter for the same conductance.

Material	*Resistivity (nΩ•m)*	*Density (g/cm³)*	*Resistivity-densityproduct (nΩ•m •g/cm³)*
Sodium	47.7	0.97	46
Lithium	92.8	0.53	49
Calcium	33.6	1.55	52
Potassium	72.0	0.89	64
Beryllium	35.6	1.85	66
Aluminium	26.50	2.70	72
Magnesium	43.90	1.74	76.3
Copper	16.78	8.96	150
Silver	15.87	10.49	166
Gold	22.14	19.30	427
Iron	96.1	7.874	757

Silver, although it is the least resistive metal known, has a high density and does poorly by this measure. Calcium and the alkali metals have the best resistivity-density products, but are rarely used for conductors due to their high reactivity with water and oxygen. Aluminium is far more stable. Two other important attributes, price and toxicity, exclude the (otherwise) best choice: Beryllium. Thus, aluminium is usually the metal of choice when the weight of some required conduction (and/or the cost of conduction) is the driving consideration.

Ionic Conductivity (Solid State)

Ionic conduction (denoted by λ-lambda) is the movement of an ion from one site to another through defects in the crystal lattice of a solid or aqueous solution.

Ionic conduction is one aspect of current. In solids, ions typically occupy fixed positions in the crystal lattice and do not move. However, ionic conduction can occur, especially, as the temperature increases.

Ionic conduction is one of the mechanisms by which microwave ovens are believed to work. Microwaves cause ions dissolved in the microwaved sample to oscillate, colliding with neighbouring molecules or atoms. These collisions cause agitation, motion or heat. This mechanism is "important when considering the heating behaviour of ionic liquids" within a microwave.

Ionic conduction in solids has been a subject of interest since the beginning of the 19th century. Michael Faraday established in 1839 that the laws of electrolysis are also obeyed in ionic solids like lead(II) fluoride (PbF_2) and silver sulfide (Ag_2S).

Silver Iodide

In 1921, Tubandt et al. found that silver iodide (AgI) has extraordinary high ionic conductivity. While measuring conductivity, they found that above 147 °C, AgI changes into a phase that has an ionic conductivity of ~ 1 –1 cm^{-1}, similar to that of its liquid phase. As a result, the low temperature phase of AgI was the first superionic conductor ever discovered. The highly conductive phase of AgI is now known as -AgI. It was shown that a sublattice cationic disorder takes place in -AgI. The liquid-like state of Ag^+ ions, as proposed by Strock (1934, 1936) and later reinforced by others (Geller, 1977; Funke, 1976), consists of a cubic unit cell of iodide ions (I^-), in which a total of 42 sites (6 octahedral, 12 tetragonal and 24 trigonal bipyramidal) are available for 2 Ag^+ ions. O' Keeffe and Hyde (1976) have argued that this phase transition in AgI is dramatic and powerful, nothing less than the melting and have also shown that the entropy change at the super-ionic transition is comparable to its value at the melting. Thus, in the -phase, I^- ions form a body-centred cubic lattice and the Ag^+ ions are distributed in such a way that 42 crystallographic equivalent interstices are available for the two Ag^+ ions.

Alpha phase crystals of various materials like Ag_2S, Ag_2Se, Ag_2Te, etc. were soon discovered (Tubandt, 1932). By the early 1930s, it was demonstrated that these fast ionically conducting solids could be treated entirely the same as aqueous electrolytes from the viewpoint of chemical reactions and thermodynamics, hence these materials were labelled *solid electrolytes.*

Intrinsic and Extrinsic Semi Conductivity

Intrinsic Semiconductor

An intrinsic semiconductor, also called an undoped semiconductor or i-type semiconductor, is a pure semiconductor without any significant dopant species present. The number of charge carriers is therefore determined by the properties of the material itself instead of the amount of impurities. In intrinsic semiconductors the number of excited electrons and the number of holes are equal: n = p.

The electrical conductivity of intrinsic semiconductors can be due to crystallographic defects or electron excitation. In an intrinsic semiconductor the number of electrons in the conduction band is equal to the number of holes in the valence band. An example is Hg

0.8Cd

0.2Te at room temperature.

An indirect band gap intrinsic semiconductor is one in which the maximum energy of the valence band occurs at a different k (k-space wave vector) than the minimum energy of the conduction band. Examples include silicon and germanium. A direct band gap intrinsic semiconductor is one where the maximum energy of the valence band occurs at the same k as the minimum energy of the conduction band. Examples include gallium arsenide.

A silicon crystal is different from an insulator because at any temperature above absolute zero temperature, there is a finite probability that an electron in the lattice will be knocked loose from its position, leaving behind an electron deficiency called a "hole".

If a voltage is applied, then both the electron and the hole can contribute to a small current flow.

The conductivity of a semiconductor can be modelled in terms of the band theory of solids. The band model of a semiconductor suggests that at ordinary temperatures there is a finite possibility that electrons can reach the conduction band and contribute to electrical conduction.

The term intrinsic here distinguishes between the properties of pure "intrinsic" silicon and the dramatically different properties of doped n-type or p-type semiconductor

Electrons and Holes

In an intrinsic semiconductor such as silicon at temperatures above absolute zero, there will be some electrons which are excited across the band gap into the conduction band and which can support current flow. When the electron in pure silicon crosses the gap, it leaves behind an electron vacancy or "hole" in the regular silicon lattice. Under the influence of an external voltage, both the electron and the hole can move across the material. In an n-type semiconductor, the dopant contributes extra electrons, dramatically increasing the conductivity. In a p-type semiconductor, the dopant produces extra vacancies or holes, which likewise increase the conductivity. It is however the behaviour of the p-n junction which is the key to the enormous variety of solid-state electronic devices.

Semiconductor Current

The current which will flow in an intrinsic semiconductor consists of both electron and hole current. That is, the electrons which have been freed from their lattice positions into the conduction band can move through the material. In addition, other electrons can hop between lattice positions to fill the vacancies left by the freed electrons. This additional mechanism is called hole conduction because it is as if the holes are migrating across the material in the direction opposite to the free electron movement. The current flow in an intrinsic semiconductor is influenced by the density of energy states which in turn influences the electron density in the conduction band. This current is highly temperature dependent.

Extrinsic Semiconductor

An extrinsic semiconductor is a semiconductor that has been *doped*, that is, into which a doping agent has been introduced, giving it different electrical properties than the intrinsic (pure) semiconductor.

Doping involves adding dopant atoms to an intrinsic semiconductor, which changes the electron and hole carrier concentrations of the semiconductor at thermal equilibrium. Dominant carrier concentrations in an extrinsic semiconductor classify it as either an n-type or p-type semiconductor. The electrical properties of extrinsic semiconductors make them essential components of many electronic devices.

Semiconductor Doping

Semiconductor doping is the process that changes an intrinsic semiconductor to an extrinsic semiconductor. During doping, impurity atoms are introduced to an intrinsic semiconductor. Impurity atoms are atoms of a different element than the atoms of the intrinsic semiconductor. Impurity atoms act as either donors or acceptors to the intrinsic semiconductor, changing the electron and hole concentrations of the semiconductor. Impurity atoms are classified as donor or acceptor atoms based on the effect they have on the intrinsic semiconductor. Donor impurity atoms have more valence electrons than the atoms they replace in the intrinsic semiconductor lattice. Donor impurities "donate" their extra valence electrons to a semiconductor's conduction band, providing excess electrons to the intrinsic semiconductor. Excess electrons increase the electron carrier concentration (n_0) of the semiconductor, making it n-type.

Acceptor impurity atoms have fewer valence electrons than the atoms they replace in the intrinsic semiconductor. They "accept" electrons from the semiconductor's valence band. This provides excess holes to the intrinsic semiconductor. Excess holes increase the hole carrier concentration (p_0) of the semiconductor, creating a p-type semiconductor. Semiconductors and dopant atoms are defined by the column of the periodic in which they fall. The column definition of the semiconductor determines how many valence electrons its atoms have and whether dopant atoms act as the semiconductor's donors or acceptors. Group IV semiconductors use group V atoms as donors and group III atoms as acceptors.

Group III-V semiconductors, the compound semiconductors, use group VI atoms as donors and group II atoms as acceptors. Group III-V semiconductors can also use group IV atoms as either donors or acceptors. When a group IV atom replaces the group III element in the semiconductor lattice, the group IV atom acts as a donor. Conversely, when a group IV atom replaces the group V element, the group IV atom acts as an acceptor. Group IV atoms can act as both donors and

	Intrinsic semiconductor	*Donor atoms*	*Acceptor atoms*
Group IV semiconductors	Silicon, Germanium	Phosphorus, Arsenic	Boron, Aluminium
Group III-V semiconductors	Aluminum phosphide, Aluminum arsenide, Gallium arsenide, Gallium nitride	Selenium, Tellurium, Silicon, Germanium	Beryllium, Zinc, Cadmium, Silicon, Germanium

The Two Types of Extrinsic Semiconductor

N-type semiconductors:

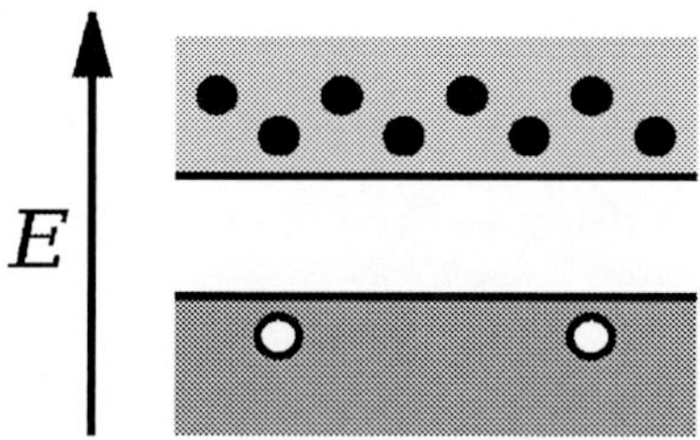

Figure: *Band structure of an n-type semiconductor. Dark circles in the conduction band are electrons and light circles in the valence band are holes. The image shows that the electrons are the majority charge carrier.*

Extrinsic semiconductors with a larger electron concentration than hole concentration are known as n-type semiconductors. The phrase 'n-type' comes from the negative charge of the electron. In n-type semiconductors, electrons are the majority carriers and holes are the minority carriers. N-type semiconductors are created by doping an intrinsic semiconductor with donor impurities. In an n-type semiconductor, the Fermi energy level is greater than that of the intrinsic semiconductor and lies closer to the conduction band than the valence band.

P-type Semiconductors

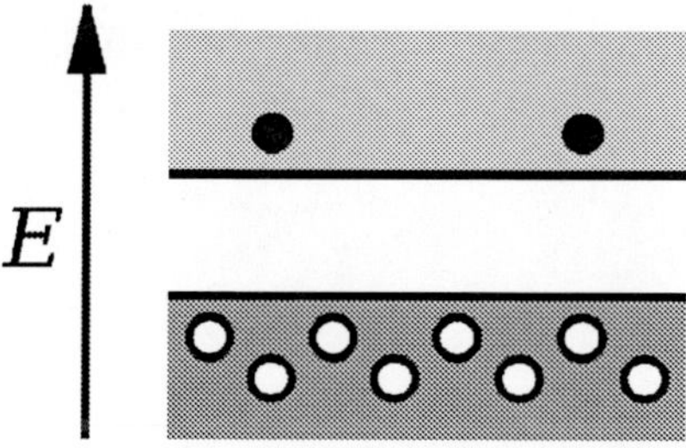

Figure: *Band structure of a p-type semiconductor. Dark circles in the conduction band are electrons and light circles in the valence band are holes. The image shows that the holes are the majority charge carrier*

As opposed to n-type semiconductors, p-type semiconductors have a larger hole concentration than electron concentration. The phrase 'p-type' refers to the positive charge of the hole. In p-type semiconductors, holes are the majority carriers and electrons are the minority carriers.

P-type semiconductors are created by doping an intrinsic semiconductor with acceptor impurities. P-type semiconductors have Fermi energy levels below the intrinsic Fermi energy level. The Fermi energy level lies closer to the valence band than the conduction band in a p-type semiconductor.

Use of Extrinsic Semiconductors

Extrinsic semiconductors are components of many common electrical devices. A semiconductor diode (devices that allow current in only one direction) consists of p-type and n-type semiconductors placed in junction with one another. Currently, most semiconductor diodes use doped silicon or germanium.

Transistors (devices that enable current switching) also make use of extrinsic semiconductors. Bipolar junction transistors (BJT) are one type of transistor.

The most common BJTs are NPN and PNP type. NPN transistors have two layers of n-type semiconductors sandwiching a p-type semiconductor. PNP transistors have two layers of p-type semiconductors sandwiching an n-type semiconductor.

Field-effect transistors (FET) are another type of transistor implementing extrinsic semiconductors. As opposed to BJTs, they are unipolar and considered either N-channel or P-channel. FETs are broken into two families, junction gate FET (JFET) and insulated gate FET (IGFET).

Other devices implementing the extrinsic semiconductor:

- Lasers
- Solar cells
- Photodetectors
- Light-emitting diodes
- Thyristors

Chapter 5

Semiconductor Devices

Semiconductor devices are electronic components that exploit the electronic properties of semiconductor materials, principally silicon, germanium, and gallium arsenide, as well as organic semiconductors. Semiconductor devices have replaced thermionic devices (vacuum tubes) in most applications. They use electronic conduction in the solid state as opposed to the gaseous state or thermionic emission in a high vacuum. Semiconductor devices are manufactured both as single discrete devices and as *integrated circuits* (ICs), which consist of a number—from a few (as low as two) to billions—of devices manufactured and interconnected on a single semiconductor substrate, or wafer.

Semiconductor materials are so useful because their behaviour can be easily manipulated by the addition of impurities, known as doping. Semiconductor conductivity can be controlled by introduction of an electric or magnetic field, by exposure to light or heat, or by mechanical deformation of a doped monocrystalline grid; thus, semiconductors can make excellent sensors.

Current conduction in a semiconductor occurs via mobile or "free" *electrons* and *holes*, collectively known as *charge carriers.* Doping a semiconductor such as silicon with a small amount of impurity atoms, such as phosphorus or boron, greatly increases the number of free electrons or holes within the semiconductor. When a doped semiconductor contains excess holes it is called "p-type", and when it contains excess free electrons it is known as "n-type", where *p* (positive for holes) or *n* (negative for electrons) is the sign of the charge of the majority mobile charge carriers. The semiconductor material used in devices is doped under highly controlled conditions in a fabrication facility, or *fab*, to precisely control the location and concentration of p- and n-type dopants. The junctions which form where n-type and p-type semiconductors join together are called p–n junctions.

Diode

In electronics, a diode is a two-terminal electronic component with asymmetric conductance, it has low (ideally zero) resistance to current flow in one direction, and high (ideally infinite) resistance in the other. A semiconductor diode, the most common type today, is a crystalline piece of semiconductor material with a p–n junction connected to two electrical terminals. A vacuum tube diode has two electrodes, a plate (anode) and a heated cathode.

The most common function of a diode is to allow an electric current to pass in one direction (called the diode's *forward* direction), while blocking current in the opposite direction (the *reverse* direction). Thus, the diode can be viewed as an electronic version of a check valve. This unidirectional behaviour is called rectification, and is used to convert alternating current to direct current, including extraction of modulation from radio signals in radio receivers—these diodes are forms of rectifiers.

However, diodes can have more complicated behaviour than this simple on–off action. Semiconductor diodes begin conducting electricity only if a certain threshold voltage or cut-in voltage is present in the forward direction (a state in which the diode is said to be *forward-biased*). The voltage drop across a forward-biased diode varies only a little with the current, and is a function of temperature; this effect can be used as a temperature sensor or voltage reference.

Semiconductor diodes' nonlinear current–voltage characteristic can be tailored by varying the semiconductor materials and doping, introducing impurities into the materials. These are exploited in special-purpose diodes that perform many different functions. For example, diodes are used to regulate voltage (Zener diodes), to protect circuits from high voltage surges (avalanche diodes), to electronically tune radio and TV receivers (varactor diodes), to generate radio frequency oscillations (tunnel diodes, Gunn diodes, IMPATT diodes), and to produce light (light emitting diodes). Tunnel diodes exhibit negative resistance, which makes them useful in some types of circuits.

Diodes were the first semiconductor electronic devices. The discovery of crystals' rectifying abilities was made by German physicist Ferdinand Braun in 1874. The first semiconductor diodes, called cat's whisker diodes, developed around 1906, were made of mineral crystals such as galena. Today most diodes are made of silicon, but other semiconductors such as selenium or germanium are sometimes used.

History

Thermionic (vacuum tube) diodes and solid state (semiconductor) diodes were developed separately, at approximately the same time, in the early 1900s, as radio receiver detectors. Until the 1950s vacuum tube diodes were more often used in radios because the early point-contact type semiconductor diodes (cat's-whisker detectors) were less stable, and because most receiving sets had vacuum tubes for amplification that could easily have diodes included in the tube (for example the 12SQ7 double-diode triode), and vacuum tube rectifiers and gas-filled rectifiers handled some high voltage/high current rectification tasks beyond the capabilities of semiconductor diodes (such as selenium rectifiers) available at the time.

Discovery of Vacuum tube Diodes

In 1873, Frederick Guthrie discovered the basic principle of operation of thermionic diodes. Guthrie discovered that a positively charged electroscope could be discharged by bringing a grounded piece of white-hot metal close to it (but not actually touching it). The same did not apply to a negatively charged electroscope, indicating that the current flow was only possible in one direction.

Thomas Edison independently rediscovered the principle on February 13, 1880. At the time, Edison was investigating why the filaments of his carbon-filament light bulbs nearly always burned out at the positive-connected end. He had a special bulb made with a metal plate sealed into the glass envelope. Using this device, he confirmed that an invisible current flowed from the glowing filament through the vacuum to the metal plate, but only when the plate was connected to the positive supply.

Edison devised a circuit where his modified light bulb effectively replaced the resistor in a DC voltmeter. Edison was awarded a patent for this invention in 1884. Since there was no apparent practical use for such a device at the time, the patent application was most likely simply a precaution in case someone else did find a use for the so-called Edison effect.

About 20 years later, John Ambrose Fleming (scientific adviser to the Marconi Company and former Edison employee) realized that the Edison effect could be used as a precision radio detector. Fleming patented the first true thermionic diode, the Fleming valve, in Britain on November 16, 1904 (followed by U.S. Patent 803,684 in November 1905).

Solid-state Diodes

In 1874 German scientist Karl Ferdinand Braun discovered the "unilateral conduction" of crystals. Braun patented the crystal rectifier in 1899. Copper oxide and selenium rectifiers were developed for power applications in the 1930s.

Indian scientist Jagadish Chandra Bose was the first to use a crystal for detecting radio waves in 1894. He also worked with microwaves in the centimeter and also the millimeter range. The crystal detector was developed into a practical device for wireless telegraphy by Greenleaf Whittier Pickard, who invented a silicon crystal detector in 1903 and received a patent for it on November 20, 1906. Other experimenters tried a variety of other substances, of which the most widely used was the mineral galena (lead sulfide). Other substances offered slightly better performance, but galena was most widely used because it had the advantage of being cheap and easy to obtain. The crystal detector in these early crystal radio sets consisted of an adjustable wire point-contact (the so-called "cat's whisker"), which could be manually moved over the face of the crystal in order to obtain optimum signal. This troublesome device was superseded by thermionic diodes by the 1920s, but after high purity semiconductor materials became available, the crystal detector returned to dominant use with the advent of inexpensive fixed-germanium diodes in the 1950s. Bell Labs also developed a germanium diode for microwave reception, and AT&T used these in their microwave towers that criss-crossed the nation starting in the late 1940s, carrying telephone and network television signals. Bell Labs did not develop a satisfactory thermionic diode for microwave reception.

Etymology

At the time of their invention, such devices were known as rectifiers. In 1919, the year tetrodes were invented, William Henry Eccles coined the term *diode* from the Greek roots *di* (from δβ), meaning "two", and *ode* (from *Aäüò*), meaning "path". (However, the word *diode* itself, as well as *triode, tetrode, penthode, hexode*, was already in use as a term of multiplex telegraphy).

Rectifiers

Although all diodes *rectify*, the term 'rectifier' is normally reserved for higher currents and voltages than would normally be found in the rectification of lower power signals; examples include:

- power supply (*half-wave* or *full-wave* or *bridge*) rectifiers
- CRT (especially TV) Extra-high voltage flyback, "damper" or "booster" diodes such as the 6AU4GTA.

Thermionic Diodes

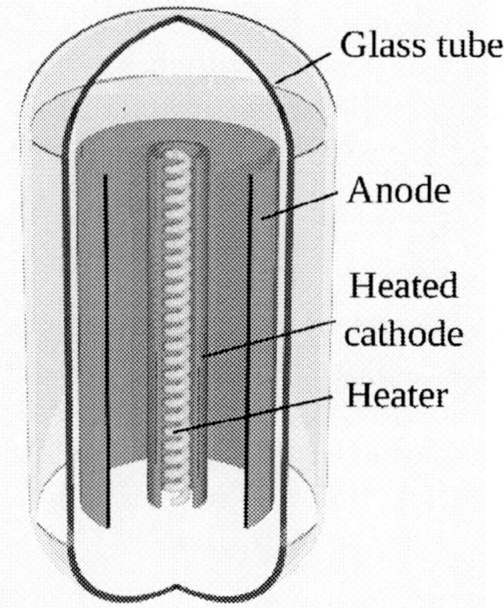

Figure: *Diode vacuum tube construction*

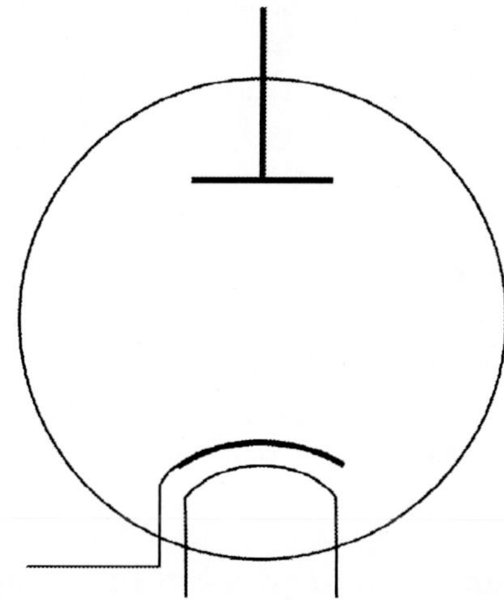

Figure: The symbol for an indirect heated vacuum-tube diode. From top to bottom, the components are the anode, the cathode, and the heater filament.

A thermionic diode is a thermionic-valve device (also known as a vacuum tube, tube, or valve), consisting of a sealed evacuated glass envelope containing two electrodes: a cathode heated by a filament, and a plate (anode). Early examples were fairly similar in appearance to incandescent light bulbs.

In operation, a separate current through the filament (heater), a high resistance wire made of nichrome, heats the cathode red hot (800-1000° C), causing it to release electrons into the vacuum, a process called thermionic emission.

The cathode is coated with oxides of alkaline earth metals such as barium and strontium oxides, which have a low work function, to increase the number of electrons emitted. (Some valves use *direct heating*, in which a tungsten filament acts as both heater and cathode.) The alternating voltage to be rectified is applied between the cathode and the concentric plate electrode. When the plate has a positive voltage with respect to the cathode, it electrostatically attracts the electrons from the cathode, so a current of electrons flows through the tube from cathode to plate. However when the polarity is reversed and the plate has a negative voltage, no current flows, because the cathode electrons are not attracted to it. The unheated plate does not emit any electrons itself. So current can only flow through the tube in one direction, from cathode to plate.

In a mercury-arc valve, an arc forms between a refractory conductive anode and a pool of liquid mercury acting as cathode. Such units were made with ratings up to hundreds of kilowatts, and were important in the development of HVDC power transmission. Some types of smaller thermionic rectifiers sometimes had mercury vapor fill to reduce their forward voltage drop and to increase current rating over thermionic hard-vacuum devices.

Throughout the vacuum tube era, valve diodes were used in analog signal applications and as rectifiers in DC power supplies in consumer electronics such as radios, televisions, and sound systems. They were replaced in power supplies beginning in the 1940s by selenium rectifiers and then by semiconductor diodes by the 1960s. Today they are still used in a few high power applications where their ability to withstand transients and their robustness gives them an advantage over semiconductor devices. The recent (2012) resurgence of interest among audiophiles and recording studios in old valve audio gear such as guitar amplifiers and home audio systems has provided a market for the legacy consumer diode valves.

Semiconductor Diodes

Electronic Symbols: The symbol used for a semiconductor diode in a circuit diagram specifies the type of diode. There are alternate symbols for some types of diodes, though the differences are minor.

Point-contact Diodes: A point-contact diode works the same as the junction diodes described below, but their construction is simpler. A block of n-type semiconductor is built, and a conducting sharp-point contact made with some group-3 metal is placed in contact with the semiconductor. Some metal migrates into the semiconductor to make a small region of p-type semiconductor near the contact. The long-popular 1N34 germanium version is still used in radio receivers as a detector and occasionally in specialized analog electronics.

Junction Diodes: Most diodes today are silicon junction diodes. A junction is formed between the p and n regions which is also called a depletion region.

p–n Junction Diode: A p–n junction diode is made of a crystal of semiconductor. Impurities are added to it to create a region on one side that contains negative charge carriers (electrons), called n-type semiconductor, and a region on the other side that contains positive charge carriers (holes), called p-type semiconductor. When two materials i.e. n-type and p-type are attached together, a momentary flow of electrons occur from n to p side resulting in a third region where no charge carriers are present. It is called Depletion region due to the absence of charge carriers (electrons and holes in this case). The diode's terminals are attached to each of these regions. The boundary between these two regions, called a p–n junction, is where the action of the diode takes place. The crystal allows electrons to flow from the N-type side (called the cathode) to the P-type side (called the anode), but not in the opposite direction.

Schottky Diode

Another type of junction diode, the Schottky diode, is formed from a metal–semiconductor junction rather than a p–n junction, which reduces capacitance and increases switching speed.

Current–voltage Characteristic: A semiconductor diode's behaviour in a circuit is given by its current–voltage characteristic, or I–V graph. The shape of the curve is determined by the transport of charge carriers through the so-called *depletion layer* or *depletion region* that exists at the p–n junction between differing semiconductors.

When a p–n junction is first created, conduction-band (mobile) electrons from the N-doped region diffuse into the P-doped region where there is a large population of holes (vacant places for electrons) with which the electrons "recombine". When a mobile electron recombines with a hole, both hole and electron vanish, leaving behind an immobile positively charged donor (dopant) on the N side and negatively charged acceptor (dopant) on the P side. The region around the p–n junction becomes depleted of charge carriers and thus behaves as an insulator.

However, the width of the depletion region (called the depletion width) cannot grow without limit. For each electron–hole pair that recombines, a positively charged dopant ion is left behind in the N-doped region, and a negatively charged dopant ion is left behind in the P-doped region. As recombination proceeds more ions are created, an increasing electric field develops through the depletion zone that acts to slow and then finally stop recombination. At this point, there is a "built-in" potential across the depletion zone.

If an external voltage is placed across the diode with the same polarity as the built-in potential, the depletion zone continues to act as an insulator, preventing any significant electric current flow (unless electron/hole pairs are actively being created in the junction by, for instance, light). This is the *reverse bias* phenomenon. However, if the polarity of the external voltage opposes the built-in potential, recombination can once again proceed, resulting in substantial electric current through the p–n junction (i.e. substantial numbers of electrons and holes recombine at the junction). For silicon diodes, the built-in potential is approximately 0.7 V (0.3 V for Germanium and 0.2 V for Schottky). Thus, if an external current is passed through the diode, about 0.7 V will be developed across the diode such that the P-doped region is positive with respect to the N-doped region and the diode is said to be "turned on" as it has a *forward bias*.

A diode's *I–V characteristic* can be approximated by four regions of operation.

At very large reverse bias, beyond the peak inverse voltage or PIV, a process called reverse breakdown occurs that causes a large increase in current (i.e., a large number of electrons and holes are created at, and move away from the p–n junction) that usually damages the device permanently. The avalanche diode is deliberately designed for use in the avalanche region. In the Zener diode, the concept of PIV is not applicable. A Zener diode contains a heavily doped p–n junction allowing electrons to tunnel from the valence band of the p-type material to the conduction band of the n-type material, such that the reverse voltage is "clamped" to a known value (called the *Zener voltage*), and avalanche does not occur.

Both devices, however, do have a limit to the maximum current and power in the clamped reverse-voltage region. Also, following the end of forward conduction in any diode, there is reverse current for a short time. The device does not attain its full blocking capability until the reverse current ceases.

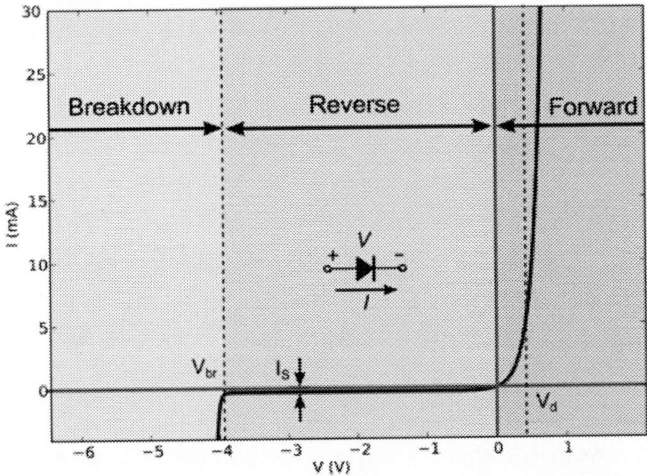

***Figure:** I–V characteristics of a p–n junction diode.*

The second region, at reverse biases more positive than the PIV, has only a very small reverse saturation current. In the reverse bias region for a normal P–N rectifier diode, the current through the device is very low (in the μA range). However, this is temperature dependent, and at sufficiently high temperatures, a substantial amount of reverse current can be observed (mA or more).

The third region is forward but small bias, where only a small forward current is conducted.

As the potential difference is increased above an arbitrarily defined "cut-in voltage" or "on-voltage" or "diode forward voltage drop (V_d)", the diode current becomes appreciable (the level of current considered "appreciable" and the value of cut-in voltage depends on the application), and the diode presents a very low resistance. The current–voltage curve is exponential. In a normal silicon diode at rated currents, the arbitrary cut-in voltage is defined as 0.6 to 0.7 volts. The value is different for other diode types—Schottky diodes can be rated as low as 0.2 V, Germanium diodes 0.25 to 0.3 V, and red or blue light-emitting diodes (LEDs) can have values of 1.4 V and 4.0 V respectively.

At higher currents the forward voltage drop of the diode increases. A drop of 1 V to 1.5 V is typical at full rated current for power diodes.

Shockley Diode Equation

The *Shockley ideal diode equation* or the *diode law* (named after transistor co-inventor William Bradford Shockley) gives the I–V characteristic of an ideal diode in either forward or reverse bias (or no bias). The *Shockley ideal diode equation* is below, where n, the ideality factor, is equal to 1 :

$$I = I_S\left(e^{V_D/(nV_T)} - 1\right),$$

where

I is the diode current,

I_S is the reverse bias saturation current (or scale current),

V_D is the voltage across the diode,

V_T is the thermal voltage, and

n is the *ideality factor*, also known as the *quality factor* or sometimes *emission coefficient*. The ideality factor n typically varies from 1 to 2 (though can in some cases be higher), depending on the fabrication process and semiconductor material and in many cases is assumed to be approximately equal to 1 (thus the notation n is omitted).

The ideality factor does not form part of the *Shockley ideal diode equation*, and was added to account for imperfect junctions as observed in real transistors. By setting n = 1 above, the equation reduces to the *Shockley ideal diode equation.*

The thermal voltage V_T is approximately 25.85 mV at 300 K, a temperature close to "room temperature" commonly used in device simulation software. At any temperature it is a known constant defined by:

$$V_T = \frac{kT}{q},$$

where k is the Boltzmann constant, T is the absolute temperature of the p–n junction, and q is the magnitude of charge of an electron (the elementary charge).

The reverse saturation current, I_S, is not constant for a given device, but varies with temperature; usually more significantly than V_T, so that V_D typically decreases as T increases.

The *Shockley ideal diode equation* or the *diode law* is derived with the assumption that the only processes giving rise to the current in the diode are drift (due to electrical field), diffusion, and thermal recombination–generation (R–G) (this equation is derived by setting n = 1 above). It also assumes that the R–G current in the depletion region is insignificant. This means that the *Shockley ideal diode equation* doesn't account for the processes involved in reverse breakdown and photon-assisted R–G. Additionally, it doesn't describe the "leveling off" of the I–V curve at high forward bias due to internal resistance. Introducing the ideality factor, n, accounts for recombination and generation of carriers.

Under *reverse bias* voltages the exponential in the diode equation is negligible, and the current is a constant (negative) reverse current value of $-I_S$. The reverse *breakdown region* is not modelled by the Shockley diode equation.

For even rather small *forward bias* voltages the exponential is very large because the thermal voltage is very small, so the subtracted '1' in the diode equation is negligible and the forward diode current is often approximated as

$$I = I_S e^{V_D/(nV_T)}$$

The use of the diode equation in circuit problems is illustrated in the article on diode modelling.

Small-signal Behaviour

For circuit design, a small-signal model of the diode behaviour often proves useful. A specific example of diode modelling is discussed in the article on small-signal circuits.

Reverse-recovery Effect

Following the end of forward conduction in a p–n type diode, a reverse current flows for a short time. The device does not attain its blocking capability until the mobile charge in the junction is depleted.

The effect can be significant when switching large currents very quickly. A certain amount of "reverse recovery time" t_r (on the order of tens of nanoseconds to a few microseconds) may be required to remove the reverse recovery charge Q_r from the diode. During this recovery time, the diode can actually conduct in the reverse direction. In certain real-world cases it can be important to consider the losses incurred by this non-ideal diode effect. However, when the slew rate of the current is not so severe (e.g. Line frequency) the effect can be safely ignored. For most applications, the effect is also negligible for Schottky diodes.

The reverse current ceases abruptly when the stored charge is depleted; this abrupt stop is exploited in step recovery diodes for generation of extremely short pulses.

Types of Semiconductor Diode

There are several types of p–n junction diodes, which either emphasize a different physical aspect of a diode often by geometric scaling, doping level, choosing the right electrodes, are just an application of a diode in a special circuit, or are really different devices like the Gunn and laser diode and the MOSFET:

Normal (p–n) diodes, which operate as described above, are usually made of doped silicon or, more rarely, germanium. Before the development of silicon power rectifier diodes, cuprous oxide and later selenium was used; its low efficiency gave it a much higher forward voltage drop (typically 1.4 to 1.7 V per "cell", with multiple cells stacked to increase the peak inverse voltage rating in high voltage rectifiers), and required a large heat sink (often an extension of the diode's metal substrate), much larger than a silicon diode of the same current ratings would require.

The vast majority of all diodes are the p–n diodes found in CMOS integrated circuits, which include two diodes per pin and many other internal diodes.

Avalanche Diodes

Diodes that conduct in the reverse direction when the reverse bias voltage exceeds the breakdown voltage. These are electrically very similar to Zener diodes, and are often mistakenly called Zener diodes, but break down by a different mechanism, the *avalanche effect*. This occurs when the reverse electric field across the p–n junction causes a wave of ionization, reminiscent of an avalanche, leading to a large current. Avalanche diodes are designed to break down at a well-defined reverse voltage without being destroyed. The difference between the avalanche diode (which has a reverse breakdown above about 6.2 V) and the Zener is that the channel length of the former exceeds the mean free path of the electrons, so there are collisions between them on the way out. The only practical difference is that the two types have temperature coefficients of opposite polarities.

Cat's Whisker or Crystal Diodes

These are a type of point-contact diode. The cat's whisker diode consists of a thin or sharpened metal wire pressed against a semiconducting crystal, typically galena or a piece of coal. The wire forms the anode and the crystal forms the cathode. Cat's whisker diodes were also called crystal diodes and found application in crystal radio receivers. Cat's whisker diodes are generally obsolete, but may be available from a few manufacturers.

Constant Current Diodes

These are actually a JFET with the gate shorted to the source, and function like a two-terminal current-limiter analog to the Zener diode, which is limiting voltage. They allow a current through them to rise to a certain value, and then level off at a specific value. Also called *CLDs*, *constant-current diodes*, *diode-connected transistors*, or *current-regulating diodes*.

Esaki or Tunnel Diodes

These have a region of operation showing negative resistance caused by quantum tunnelling, allowing amplification of signals and very simple bistable circuits. Due to the high carrier concentration, tunnel diodes are very fast, may be used at low (mK) temperatures, high magnetic fields, and in high radiation environments. Because of these properties, they are often used in spacecraft.

Gunn Diodes

These are similar to tunnel diodes in that they are made of materials such as GaAs or InP that exhibit a region of negative differential resistance. With appropriate biasing, dipole domains form and travel across the diode, allowing high frequency microwave oscillators to be built.

Light-emitting Diodes (LEDs)

In a diode formed from a direct band-gap semiconductor, such as gallium arsenide, carriers that cross the junction emit photons when they recombine with the majority carrier on the other side. Depending on the material, wavelengths (or colours) from the infrared to the near ultraviolet may be produced. The forward potential of these diodes depends on the wavelength of the emitted photons: 2.1 V corresponds to red, 4.0 V to violet. The first LEDs were red and yellow, and higher-frequency diodes have been developed over time. All LEDs produce incoherent, narrow-spectrum light; "white" LEDs are actually combinations of three LEDs of a different colour, or a blue LED with a yellow scintillator coating.

LEDs can also be used as low-efficiency photodiodes in signal applications. An LED may be paired with a photodiode or phototransistor in the same package, to form an opto-isolator.

Laser Diodes

When an LED-like structure is contained in a resonant cavity formed by polishing the parallel end faces, a laser can be formed. Laser diodes are commonly used in optical storage devices and for high speed optical communication.

Thermal Diodes

This term is used both for conventional p–n diodes used to monitor temperature due to their varying forward voltage with temperature, and for Peltier heat pumps for thermoelectric heating and cooling. Peltier heat pumps may be made from semiconductor, though they do not have any rectifying junctions, they use the differing behaviour of charge carriers in N and P type semiconductor to move heat.

Photodiodes

All semiconductors are subject to optical charge carrier generation. This is typically an undesired effect, so most semiconductors are packaged in light blocking material. Photodiodes are intended to sense light (photodetector), so they are packaged in materials that allow light to pass, and are usually PIN (the kind of diode most sensitive to light). A photodiode can be used in solar cells, in photometry, or in optical communications. Multiple photodiodes may be packaged in a single device, either as a linear array or as a two-dimensional array. These arrays should not be confused with charge-coupled devices.

PIN Diodes

A PIN diode has a central un-doped, or *intrinsic*, layer, forming a p-type/intrinsic/n-type structure. They are used as radio frequency switches and attenuators. They are also used as large volume ionizing radiation detectors and as photodetectors. PIN diodes are also used in power electronics, as their central layer can withstand high voltages.

Furthermore, the PIN structure can be found in many power semiconductor devices, such as IGBTs, power MOSFETs, and thyristors.

Schottky Diodes

Schottky diodes are constructed from a metal to semiconductor contact. They have a lower forward voltage drop than p–n junction diodes. Their forward voltage drop at forward currents of about 1 mA is in the range 0.15 V to 0.45 V, which makes them useful in voltage clamping applications and prevention of transistor saturation. They can also be used as low loss rectifiers, although their reverse leakage current is in general higher than that of other diodes. Schottky diodes are majority carrier devices and so do not suffer from minority carrier storage problems that slow down many other diodes—so they have a faster reverse recovery than p–n junction diodes. They also tend to have much lower junction capacitance than p–n diodes, which provides for high switching speeds and their use in high-speed circuitry and RF devices such as switched-mode power supply, mixers, and detectors.

Super Barrier Diodes

Super barrier diodes are rectifier diodes that incorporate the low forward voltage drop of the Schottky diode with the surge-handling capability and low reverse leakage current of a normal p–n junction diode.

Gold-doped Diodes

As a dopant, gold (or platinum) acts as recombination centres, which helps a fast recombination of minority carriers. This allows the diode to operate at signal frequencies, at the expense of a higher forward voltage drop. Gold-doped diodes are faster than other p–n diodes (but not as fast as Schottky diodes). They also have less reverse-current leakage than Schottky diodes (but not as good as other p–n diodes). A typical example is the 1N914.

Snap-off or Step Recovery Diodes

The term *step recovery* relates to the form of the reverse recovery characteristic of these devices. After a forward current has been passing in an SRD and the current is interrupted or reversed, the reverse conduction will cease very abruptly (as in a step waveform). SRDs can, therefore, provide very fast voltage transitions by the very sudden disappearance of the charge carriers.

Stabistors or Forward Reference Diodes

The term *stabistor* refers to a special type of diodes featuring extremely stable forward voltage characteristics. These devices are specially designed for low-voltage stabilization applications requiring a guaranteed voltage over a wide current range and highly stable over temperature.

Transient Voltage Suppression Diode (TVS)

These are avalanche diodes designed specifically to protect other semiconductor devices from high-voltage transients. Their p–n junctions have a much larger cross-sectional area than those of a normal diode, allowing them to conduct large currents to ground without sustaining damage.

Varicap or Varactor Diodes

These are used as voltage-controlled capacitors. These are important in PLL (phase-locked loop) and FLL (frequency-locked loop) circuits, allowing tuning circuits, such as those in television receivers, to lock quickly. They also enabled tunable oscillators in early discrete tuning of radios, where a cheap and stable, but fixed-frequency, crystal oscillator provided the reference frequency for a voltage-controlled oscillator.

Zener Diodes

Diodes that can be made to conduct backward, correctly termed reverse breakdown diodes. This effect, called Zener breakdown, occurs at a precisely defined voltage, allowing the diode to be used as a precision voltage reference. The term Zener diode is colloquially applied to several types of breakdown diodes, but strictly speaking Zener diodes have a breakdown voltage of below 5 volts, whilst those above that value are usually avalanche diodes.

In practical voltage reference circuits, Zener and switching diodes are connected in series and opposite directions to balance the temperature coefficient to near-zero. Some devices labelled as high-voltage Zener diodes are actually avalanche diodes. Two (equivalent) Zeners in series and in reverse order, in the same package, constitute a transient absorber (or Transorb, a registered trademark). The Zener diode is named for Dr. Clarence Melvin Zener of Carnegie Mellon University, inventor of the device.

Other uses for semiconductor diodes include sensing temperature, and computing analog logarithms.

Numbering and Coding Schemes

There are a number of common, standard and manufacturer-driven numbering and coding schemes for diodes; the two most common being the EIA/JEDEC standard and the European Pro Electron standard:

EIA/JEDEC

The standardized 1N-series numbering *EIA370* system was introduced in the US by EIA/JEDEC (Joint Electron Device Engineering Council) about 1960. Among the most popular in this series were: 1N34A/1N270 (Germanium signal), 1N914/1N4148 (Silicon signal), 1N4001-1N4007 (Silicon 1A power rectifier) and 1N54xx (Silicon 3A power rectifier)

Jis

The JIS semiconductor designation system has all semiconductor diode designations starting with "1S".

Pro Electron

The European Pro Electron coding system for active components was introduced in 1966 and comprises two letters followed by the part code. The first letter represents the semiconductor material used for the component (A = Germanium and B = Silicon) and the second letter represents the general function of the part (for diodes: A = low-power/ signal, B = Variable capacitance, X = Multiplier, Y = Rectifier and Z = Voltage reference), for example:

- AA-series germanium low-power/signal diodes (e.g.: AA119)
- BA-series silicon low-power/signal diodes (e.g.: BAT18 Silicon RF Switching Diode)
- BY-series silicon rectifier diodes (e.g.: BY127 1250V, 1A rectifier diode)
- BZ-series silicon Zener diodes (e.g.: BZY88C4V7 4.7V Zener diode)

Other common numbering / coding systems (generally manufacturer-driven) include:

- GD-series germanium diodes (e.g.: GD9) – this is a very old coding system
- OA-series germanium diodes (e.g.: OA47) – a coding sequence developed by Mullard, a UK company

As well as these common codes, many manufacturers or organisations have their own systems too – for example:

- HP diode 1901-0044 = JEDEC 1N4148
- UK military diode CV448 = Mullard type OA81 = GEC type GEX23

Related Devices

- Rectifier
- Transistor
- Thyristor or silicon controlled rectifier (SCR)
- TRIAC
- Diac
- Varistor

In optics, an equivalent device for the diode but with laser light would be the Optical isolator, also known as an Optical Diode, that allows light to only pass in one direction. It uses a Faraday rotator as the main component.

Applications

Radio Demodulation: The first use for the diode was the demodulation of amplitude modulated (AM) radio broadcasts. The history of this discovery is treated in depth in the radio article. In summary, an AM signal consists of alternating positive and negative peaks of a radio carrier wave, whose amplitude or envelope is proportional to the original audio signal.

The diode (originally a crystal diode) rectifies the AM radio frequency signal, leaving only the positive peaks of the carrier wave. The audio is then extracted from the rectified carrier wave using a simple filter and fed into an audio amplifier or transducer, which generates sound waves.

Power Conversion

Rectifiers are constructed from diodes, where they are used to convert alternating current (AC) electricity into direct current (DC). Automotive alternators are a common example, where the diode, which rectifies the AC into DC, provides better performance than the commutator or earlier, dynamo. Similarly, diodes are also used in *Cockcroft–Walton voltage multipliers* to convert AC into higher DC voltages.

Over-voltage Protection

Diodes are frequently used to conduct damaging high voltages away from sensitive electronic devices. They are usually reverse-biased (non-conducting) under normal circumstances.

When the voltage rises above the normal range, the diodes become forward-biased (conducting). For example, diodes are used in (stepper motor and H-bridge) motor controller and relay circuits to de-energize coils rapidly without the damaging voltage spikes that would otherwise occur. (Any diode used in such an application is called a flyback diode). Many integrated circuits also incorporate diodes on the connection pins to prevent external voltages from damaging their sensitive transistors. Specialized diodes are used to protect from over-voltages at higher power.

Logic Gates

Diodes can be combined with other components to construct AND and OR logic gates. This is referred to as diode logic.

Ionizing Radiation Detectors

In addition to light, mentioned above, semiconductor diodes are sensitive to more energetic radiation. In electronics, cosmic rays and other sources of ionizing radiation cause noise pulses and single and multiple bit errors. This effect is sometimes exploited by particle detectors to detect radiation.

A single particle of radiation, with thousands or millions of electron volts of energy, generates many charge carrier pairs, as its energy is deposited in the semiconductor material. If the depletion layer is large enough to catch the whole shower or to stop a heavy particle, a fairly accurate measurement of the particle's energy can be made, simply by measuring the charge conducted and without the complexity of a magnetic spectrometer, etc.

These semiconductor radiation detectors need efficient and uniform charge collection and low leakage current. They are often cooled by liquid nitrogen. For longer-range (about a centimetre) particles, they need a very large depletion depth and large area. For short-range particles, they need any contact or un-depleted semiconductor on at least one surface to be very thin. The back-bias voltages are near breakdown (around a thousand volts per centimetre). Germanium and silicon are common materials. Some of these detectors sense position as well as energy. They have a finite life, especially when detecting heavy particles, because of radiation damage. Silicon and germanium are quite different in their ability to convert gamma rays to electron showers.

Semiconductor detectors for high-energy particles are used in large numbers. Because of energy loss fluctuations, accurate measurement of the energy deposited is of less use.

Temperature Measurements

A diode can be used as a temperature measuring device, since the forward voltage drop across the diode depends on temperature, as in a silicon bandgap temperature sensor. From the Shockley ideal diode equation given above, it might *appear* that the voltage has a *positive* temperature coefficient (at a constant current), but usually the variation of the reverse saturation current term is more significant than the variation in the thermal voltage term. Most diodes therefore have a *negative* temperature coefficient, typically "2 mV/ÚC for silicon diodes at room temperature. This is approximately linear for temperatures above about 20 kelvins. Some graphs are given for: 1N400x series, and CY7 cryogenic temperature sensor.

Current Steering

Diodes will prevent currents in unintended directions. To supply power to an electrical circuit during a power failure, the circuit can draw current from a battery. An uninterruptible power supply may use diodes in this way to ensure that current is only drawn from the battery when necessary.

Likewise, small boats typically have two circuits each with their own battery/batteries: one used for engine starting; one used for domestics. Normally, both are charged from a single alternator, and a heavy-duty split-charge diode is used to prevent the higher-charge battery (typically the engine battery) from discharging through the lower-charge battery when the alternator is not running.

Diodes are also used in electronic musical keyboards. To reduce the amount of wiring needed in electronic musical keyboards, these instruments often use keyboard matrix circuits. The keyboard controller scans the rows and columns to determine which note the player has pressed. The problem with matrix circuits is that, when several notes are pressed at once, the current can flow backwards through the circuit and trigger "phantom keys" that cause "ghost" notes to play. To avoid triggering unwanted notes, most keyboard matrix circuits have diodes soldered with the switch under each key of the musical keyboard. The same principle is also used for the switch matrix in solid-state pinball machines.

Abbreviations

Diodes are usually referred to as *D* for diode on PCBs. Sometimes the abbreviation *CR* for *crystal rectifier* is used.

Two-terminal Nonlinear Devices

Many other two-terminal nonlinear devices exist, for example a neon lamp has two terminals in a glass envelope and has interesting and useful nonlinear properties. Lamps including arc-discharge lamps, incandescent lamps, fluorescent lamps and mercury vapor lamps have two terminals and display nonlinear current–voltage characteristics.

Transistor

A transistor is a semiconductor device used to amplify and switch electronic signals and electrical power. It is composed of semiconductor material with at least three terminals for connection to an external circuit. A voltage or current applied to one pair of the transistor's terminals changes the current through another pair of terminals. Because the controlled (output) power can be higher than the controlling (input) power, a transistor can amplify a signal. Today, some transistors are packaged individually, but many more are found embedded in integrated circuits.

The transistor is the fundamental building block of modern electronic devices, and is ubiquitous in modern electronic systems. Following its development in the early 1950s, the transistor revolutionized the field of electronics, and paved the way for smaller and cheaper radios, calculators, and computers, among other things.

History

The thermionic triode, a vacuum tube invented in 1907, propelled the electronics age forward, enabling amplified radio technology and long-distance telephony. The triode, however, was a fragile device that consumed a lot of power. Physicist Julius Edgar Lilienfeld filed a patent for a field-effect transistor (FET) in Canada in 1925, which was intended to be a solid-state replacement for the triode. Lilienfeld also filed identical patents in the United States in 1926 and 1928. However, Lilienfeld did not publish any research articles about his devices nor did his patents cite any specific examples of a working prototype. Because the production of high-quality semiconductor materials was still decades away, Lilienfeld's solid-state amplifier ideas would not have found practical use in the 1920s and 1930s, even if such a device had been built. In 1934, German inventor Oskar Heil patented a similar device.

From November 17, 1947 to December 23, 1947, John Bardeen and Walter Brattain at AT&T's Bell Labs in the United States, performed experiments and observed that when two gold point contacts were applied to a crystal of germanium, a signal was produced with the output power greater than the input. Solid State Physics Group leader William Shockley saw the potential in this, and over the next few months worked to greatly expand the knowledge of semiconductors. The term *transistor* was coined by John R. Pierce as a portmanteau of the term "transfer resistor".

According to Lillian Hoddeson and Vicki Daitch, authors of a biography of John Bardeen, Shockley had proposed that Bell Labs' first patent for a transistor should be based on the field-effect and that he be named as the inventor. Having unearthed Lilienfeld's patents that went into obscurity years earlier, lawyers at Bell Labs advised against Shockley's proposal because the idea of a field-effect transistor that used an electric field as a "grid" was not new. Instead, what Bardeen, Brattain, and Shockley invented in 1947 was the first point-contact transistor. In acknowledgement of this accomplishment, Shockley, Bardeen, and Brattain were jointly awarded the 1956 Nobel Prize in Physics "for their researches on semiconductors and their discovery of the transistor effect."

In 1948, the point-contact transistor was independently invented by German physicists Herbert Mataré and Heinrich Welker while working at the Compagnie des Freins et Signaux, a Westinghouse subsidiary located in Paris. Mataré had previous experience in developing crystal rectifiers from silicon and germanium in the German radar effort during World War II.

Using this knowledge, he began researching the phenomenon of "interference" in 1947. By witnessing currents flowing through point-contacts, similar to what Bardeen and Brattain had accomplished earlier in December 1947, Mataré by June 1948, was able to produce consistent results by using samples of germanium produced by Welker. Realizing that Bell Labs' scientists had already invented the transistor before them, the company rushed to get its "transistron" into production for amplified use in France's telephone network.

The first high-frequency transistor was the surface-barrier germanium transistor developed by Philco in 1953, capable of operating up to 60 MHz.

These were made by etching depressions into an N-type germanium base from both sides with jets of indium sulphate until it was a few ten-thousandths of an inch thick. Indium electroplated into the depressions formed the collector and emitter. The first all-transistor car radio, which was produced in 1955 by Chrysler and Philco, used these transistors in its circuitry and also they were the first suitable for high-speed computers.

The first working silicon transistor was developed at Bell Labs on January 26, 1954 by Morris Tanenbaum. The first commercial silicon transistor was produced by Texas Instruments in 1954. This was the work of Gordon Teal, an expert in growing crystals of high purity, who had previously worked at Bell Labs. The first MOS transistor actually built was by Kahng and Atalla at Bell Labs in 1960.

In March 2013 American researchers at Stanford University announced they had built a transistor out of DNA and RNA molecules.

Importance

The transistor is the key active component in practically all modern electronics. Many consider it to be one of the greatest inventions of the 20th century. Its importance in today's society rests on its ability to be mass-produced using a highly automated process (semiconductor device fabrication) that achieves astonishingly low per-transistor costs. The invention of the first transistor at Bell Labs was named an IEEE Milestone in 2009.

Although several companies each produce over a billion individually packaged (known as *discrete*) transistors every year, the vast majority of transistors are now produced in integrated circuits (often shortened to *IC*, *microchips* or simply *chips*), along with diodes, resistors, capacitors and other electronic components, to produce complete electronic circuits. A logic gate consists of up to about twenty transistors whereas an advanced microprocessor, as of 2009, can use as many as 3 billion transistors (MOSFETs). "About 60 million transistors were built in 2002 ... for [each] man, woman, and child on Earth."

The transistor's low cost, flexibility, and reliability have made it a ubiquitous device. Transistorized mechatronic circuits have replaced electromechanical devices in controlling appliances and machinery. It is often easier and cheaper to use a standard microcontroller and write a computer programme to carry out a control function than to design an equivalent mechanical control function.

Simplified Operation

The essential usefulness of a transistor comes from its ability to use a small signal applied between one pair of its terminals to control a much larger signal at another pair of terminals. This property is called gain. A transistor can control its output in proportion to the input signal; that is, it can act as an amplifier. Alternatively, the transistor can be used to turn current on or off in a circuit as an electrically controlled switch, where the amount of current is determined by other circuit elements. There are two types of transistors, which have slight differences in how they are used in a circuit. A *bipolar transistor* has terminals labelled base, collector, and emitter. A small current at the base terminal (that is, flowing between the base and the emitter) can control or switch a much larger current between the collector and emitter terminals. For a *field-effect transistor*, the terminals are labelled gate, source, and drain, and a voltage at the gate can control a current between source and drain.

The image to the right represents a typical bipolar transistor in a circuit. Charge will flow between emitter and collector terminals depending on the current in the base. Because internally the base and emitter connections behave like a semiconductor diode, a voltage drop develops between base and emitter while the base current exists. The amount of this voltage depends on the material the transistor is made from, and is referred to as V_{BE}.

Transistor as a Switch

Transistors are commonly used as electronic switches, both for high-power applications such as switched-mode power supplies and for low-power applications such as logic gates.

In a grounded-emitter transistor circuit, such as the light-switch circuit shown, as the base voltage rises, the emitter and collector currents rise exponentially. The collector voltage drops because of reduced resistance from collector to emitter. If the voltage difference between the collector and emitter were zero (or near zero), the collector current would be limited only by the load resistance (light bulb) and the supply voltage. This is called *saturation* because current is flowing from collector to emitter freely. When saturated the switch is said to be *on*.

Providing sufficient base drive current is a key problem in the use of bipolar transistors as switches. The transistor provides current gain, allowing a relatively large current in the collector to be switched by a much smaller current into the base terminal.

The ratio of these currents varies depending on the type of transistor, and even for a particular type, varies depending on the collector current. In the example light-switch circuit shown, the resistor is chosen to provide enough base current to ensure the transistor will be saturated.

In any switching circuit, values of input voltage would be chosen such that the output is either completely off, or completely on. The transistor is acting as a switch, and this type of operation is common in digital circuits where only "on" and "off" values are relevant.

Transistor as an Amplifier

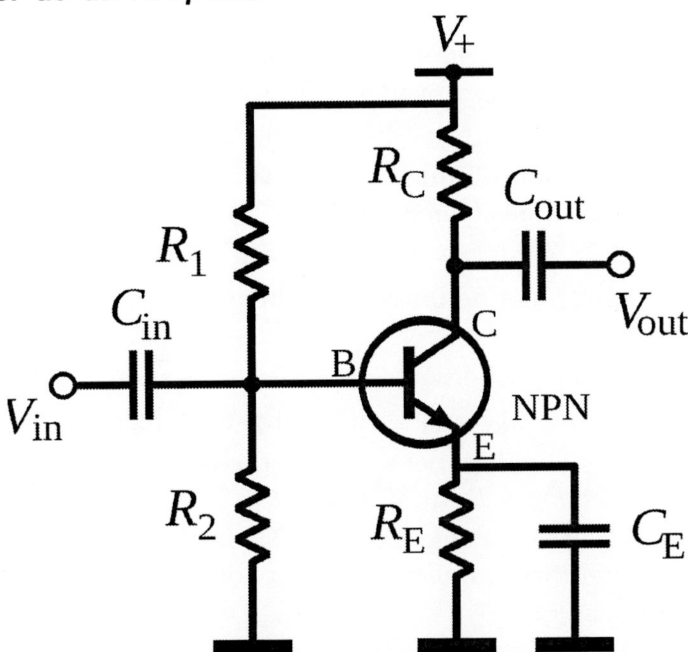

Figure: *Amplifier circuit, common-emitter configuration with a voltage-divider bias circuit.*

The common-emitter amplifier is designed so that a small change in voltage (V_{in}) changes the small current through the base of the transistor; the transistor's current amplification combined with the properties of the circuit mean that small swings in V_{in} produce large changes in V_{out}.

Various configurations of single transistor amplifier are possible, with some providing current gain, some voltage gain, and some both. From mobile phones to televisions, vast numbers of products include amplifiers for sound reproduction, radio transmission, and signal processing. The first discrete-transistor audio amplifiers barely supplied a few hundred milliwatts, but power and audio fidelity gradually increased as better transistors became available and amplifier architecture evolved. Modern transistor audio amplifiers of up to a few hundred watts are common and relatively inexpensive.

Comparison with Vacuum Trubes

Prior to the development of transistors, vacuum (electron) tubes (or in the UK "thermionic valves" or just "valves") were the main active components in electronic equipment.

Advantages

The key advantages that have allowed transistors to replace their vacuum tube predecessors in most applications are

- No power consumption by a cathode heater.
- Small size and minimal weight, allowing the development of miniaturized electronic devices.
- Low operating voltages compatible with batteries of only a few cells.
- No warm-up period for cathode heaters required after power application.
- Lower power dissipation and generally greater energy efficiency.
- Higher reliability and greater physical ruggedness.
- Extremely long life. Some transistorized devices have been in service for more than 50 years.
- Complementary devices available, facilitating the design of complementary-symmetry circuits, something not possible with vacuum tubes.
- Insensitivity to mechanical shock and vibration, thus avoiding the problem of microphonics in audio applications.

Limitations

- Silicon transistors can age and fail.
- High-power, high-frequency operation, such as that used in over-the-air television broadcasting, is better achieved in vacuum tubes due to improved electron mobility in a vacuum.

- Solid-state devices are more vulnerable to Electrostatic discharge in handling and operation
- A vacuum tube momentarily overloaded will just get a little hotter; solid-state devices have less mass to absorb the heat due to overloads, in proportion to their rating
- Sensitivity to radiation and cosmic rays (special radiation hardened chips are used for spacecraft devices).
- Vacuum tubes create a distortion, the so-called tube sound, that some people find to be more tolerable to the ear.

Bipolar junction transistor (BJT)

Bipolar transistors are so named because they conduct by using both majority and minority carriers. The bipolar junction transistor, the first type of transistor to be mass-produced, is a combination of two junction diodes, and is formed of either a thin layer of p-type semiconductor sandwiched between two n-type semiconductors (an n–p–n transistor), or a thin layer of n-type semiconductor sandwiched between two p-type semiconductors (a p–n–p transistor).

This construction produces two p–n junctions: a base–emitter junction and a base–collector junction, separated by a thin region of semiconductor known as the base region (two junction diodes wired together without sharing an intervening semiconducting region will not make a transistor).

BJTs have three terminals, corresponding to the three layers of semiconductor—an *emitter*, a *base*, and a *collector*. They are useful in amplifiers because the currents at the emitter and collector are controllable by a relatively small base current." In an n–p–n transistor operating in the active region, the emitter–base junction is forward biased (electrons and holes recombine at the junction), and electrons are injected into the base region.

Because the base is narrow, most of these electrons will diffuse into the reverse-biased (electrons and holes are formed at, and move away from the junction) base–collector junction and be swept into the collector; perhaps one-hundredth of the electrons will recombine in the base, which is the dominant mechanism in the base current. By controlling the number of electrons that can leave the base, the number of electrons entering the collector can be controlled. Collector current is approximately β (common-emitter current gain) times the base current. It is typically greater than 100 for small-signal transistors but can be smaller in transistors designed for high-power applications. Unlike the field-effect transistor, the BJT is a low–input-impedance device. Also, as the base–emitter voltage (V_{be}) is increased the base–emitter current and hence the collector–emitter current (I_{ce}) increase exponentially according to the Shockley diode model and the Ebers-Moll model. Because of this exponential relationship, the BJT has a higher transconductance than the FET.

Bipolar transistors can be made to conduct by exposure to light, because absorption of photons in the base region generates a photocurrent that acts as a base current; the collector current is approximately β times the photocurrent. Devices designed for this purpose have a transparent window in the package and are called phototransistors.

Field-effect Transistor (FET)

The *field-effect transistor*, sometimes called a *unipolar transistor*, uses either electrons (in *n-channel FET*) or holes (in *p-channel FET*) for conduction. The four terminals of the FET are named *source*, *gate*, *drain*, and *body* (*substrate*). On most FETs, the body is connected to the source inside the package, and this will be assumed for the following description.

In a FET, the drain-to-source current flows via a conducting channel that connects the *source* region to the *drain* region. The conductivity is varied by the electric field that is produced when a voltage is applied between the gate and source terminals; hence the current flowing between the drain and source is controlled by the voltage applied between the gate and source. As the gate–source voltage (V_{gs}) is increased, the drain–source current (I_{ds}) increases exponentially for V_{gs} below threshold, and then at a roughly quadratic rate ($I_{ds} \propto (V_{gs} - V_T)^2$) (where V_T is the threshold voltage at which drain current begins) in the "space-charge-limited" region above threshold. A quadratic behaviour is not observed in modern devices, for example, at the 65 nm technology node.

For low noise at narrow bandwidth the higher input resistance of the FET is advantageous.

FETs are divided into two families: *junction FET* (JFET) and *insulated gate FET* (IGFET). The IGFET is more commonly known as a *metal–oxide–semiconductor FET* (MOSFET), reflecting its original construction from layers of metal (the gate), oxide (the insulation), and semiconductor. Unlike IGFETs, the JFET gate forms a p–n diode with the channel which lies between the source and drain. Functionally, this makes the n-channel JFET the solid-state equivalent of the vacuum tube triode which, similarly, forms a diode between its grid and cathode. Also, both devices operate in the *depletion mode*, they both have a high input impedance, and they both conduct current under the control of an input voltage.

Metal–semiconductor FETs (MESFETs) are JFETs in which the reverse biased p–n junction is replaced by a metal–semiconductor junction. These, and the HEMTs (high-electron-mobility transistors, or HFETs), in which a two-dimensional electron gas with very high carrier mobility is used for charge transport, are especially suitable for use at very high frequencies (microwave frequencies; several GHz).

Unlike bipolar transistors, FETs do not inherently amplify a photocurrent. Nevertheless, there are ways to use them, especially JFETs, as light-sensitive devices, by exploiting the photocurrents in channel–gate or channel–body junctions.

FETs are further divided into *depletion-mode* and *enhancement-mode* types, depending on whether the channel is turned on or off with zero gate-to-source voltage.

For enhancement mode, the channel is off at zero bias, and a gate potential can "enhance" the conduction. For the depletion mode, the channel is on at zero bias, and a gate potential (of the opposite polarity) can "deplete" the channel, reducing conduction. For either mode, a more positive gate voltage corresponds to a higher current for n-channel devices and a lower current for p-channel devices. Nearly all JFETs are depletion-mode because the diode junctions would forward bias and conduct if they were enhancement-mode devices; most IGFETs are enhancement-mode types.

Usage of Bipolar and Field-effect Transistors

The bipolar junction transistor (BJT) was the most commonly used transistor in the 1960s and 70s. Even after MOSFETs became widely available, the BJT remained the transistor of choice for many analog circuits such as amplifiers because of their greater linearity and ease of manufacture. In integrated circuits, the desirable properties of MOSFETs allowed them to capture nearly all market share for digital circuits. Discrete MOSFETs can be applied in transistor applications, including analog circuits, voltage regulators, amplifiers, power transmitters and motor drivers.

Other Transistor Types

Figure: *Transistor symbol drawn on Portuguese pavement in the University of Aveiro.*

- Bipolar junction transistor
    - Heterojunction bipolar transistor, up to several hundred GHz, common in modern ultrafast and RF circuits
    - Schottky transistor
    - Avalanche transistor
    - Darlington transistors are two BJTs connected together to provide a high current gain equal to the product of the current gains of the two transistors.
    - Insulated-gate bipolar transistors (IGBTs) use a medium-power IGFET, similarly connected to a power BJT, to give a high input impedance. Power diodes are often connected between certain terminals depending on specific use. IGBTs are particularly suitable for heavy-duty industrial applications. The Asea Brown Boveri (ABB) *5SNA2400-E170100* illustrates just how far power semiconductor technology has advanced. Intended for three-phase power supplies, this device houses three n–p–n IGBTs in a case measuring 38 by 140 by 190 mm and weighing 1.5 kg. Each IGBT is rated at 1,700 volts and can handle 2,400 amperes.

  - o Photo transistor
  - o Multiple-emitter transistor, used in transistor–transistor logic
  - o Multiple-base transistor, used to amplify very-low-level signals in noisy environments such as the pickup of a record player or radio front ends. Effectively, it is a very large number of transistors in parallel where, at the output, the signal is added constructively, but random noise is added only stochastically.
- Field-effect transistor
  - o Carbon nanotube field-effect transistor (CNFET)
  - o JFET, where the gate is insulated by a reverse-biased p–n junction
  - o MESFET, similar to JFET with a Schottky junction instead of a p–n junction
    - – High-electron-mobility transistor (HEMT, HFET, MODFET)
  - o MOSFET, where the gate is insulated by a shallow layer of insulator
  - o Inverted-T field-effect transistor (ITFET)
  - o FinFET, source/drain region shapes fins on the silicon surface.
  - o FREDFET, fast-reverse epitaxial diode field-effect transistor
  - o Thin-film transistor, in LCDs.
  - o Organic field-effect transistor (OFET), in which the semiconductor is an organic compound
  - o Ballistic transistor
  - o Floating-gate transistor, for non-volatile storage.
  - o FETs used to sense environment
    - – Ion-sensitive field effect transistor (IFSET), to measure ion concentrations in solution.
    - – EOSFET, electrolyte-oxide-semiconductor field-effect transistor (Neurochip)
    - – DNAFET, deoxyribonucleic acid field-effect transistor
- Diffusion transistor, formed by diffusing dopants into semiconductor substrate; can be both BJT and FET

- Unijunction transistors can be used as simple pulse generators. They comprise a main body of either P-type or N-type semiconductor with ohmic contacts at each end (terminals *Base1* and *Base2*). A junction with the opposite semiconductor type is formed at a point along the length of the body for the third terminal (*Emitter*).
- Single-electron transistors (SET) consist of a gate island between two tunnelling junctions. The tunnelling current is controlled by a voltage applied to the gate through a capacitor.
- Nanofluidic transistor, controls the movement of ions through sub-microscopic, water-filled channels.
- Multigate devices
    - Tetrode transistor
    - Pentode transistor
    - Trigate transistors (Prototype by Intel)
    - Dual-gate FETs have a single channel with two gates in cascode; a configuration optimized for *high-frequency amplifiers*, *mixers*, and oscillators.
- Junctionless nanowire transistor (JNT), developed at Tyndall National Institute in Ireland, was the first transistor successfully fabricated without junctions. (Even MOSFETs have junctions, although its gate is electrically insulated from the region the gate controls.) Junctions are difficult and expensive to fabricate, and, because they are a significant source of current leakage, they waste significant power and generate significant waste heat. Eliminating them held the promise of cheaper and denser microchips. The JNT uses a simple nanowire of silicon surrounded by an electrically isolated "wedding ring" that acts to gate the flow of electrons through the wire. This method has been described as akin to squeezing a garden hose to gate the flow of water through the hose. The nanowire is heavily n-doped, making it an excellent conductor. Crucially the gate, comprising silicon, is heavily p-doped; and its presence depletes the underlying silicon nanowire thereby preventing carrier flow past the gate.
- Vacuum-channel transistor: In 2012, NASA and the National Nanofab Centre in South Korea were reported to have built a prototype vacuum-channel transistor in only 150 nanometers in size, can be manufactured cheaply using standard silicon

semiconductor processing, can operate at high speeds even in hostile environments, and could consume just as much power as a standard transistor.

Part Numbering Standards/Specifications

The types of some transistors can be parsed from the part number. There are three major semiconductor naming standards; in each the alphanumeric prefix provides clues to type of the device.

Japanese Industrial Standard (JIS)

JIS	***Transistor Prefix Table***
Prefix	Type of transistor
2SA	high-frequency p–n–p BJTs
2SB	audio-frequency p–n–p BJTs
2SC	high-frequency n–p–n BJTs
2SD	audio-frequency n–p–n BJTs
2SJ	P-channel FETs (both JFETs and MOSFETs)
2SK	N-channel FETs (both JFETs and MOSFETs)

The *JIS-C-7012* specification for transistor part numbers starts with "2S", e.g. 2SD965, but sometimes the "2S" prefix is not marked on the package – a 2SD965 might only be marked "D965"; a 2SC1815 might be listed by a supplier as simply "C1815".

This series sometimes has suffixes (such as "R", "O", "BL"... standing for "Red", "Orange", "Blue" etc.) to denote variants, such as tighter h_{FE} (gain) groupings.

European Electronic Component Manufacturers Association (EECA)

The Pro Electron standard, the European Electronic Component Manufacturers Association part numbering scheme, begins with two letters: the first gives the semiconductor type (A for germanium, B for silicon, and C for materials like GaAs); the second letter denotes the intended use (A for diode, C for general-purpose transistor, etc.). A 3-digit sequence number (or one letter then 2 digits, for industrial types) follows.

With early devices this indicated the case type. Suffixes may be used, with a letter (e.g. "C" often means high h_{FE}, such as in: BC549C) or other codes may follow to show gain (e.g. BC327-25) or voltage rating (e.g. BUK854-800A). The more common prefixes are:

Table: *Pro Electron / EECA Transistor Prefix Table*

Prefix class	***Type and usage***	***Example***	***Equivalent***	***Reference***
AC	Germanium small-signal AF transistor	AC126	NTE102A	Datasheet
AD	Germanium AF power transistor	AD133	NTE179	Datasheet
AF	Germanium small-signal RF transistor	AF117	NTE160	Datasheet
AL	Germanium RF power transistor	ALZ10	NTE100	Datasheet
AS	Germanium switching transistor	ASY28	NTE101	Datasheet
AU	Germanium power switching transistor	AU103	NTE127	Datasheet
BC	Silicon, small-signal transistor ("general purpose")	BC548	2N3904	Datasheet
BD	Silicon, power transistor	BD139	NTE375	Datasheet
BF	Silicon, RF (high frequency) BJT or FET	BF245	NTE133	Datasheet
BS	Silicon, switching transistor (BJT or MOSFET)	BS170	2N7000	Datasheet
BL	Silicon, high frequency, high power (for transmitters)	BLW60	NTE325	Datasheet
BU	Silicon, high voltage (for CRT horizontal deflection circuits)	BU2520A	NTE2354	Datasheet
CF	Gallium Arsenide small-signal Microwave transistor (MESFET)	CF739	—	Datasheet
CL	Gallium Arsenide Microwave power transistor (FET)	CLY10	—	Datasheet

Joint Electron Devices Engineering Council (JEDEC)

The JEDEC *EIA370* transistor device numbers usually start with "2N", indicating a three-terminal device (dual-gate field-effect transistors are four-terminal devices, so begin with 3N), then a 2, 3 or 4-digit sequential number with no significance as to device properties (although early devices with low numbers tend to be germanium). For example 2N3055 is a silicon n–p–n power transistor, 2N1301 is a p–n–p germanium switching transistor. A letter suffix (such as "A") is sometimes used to indicate a newer variant, but rarely gain groupings.

Proprietary

Manufacturers of devices may have their own proprietary numbering system, for example CK722. Note that a manufacturer's prefix (like "MPF" in MPF102, which originally would denote a Motorola FET) now is an unreliable indicator of who made the device.

Some proprietary naming schemes adopt parts of other naming schemes, for example a PN2222A is a (possibly Fairchild Semiconductor) 2N2222A in a plastic case (but a PN108 is a plastic version of a BC108, not a 2N108, while the PN100 is unrelated to other xx100 devices).

Military part numbers sometimes are assigned their own codes, such as the British Military CV Naming System.

Manufacturers buying large numbers of similar parts may have them supplied with "house numbers", identifying a particular purchasing specification and not necessarily a device with a standardized registered number. For example, an HP part 1854,0053 is a (JEDEC) 2N2218 transistor which is also assigned the CV number: CV7763

Naming Problems

With so many independent naming schemes, and the abbreviation of part numbers when printed on the devices, ambiguity sometimes occurs. For example two different devices may be marked "J176" (one the J176 low-power Junction FET, the other the higher-powered MOSFET 2SJ176).

As older "through-hole" transistors are given surface-mount packaged counterparts, they tend to be assigned many different part numbers because manufacturers have their own systems to cope with the variety in pinout arrangements and options for dual or matched n–p–n+p–n–p devices in one pack.

So even when the original device (such as a 2N3904) may have been assigned by a standards authority, and well known by engineers over the years, the new versions are far from standardized in their naming.

Construction

Semiconductor Material

Table: *Semiconductor material characteristics*

Semiconductor material	*Junction forward voltageV @ 25 °C*	*Electron mobilitym²/ (V·s) @ 25 °C*	*Hole mobilitym²/ (V·s) @ 25 °C*	*Max.junction temp.°C*
Ge	0.27	0.39	0.19	70 to 100
Si	0.71	0.14	0.05	150 to 200
GaAs	1.03	0.85	0.05	150 to 200
Al-Si junction	0.3	—	—	150 to 200

The first BJTs were made from germanium (Ge). Silicon (Si) types currently predominate but certain advanced microwave and high-performance versions now employ the *compound semiconductor* material gallium arsenide (GaAs) and the *semiconductor alloy* silicon germanium (SiGe). Single element semiconductor material (Ge and Si) is described as *elemental.*

Rough parameters for the most common semiconductor materials used to make transistors are given in the table to the right; these parameters will vary with increase in temperature, electric field, impurity level, strain, and sundry other factors.

The *junction forward voltage* is the voltage applied to the emitter–base junction of a BJT in order to make the base conduct a specified current. The current increases exponentially as the junction forward voltage is increased. The values given in the table are typical for a current of 1 mA (the same values apply to semiconductor diodes). The lower the junction forward voltage the better, as this means that less power is required to "drive" the transistor. The junction forward voltage for a given current decreases with increase in temperature. For a typical silicon junction the change is "2.1 mV/°C. In some circuits special compensating elements (sensistors) must be used to compensate for such changes. The density of mobile carriers in the channel of a MOSFET is a function of the electric field forming the channel and of various other phenomena such as the impurity level in the channel. Some impurities, called dopants, are introduced deliberately in making a MOSFET, to control the MOSFET electrical behaviour. The *electron mobility* and *hole mobility* columns show the average speed that electrons and holes diffuse through the semiconductor material with an electric field of 1 volt per metre applied across the material. In general, the higher the electron mobility the faster the transistor can operate. The table indicates that Ge is a better material than Si in this respect. However, Ge has four major shortcomings compared to silicon and gallium arsenide:

- Its maximum temperature is limited;
- it has relatively high leakage current;
- it cannot withstand high voltages;
- it is less suitable for fabricating integrated circuits.

Because the electron mobility is higher than the hole mobility for all semiconductor materials, a given bipolar n–p–n transistor tends to be swifter than an equivalent p–n–p transistor. GaAs has the highest electron mobility of the three semiconductors. It is for this reason that GaAs is used in high-frequency applications. A relatively recent FET development, the *high-electron-mobility transistor* (HEMT), has a heterostructure (junction between different semiconductor materials) of aluminium gallium arsenide (AlGaAs)-gallium arsenide (GaAs) which has twice the electron mobility of a GaAs-metal barrier junction. Because of their high speed and low noise, HEMTs are used in satellite receivers working at frequencies around 12 GHz.

Max. junction temperature values represent a cross section taken from various manufacturers' data sheets. This temperature should not be exceeded or the transistor may be damaged. Al–Si junction refers to the high-speed (aluminum–silicon) metal–semiconductor barrier diode, commonly known as a Schottky diode. This is included in the table because some silicon power IGFETs have a *parasitic* reverse Schottky diode formed between the source and drain as part of the fabrication process. This diode can be a nuisance, but sometimes it is used in the circuit.

Packaging

Discrete transistors are individually packaged transistors. Transistors come in many different semiconductor packages. The two main categories are *through-hole* (or *leaded*), and *surface-mount*, also known as *surface-mount device* (SMD). The *ball grid array* (BGA) is the latest surface-mount package (currently only for large integrated circuits). It has solder "balls" on the underside in place of leads. Because they are smaller and have shorter interconnections, SMDs have better high-frequency characteristics but lower power rating.

Figure: *Assorted discrete transistors*

Transistor packages are made of glass, metal, ceramic, or plastic. The package often dictates the power rating and frequency characteristics. Power transistors have larger packages that can be clamped to heat sinks for enhanced cooling. Additionally, most power transistors have the collector or drain physically connected to the metal enclosure. At the other extreme, some surface-mount *microwave* transistors are as small as grains of sand.

Often a given transistor type is available in several packages. Transistor packages are mainly standardized, but the assignment of a transistor's functions to the terminals is not: other transistor types can assign other functions to the package's terminals. Even for the same transistor type the terminal assignment can vary (normally indicated by a suffix letter to the part number, q.e. BC212L and BC212K).

Semiconductor Device Materials

By far, silicon (Si) is the most widely used material in semiconductor devices. Its combination of low raw material cost, relatively simple processing, and a useful temperature range make it currently the best compromise among the various competing materials. Silicon used in semiconductor device manufacturing is currently fabricated into boules that are large enough in diameter to allow the production of 300 mm (12 in.) wafers.

Germanium (Ge) was a widely used early semiconductor material but its thermal sensitivity makes it less useful than silicon. Today, germanium is often alloyed with silicon for use in very-high-speed SiGe devices; IBM is a major producer of such devices.

Gallium arsenide (GaAs) is also widely used in high-speed devices but so far, it has been difficult to form large-diameter boules of this material, limiting the wafer diameter to sizes significantly smaller than silicon wafers thus making mass production of GaAs devices significantly more expensive than silicon.

Other less common materials are also in use or under investigation.

Silicon carbide (SiC) has found some application as the raw material for blue light-emitting diodes (LEDs) and is being investigated for use in semiconductor devices that could withstand very high operating temperatures and environments with the presence of significant levels of ionizing radiation. IMPATT diodes have also been fabricated from SiC.

Various indium compounds (indium arsenide, indium antimonide, and indium phosphide) are also being used in LEDs and solid state laser diodes. Selenium sulfide is being studied in the manufacture of photovoltaic solar cells.

The most common use for organic semiconductors is Organic light-emitting diodes.

Dielectric Properties

A dielectric is an electrical insulator that can be polarized by an applied electric field. When a dielectric is placed in an electric field, electric charges do not flow through the material as they do in a conductor, but only slightly shift from their average equilibrium positions causing dielectric polarization. Because of dielectric polarization, positive charges are displaced toward the field and negative charges shift in the opposite direction. This creates an internal electric field which reduces the overall field within the dielectric itself. If a dielectric is composed of weakly bonded molecules, those molecules not only become polarized, but also reorient so that their symmetry axis aligns to the field.

While the term "insulator" implies low electrical conduction, "dielectric" is typically used to describe materials with a high polarizability. The latter is expressed by a number called the relative permittivity (also known in older texts as dielectric constant). The term insulator is generally used to indicate electrical obstruction while the term dielectric is used to indicate the energy storing capacity of the material (by means of polarization). A common example of a dielectric is the electrically insulating material between the metallic plates of a capacitor. The polarization of the dielectric by the applied electric field increases the capacitor's surface charge. The study of dielectric properties is concerned with the storage and dissipation of electric and magnetic energy in materials. It is important to explain various phenomena in electronics, optics, and solid-state physics.

The term "dielectric" was coined by William Whewell (from "dia-electric") in response to a request from Michael Faraday.

Permittivity

In electromagnetism, absolute permittivity is the measure of the resistance that is encountered when forming an electric field in a medium. In other words, permittivity is a measure of how an electric field affects, and is affected by, a dielectric medium.

The permittivity of a medium describes how much electric field (more correctly, flux) is 'generated' per unit charge in that medium. More electric flux exists in a medium with a high permittivity (per unit charge) because of polarization effects.

Permittivity is directly related to electric susceptibility, which is a measure of how easily a dielectric polarizes in response to an electric field. Thus, permittivity relates to a material's ability to transmit (or "permit") an electric field.

In SI units, permittivity ε is measured in farads per metre (F/m); electric susceptibility χ is dimensionless. They are related to each other through

$$\varepsilon = \varepsilon_r \varepsilon_0 = (1 + \chi)\varepsilon_0$$

where ε_r is the relative permittivity of the material, and ε_0 = 8.854187817.. $\times 10^{-12}$ F/m is the vacuum permittivity.

Explanation

In electromagnetism, the electric displacement field D represents how an electric field E influences the organization of electrical charges in a given medium, including charge migration and electric dipole reorientation. Its relation to permittivity in the very simple case of *linear, homogeneous, isotropic* materials with *"instantaneous" response* to changes in electric field is

$$D = \varepsilon E$$

where the permittivity ε is a scalar. If the medium is anisotropic, the permittivity is a second rank tensor.

In general, permittivity is not a constant, as it can vary with the position in the medium, the frequency of the field applied, humidity, temperature, and other parameters. In a nonlinear medium, the permittivity can depend on the strength of the electric field. Permittivity as a function of frequency can take on real or complex values.

In SI units, permittivity is measured in farads per metre (F/m or $A^2 \cdot s^4 \cdot kg^{-1} \cdot m^{-3}$). The displacement field D is measured in units of coulombs per square metre (C/m^2), while the electric field E is measured in volts per metre (V/m). D and E describe the interaction between charged objects. D is related to the *charge densities* associated with this interaction, while E is related to the *forces* and *potential differences.*

Vacuum Permittivity

The vacuum permittivity ε_0 (also called permittivity of free space or the electric constant) is the ratio D/E in free space. It also appears in the Coulomb force constant, $k_e = 1/(4\pi\varepsilon_0)$.

Its value is

$$\varepsilon_0 \overset{\text{def}}{=} \frac{1}{c_0^2 \mu_0} = \frac{1}{35950207149.4727056\pi} \frac{\text{F}}{\text{m}} \approx 8.8541878176\ldots \times 10^{-12} \text{ F/m}$$

where

c_0 is the speed of light in free space,

μ_0 is the vacuum permeability.

Constants c_0 and μ_0 are defined in SI units to have exact numerical values, shifting responsibility of experiment to the determination of the metre and the ampere. (The approximation in the second value of ε_0 above stems from π being an irrational number.)

Relative Permittivity

The linear permittivity of a homogeneous material is usually given relative to that of free space, as a relative permittivity ε_r (also called dielectric constant, although this sometimes only refers to the static, zero-frequency relative permittivity). In an anisotropic material, the relative permittivity may be a tensor, causing birefringence. The actual permittivity is then calculated by multiplying the relative permittivity by ε_0:

$$\varepsilon = \varepsilon_r \varepsilon_0 = (1 + \chi)\varepsilon_0,$$

where χ (frequently writtenχ_e) is the electric susceptibility of the material.

The susceptibility is defined as the constant of proportionality (which may be a tensor) relating an electric field E to the induced dielectric polarization density P such that

$$P = \varepsilon_0 \chi E,$$

where ε_0 is the electric permittivity of free space.

The susceptibility of a medium is related to its relative permittivity ε_r by

$$\chi = \varepsilon_r - 1.$$

So in the case of a vacuum,

$$\chi = 0.$$

The susceptibility is also related to the polarizability of individual particles in the medium by the Clausius-Mossotti relation.

The electric displacement D is related to the polarization density P by

$$D = \varepsilon_0 E + P = \varepsilon_0 (1 + \chi) E = \varepsilon_r \varepsilon_0 E.$$

The permittivity ε and permeability μ of a medium together determine the phase velocity $v = c/n$ of electromagnetic radiation through that medium:

$$\varepsilon\mu = \frac{1}{v^2}.$$

Dispersion and Causality

In general, a material cannot polarize instantaneously in response to an applied field, and so the more general formulation as a function of time is

$$P(t) = \varepsilon_0 \int_{-\infty}^{t} \chi(t - t') E(t') dt'.$$

That is, the polarization is a convolution of the electric field at previous times with time-dependent susceptibility given by $\chi(\Delta t)$. The upper limit of this integral can be extended to infinity as well if one defines $\chi(\Delta t) = 0$ for $\Delta t < 0$. An instantaneous response corresponds to Dirac delta function susceptibility $\chi(\Delta t) = \chi\, \delta(\Delta t)$.

It is more convenient in a linear system to take the Fourier transform and write this relationship as a function of frequency. Because of the convolution theorem, the integral becomes a simple product,

$$P(\omega) = \varepsilon_0 \chi(\omega) E(\omega).$$

This frequency dependence of the susceptibility leads to frequency dependence of the permittivity. The shape of the susceptibility with respect to frequency characterizes the dispersion properties of the material.

Moreover, the fact that the polarization can only depend on the electric field at previous times (i.e. $\chi(\Delta t) = 0$ for $\Delta t < 0$), a consequence of causality, imposes Kramers–Kronig constraints on the susceptibility $\chi(0)$.

Complex Permittivity

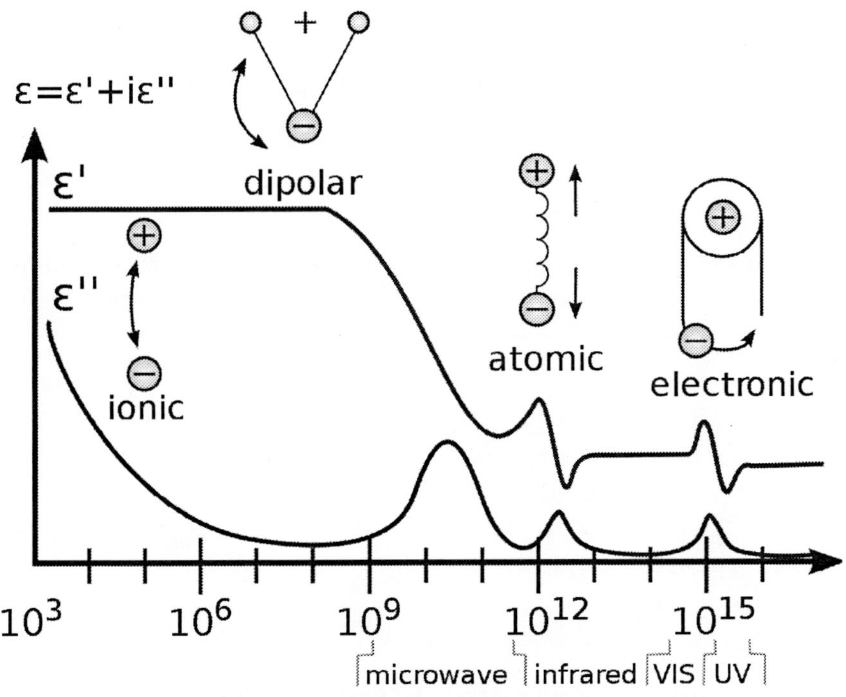

Figure: *A dielectric permittivity spectrum over a wide range of frequencies. ε2 and ε3 denote the real and the imaginary part of the permittivity, respectively. Various processes are labelled on the image: ionic and dipolar relaxation, and atomic and electronic resonances at higher energies.*

As opposed to the response of a vacuum, the response of normal materials to external fields generally depends on the frequency of the field. This frequency dependence reflects the fact that a material's polarization does not respond instantaneously to an applied field. The response must always be *causal* (arising after the applied field) which can be represented by a phase difference. For this reason, permittivity is often treated as a complex function of the (angular) frequency of the applied field ω: $\varepsilon \rightarrow \hat{\varepsilon}(\omega)$ (since complex numbers allow specification of magnitude and phase). The definition of permittivity therefore becomes

$$D_0 e^{-i\omega t} = \hat{\varepsilon}(\omega) E_0 e^{-i\omega t},$$

where

D_0 and E_0 are the amplitudes of the displacement and electrical fields, respectively,

i is the imaginary unit, $i^2 = -1$.

The response of a medium to static electric fields is described by the low-frequency limit of permittivity, also called the static permittivity ε_s (also ε_{DC}):

$$\varepsilon_s = \lim_{\omega \to 0} \hat{\varepsilon}(\omega).$$

At the high-frequency limit, the complex permittivity is commonly referred to as ε∞ At the plasma frequency and above, dielectrics behave as ideal metals, with electron gas behaviour.

The static permittivity is a good approximation for alternating fields of low frequencies, and as the frequency increases a measurable phase difference δ emerges between D and E. The frequency at which the phase shift becomes noticeable depends on temperature and the details of the medium. For moderate fields strength (E_0), D and E remain proportional, and

$$\hat{\varepsilon} = \frac{D_0}{E_0} = |\varepsilon| e^{i\delta}.$$

Since the response of materials to alternating fields is characterized by a complex permittivity, it is natural to separate its real and imaginary parts, which is done by convention in the following way:

$$\hat{\varepsilon}(\omega) = \varepsilon'(\omega) + i\varepsilon''(\omega) = \frac{D_0}{E_0}\left(\cos\delta + i\sin\delta\right).$$

where

ε' is the real part of the permittivity, which is related to the stored energy within the medium;

ε" is the imaginary part of the permittivity, which is related to the dissipation (or loss) of energy within the medium.

It is important to realize that the choice of sign for time-dependence, exp(-$i\omega t$), dictates the sign convention for the imaginary part of permittivity. The signs used here correspond to those commonly used in physics, whereas for the engineering convention one should reverse all imaginary quantities. The complex permittivity is usually a complicated function of frequency ω, since it is a superimposed description of dispersion phenomena occurring at multiple frequencies. The dielectric function $\varepsilon(\omega)$ must have poles only for frequencies with positive imaginary parts, and therefore satisfies the Kramers–Kronig relations. However, in the narrow frequency ranges that are often studied in practice, the permittivity can be approximated as frequency-independent or by model functions.

At a given frequency, the imaginary part of $\hat{\varepsilon}$ leads to absorption loss if it is positive (in the above sign convention) and gain if it is negative. More generally, the imaginary parts of the eigenvalues of the anisotropic dielectric tensor should be considered.

In the case of solids, the complex dielectric function is intimately connected to band structure. The primary quantity that characterizes the electronic structure of any crystalline material is the probability of photon absorption, which is directly related to the imaginary part of the optical dielectric function $\varepsilon(\omega)$. The optical dielectric function is given by the fundamental expression:

$$\varepsilon(\omega) = 1 + \frac{8\pi^2 e^2}{m^2} \sum_{c,v} \int W_{c,v}(E)\left[\varphi(\hbar\omega - E) - \varphi(\hbar\omega + E)\right]dx.$$

In this expression, $W_{c,v}(E)$ represents the product of the Brillouin zone-averaged transition probability at the energy E with the joint density of states, $J_{c,v}(E)$; φ is a broadening function, representing the role of scattering in smearing out the energy levels. In general, the broadening is intermediate between Lorentzian and Gaussian; for an alloy it is somewhat closer to Gaussian because of strong scattering from statistical fluctuations in the local composition on a nanometer scale.

Tensorial Permittivity

According to the Drude model of magnetized plasma, a more general expression which takes into account the interaction of the carriers with an alternating electric field at millimeter and microwave frequencies in an axially magnetized semiconductor requires the expression of the permittivity as a non-diagonal tensor.

$$D(\omega) = \begin{vmatrix} \varepsilon_1 & -i\varepsilon_2 & 0 \\ i\varepsilon_2 & \varepsilon_1 & 0 \\ 0 & 0 & \varepsilon_z \end{vmatrix} E(\omega)$$

If ε_2 vanishes, then the tensor is diagonal but not proportional to the identity and the medium is said an uniaxial medium.

Classification of Materials

Materials can be classified according to their permittivity and conductivity, σ. Materials with a large amount of loss inhibit the propagation of electromagnetic waves. In this case, generally when $\sigma/(\omega\varepsilon') >> 1$, we consider the material to be a good conductor. Dielectrics are associated with lossless or low-loss materials, where $\sigma/(\omega\varepsilon') << 1$. Those that do not fall under either limit are considered to be general media. A *perfect dielectric* is a material that has no conductivity, thus exhibiting only a displacement current. Therefore it stores and returns electrical energy as if it were an ideal capacitor.

Lossy Medium

In the case of lossy medium, i.e. when the conduction current is not negligible, the total current density flowing is:

$$J_{\text{tot}} = J_{\text{c}} + J_{\text{d}} = \sigma E - i\omega\varepsilon' E = -i\omega\hat{\varepsilon} E$$

where

σ is the conductivity of the medium;

ε' is the real part of the permittivity.

$\hat{\varepsilon}$ is the complex permittivity

The size of the displacement current is dependent on the frequency ω of the applied field *E*; there is no displacement current in a constant field.

In this formalism, the complex permittivity is defined as:

$$\hat{\varepsilon} = \varepsilon' - i\frac{\sigma}{\omega}$$

In general, the absorption of electromagnetic energy by dielectrics is covered by a few different mechanisms that influence the shape of the permittivity as a function of frequency:

- First, are the relaxation effects associated with permanent and induced molecular dipoles. At low frequencies the field changes slowly enough to allow dipoles to reach equilibrium before the field has measurably changed. For frequencies at which dipole orientations cannot follow the applied field because of the viscosity of the medium, absorption of the field's energy leads to energy dissipation. The mechanism of dipoles relaxing is called dielectric relaxation and for ideal dipoles is described by classic Debye relaxation.

- Second are the resonance effects, which arise from the rotations or vibrations of atoms, ions, or electrons. These processes are observed in the neighbourhood of their characteristic absorption frequencies.

The above effects often combine to cause non-linear effects within capacitors. For example, dielectric absorption refers to the inability of a capacitor that has been charged for a long time to completely discharge when briefly discharged. Although an ideal capacitor would remain at zero volts after being discharged, real capacitors will develop a small voltage, a phenomenon that is also called *soakage* or *battery action*. For some dielectrics, such as many polymer films, the resulting voltage may be less than 1-2% of the original voltage. However, it can be as much as 15 - 25% in the case of electrolytic capacitors or supercapacitors.

Quantum-mechanical Interpretation

In terms of quantum mechanics, permittivity is explained by atomic and molecular interactions.

At low frequencies, molecules in polar dielectrics are polarized by an applied electric field, which induces periodic rotations. For example, at the microwave frequency, the microwave field causes the periodic rotation of water molecules, sufficient to break hydrogen bonds. The field does work against the bonds and the energy is absorbed by the material as heat. This is why microwave ovens work very well for materials containing water. There are two maxima of the imaginary component (the absorptive index) of water, one at the microwave frequency, and the other at far ultraviolet (UV) frequency. Both of these resonances are at higher frequencies than the operating frequency of microwave ovens.

At moderate frequencies, the energy is too high to cause rotation, yet too low to affect electrons directly, and is absorbed in the form of resonant molecular vibrations. In water, this is where the absorptive index starts to drop sharply, and the minimum of the imaginary permittivity is at the frequency of blue light (optical regime).

At high frequencies (such as UV and above), molecules cannot relax, and the energy is purely absorbed by atoms, exciting electron energy levels. Thus, these frequencies are classified as ionizing radiation. While carrying out a complete *ab initio* (that is, first-principles) modelling is now computationally possible, it has not been widely applied yet. Thus, a phenomenological model is accepted as being an adequate method of capturing experimental behaviours. The Debye model and the Lorentz model use a 1st-order and 2nd-order (respectively) lumped system parameter linear representation (such as an RC and an LRC resonant circuit).

Measurement

The dielectric constant of a material can be found by a variety of static electrical measurements. The complex permittivity is evaluated over a wide range of frequencies by using different variants of dielectric spectroscopy, covering nearly 21 orders of magnitude from 10^{-6} to 10^{15} Hz. Also, by using cryostats and ovens, the dielectric properties of a medium can be characterized over an array of temperatures. In order to study systems for such diverse excitation fields, a number of measurement setups are used, each adequate for a special frequency range. Various microwave measurement techniques are outlined in Chen *et al.*. Typical errors for the Hakki-Coleman method employing a puck of material between conducting planes are about 0.3%.

- Low-frequency time domain measurements (10^{-6}-10^{3} Hz)
- Low-frequency frequency domain measurements (10^{-5}-10^{6} Hz)
- Reflective coaxial methods (10^{6}-10^{10} Hz)
- Transmission coaxial method (10^{8}-10^{11} Hz)
- Quasi-optical methods (10^{9}-10^{10} Hz)
- Terahertz time-domain spectroscopy (10^{11}-10^{13} Hz)
- Fourier-transform methods (10^{11}-10^{15} Hz)

At infrared and optical frequencies, a common technique is ellipsometry. Dual polarisation interferometry is also used to measure the complex refractive index for very thin films at optical frequencies.

Dielectric Polarization

Basic atomic model:

In the classical approach to the dielectric model, a material is made up of atoms. Each atom consists of a cloud of negative charge (Electrons) bound to and surrounding a positive point charge at its centre. In the presence of an electric field the charge cloud is distorted.

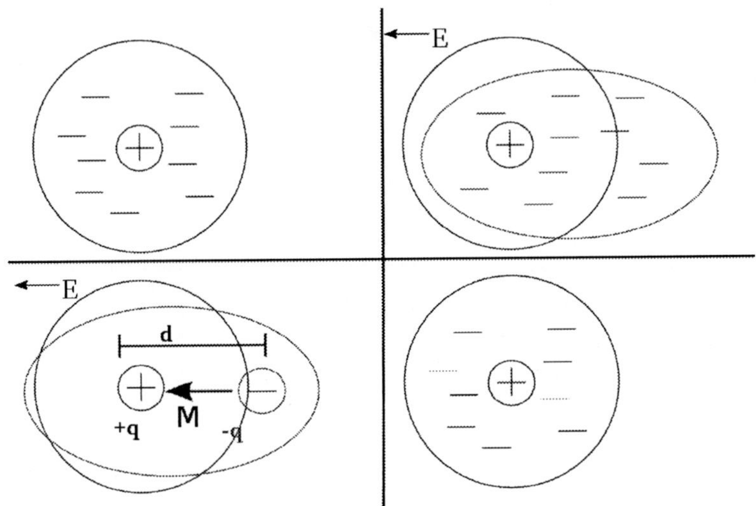

Figure: *Electric field interaction with an atom under the classical dielectric model.*

This can be reduced to a simple dipole using the superposition principle. A dipole is characterized by its dipole moment. It is the relationship between the electric field and the dipole moment that gives rise to the behaviour of the dielectric. (Note that the dipole moment is shown to be pointing in the same direction as the electric field. This isn't always correct, and it is a major simplification, but it is suitable for many materials.) When the electric field is removed the atom returns to its original state. The time required to do so is the so-called relaxation time; an exponential decay.

This is the essence of the model in physics. The behaviour of the dielectric now depends on the situation. The more complicated the situation the richer the model has to be in order to accurately describe the behaviour. Important questions are:

- Is the electric field constant or does it vary with time?
    - If the electric field does vary, at what rate?
- What are the characteristics of the material?
    - Is the direction of the field important (isotropy)?
    - Is the material the same all the way through (homogeneous)?
    - Are there any boundaries/interfaces that have to be taken into account?
- Is the system linear or do nonlinearities have to be taken into account?

The relationship between the electric field E and the dipole moment M gives rise to the behaviour of the dielectric, which, for a given material, can be characterized by the function F defined by the equation:

$$M = F(E).$$

When both the type of electric field and the type of material have been defined, one then chooses the simplest function *F* that correctly predicts the phenomena of interest. Examples of phenomena that can be so modelled include:

- Refractive index
- Group velocity dispersion
- Birefringence
- Self-focusing
- Harmonic generation

Dipolar Polarization

Dipolar polarization is a polarization that is either inherent to polar molecules (orientation polarization), or can be induced in any molecule in which the asymmetric distortion of the nuclei is possible (distortion polarization). Orientation polarization results from a permanent dipole, e.g. that arising from the 104.45° angle between the asymmetric bonds between oxygen and hydrogen atoms in the water molecule, which retains polarization in the absence of an external electric field. The assembly of these dipoles forms a macroscopic polarization.

When an external electric field is applied, the distance between charges, which is related to chemical bonding, remains constant in orientation polarization; however, the polarization itself rotates. This rotation occurs on a timescale which depends on the torque and the surrounding local viscosity of the molecules. Because the rotation is not instantaneous, dipolar polarizations lose the response to electric fields at the lowest frequency in polarizations. A molecule rotates about 1ps per radian in a fluid, thus this loss occurs at about 10^{11} Hz (in the microwave region). The delay of the response to the change of the electric field causes friction and heat.

When an external electric field is applied in the infrared, a molecule is bent and stretched by the field and the molecular moment changes in response. The molecular vibration frequency is approximately the inverse of the time taken for the molecule to bend, and the distortion polarization disappears above the infrared.

Ionic Polarization

Ionic polarization is polarization which is caused by relative displacements between positive and negative ions in ionic crystals (for example, NaCl).

If crystals or molecules do not consist of only atoms of the same kind, the distribution of charges around an atom in the crystals or molecules leans to positive or negative. As a result, when lattice vibrations or molecular vibrations induce relative displacements of the atoms, the centres of positive and negative charges might be in different locations. These centre positions are affected by the symmetry of the displacements. When the centres don't correspond, polarizations arise in molecules or crystals. This polarization is called ionic polarization.

Ionic polarization causes ferroelectric transition as well as dipolar polarization. The transition, which is caused by the order of the directional orientations of permanent dipoles along a particular direction, is called order-disorder phase transition. The transition which is caused by ionic polarizations in crystals is called displacive phase transition.

Dielectric Dispersion

In physics, dielectric dispersion is the dependence of the permittivity of a dielectric material on the frequency of an applied electric field. Because there is always a lag between changes in polarization and changes in an electric field, the permittivity of the dielectric is a complicated, complex-valued function of frequency of the electric field. It is very important for the application of dielectric materials and the analysis of polarization systems.

This is one instance of a general phenomenon known as material dispersion: a frequency-dependent response of a medium for wave propagation.

When the frequency becomes higher:

1. it becomes impossible for dipolar polarization to follow the electric field in the microwave region around 10^{10} Hz;
2. in the infrared or far-infrared region around 10^{13} Hz, ionic polarization and molecular distortion polarization lose the response to the electric field;
3. electronic polarization loses its response in the ultraviolet region around 10^{15} Hz.

In the frequency region above ultraviolet, permittivity approaches the constant ε_0 in every substance, where ε_0 is the permittivity of the free space. Because permittivity indicates the strength of the relation between an electric field and polarization, if a polarization process loses its response, permittivity decreases.

Dielectric Relaxation

Dielectric relaxation is the momentary delay (or lag) in the dielectric constant of a material. This is usually caused by the delay in molecular polarization with respect to a changing electric field in a dielectric medium (e.g. inside capacitors or between two large conducting surfaces). Dielectric relaxation in changing electric fields could be considered analogous to hysteresis in changing magnetic fields (for inductors or transformers). Relaxation in general is a delay or lag in the response of a linear system, and therefore dielectric relaxation is measured relative to the expected linear steady state (equilibrium) dielectric values. The time lag between electrical field and polarization implies an irreversible degradation of free energy(G).

In physics, dielectric relaxation refers to the relaxation response of a dielectric medium to an external electric field of microwave frequencies. This relaxation is often described in terms of permittivity as a function of frequency, which can, for ideal systems, be described by the Debye equation. On the other hand, the distortion related to ionic and electronic polarization shows behaviour of the resonance or oscillator type. The character of the distortion process depends on the structure, composition, and surroundings of the sample.

The number of possible wavelengths of emitted radiation due to dielectric relaxation can be equated using Hemmings' first law (named after Mark Hemmings)

$$n = \frac{l^2 - l}{2}$$

where

n is the number of different possible wavelengths of emitted radiation

l is the number of energy levels (including ground level).

Debye Relaxation

Debye relaxation is the dielectric relaxation response of an ideal, noninteracting population of dipoles to an alternating external electric field. It is usually expressed in the complex permittivity $\varepsilon$of a medium as a function of the field's frequency ω:

$$\hat{\varepsilon}(\omega) = \varepsilon_\infty + \frac{\Delta\varepsilon}{1 + i\omega\tau},$$

where ε_∞ is the permittivity at the high frequency limit, $\Delta\varepsilon = \varepsilon_s - \varepsilon_\infty$ where ε_s is the static, low frequency permittivity, and τ is the characteristic relaxation time of the medium.

This relaxation model was introduced by and named after the physicist Peter Debye (1913).

Variants of the Debye Equation

- Cole–Cole equation
- Cole–Davidson equation
- Havriliak–Negami relaxation
- Kohlrausch–Williams–Watts function (Fourier transform of stretched exponential function)

Paraelectricity

Paraelectricity is the ability of many materials (specifically ceramic crystals) to become polarized under an applied electric field. Unlike ferroelectricity, this can happen even if there is no permanent electric dipole that exists in the material, and removal of the fields results in the polarization in the material returning to zero. The mechanisms which give rise to paraelectric behaviour are the distortion of individual ions (displacement of the electron cloud from the nucleus) and the polarization of molecules or combinations of ions or defects.

Paraelectricity occurs in crystal phases in which electric dipoles are unaligned (i.e., unordered domains that are electrically charged) and thus have the potential to align in an external electric field and strengthen it. In comparison to the ferroelectric phase, the domains are unordered and the internal field is weak.

The $LiNbO_3$ crystal is ferroelectric below 1430 K, and above this temperature it transforms into a disordered paraelectric phase. Similarly, other perovskites also exhibit paraelectricity at high temperatures.

Paraelectricity has been explored as a possible refrigeration mechanism; polarizing a paraelectric by applying an electric field under adiabatic process conditions raises the temperature, while removing the field lowers the temperature. A heat pump that polarizes the paraelectric, allows it to return to ambient temperature, then brings it into contact with the object to be cooled, and depolarizes it, would result in refrigeration.

Applications

Capacitors:

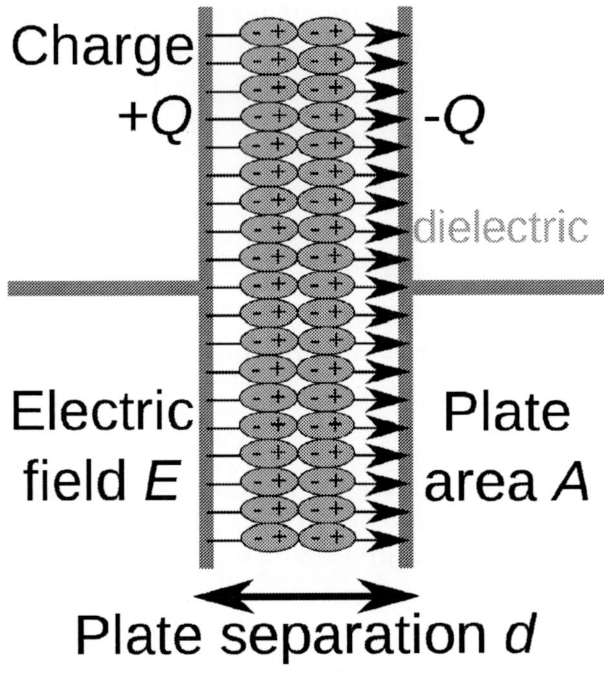

Figure: *Charge separation in a parallel-plate capacitor causes an internal electric field. A dielectric (orange) reduces the field and increases the capacitance.*

Commercially manufactured capacitors typically use a solid dielectric material with high permittivity as the intervening medium between the stored positive and negative charges. This material is often referred to in technical contexts as the "capacitor dielectric".

The most obvious advantage to using such a dielectric material is that it prevents the conducting plates on which the charges are stored from coming into direct electrical contact. More significant, however, a high permittivity allows a greater charge to be stored at a given voltage. This can be seen by treating the case of a linear dielectric with permittivity ε and thickness d between two conducting plates with uniform charge density σ_ε. In this case the charge density is given by

$$\sigma_\varepsilon = \varepsilon \frac{V}{d}$$

and the capacitance per unit area by

$$c = \frac{\sigma_\varepsilon}{V} = \frac{\varepsilon}{d}$$

From this, it can easily be seen that a larger ε leads to greater charge stored and thus greater capacitance.

Dielectric materials used for capacitors are also chosen such that they are resistant to ionization. This allows the capacitor to operate at higher voltages before the insulating dielectric ionizes and begins to allow undesirable current.

Dielectric Resonator

A *dielectric resonator oscillator* (DRO) is an electronic component that exhibits resonance for a narrow range of frequencies, generally in the microwave band. It consists of a "puck" of ceramic that has a large dielectric constant and a low dissipation factor. Such resonators are often used to provide a frequency reference in an oscillator circuit. An unshielded dielectric resonator can be used as a Dielectric Resonator Antenna (DRA).

Some Practical Dielectrics

Dielectric materials can be solids, liquids, or gases. In addition, a high vacuum can also be a useful, lossless dielectric even though its relative dielectric constant is only unity.

Solid dielectrics are perhaps the most commonly used dielectrics in electrical engineering, and many solids are very good insulators. Some examples include porcelain, glass, and most plastics. Air, nitrogen and sulphur hexafluoride are the three most commonly used gaseous dielectrics.

- Industrial coatings such as parylene provide a dielectric barrier between the substrate and its environment.
- Mineral oil is used extensively inside electrical transformers as a fluid dielectric and to assist in cooling. Dielectric fluids with higher dielectric constants, such as electrical grade castor oil, are often used in high voltage capacitors to help prevent corona discharge and increase capacitance.
- Because dielectrics resist the flow of electricity, the surface of a dielectric may retain *stranded* excess electrical charges. This

may occur accidentally when the dielectric is rubbed (the triboelectric effect). This can be useful, as in a Van de Graaff generator or electrophorus, or it can be potentially destructive as in the case of electrostatic discharge.

- Specially processed dielectrics, called electrets (which should not be confused with ferroelectrics), may retain excess internal charge or "frozen in" polarization. Electrets have a semipermanent external electric field, and are the electrostatic equivalent to magnets. Electrets have numerous practical applications in the home and industry.
- Some dielectrics can generate a potential difference when subjected to mechanical stress, or change physical shape if an external voltage is applied across the material. This property is called piezoelectricity. Piezoelectric materials are another class of very useful dielectrics.
- Some ionic crystals and polymer dielectrics exhibit a spontaneous dipole moment which can be reversed by an externally applied electric field. This behaviour is called the ferroelectric effect. These materials are analogous to the way ferromagnetic materials behave within an externally applied magnetic field. Ferroelectric materials often have very high dielectric constants, making them quite useful for capacitors.

Piezo-electricity

Piezoelectricity is the electric charge that accumulates in certain solid materials (notably crystals, certain ceramics, and biological matter such as bone, DNA and various proteins) in response to applied mechanical stress. The word *piezoelectricity* means electricity resulting from pressure. It is derived from the Greek *piezo* or *piezein*, which means to squeeze or press, and *electric* or *electron*, which stands for amber, an ancient source of electric charge. Piezoelectricity was discovered in 1880 by French physicists Jacques and Pierre Curie.

The piezoelectric effect is understood as the linear electromechanical interaction between the mechanical and the electrical state in crystalline materials with no inversion symmetry. The piezoelectric effect is a reversible process in that materials exhibiting the direct piezoelectric effect (the internal generation of electrical charge resulting from an applied mechanical force) also exhibit the reverse piezoelectric effect (the internal generation of a mechanical strain resulting from an applied electrical field). For example, lead zirconate titanate crystals will generate measurable piezoelectricity when their static structure is deformed by about 0.1% of the original dimension. Conversely, those same crystals will change about 0.1% of their static dimension when an external electric field is applied to the material. The inverse piezoelectric effect is used in production of ultrasonic sound waves.

Piezoelectricity is found in useful applications such as the production and detection of sound, generation of high voltages, electronic frequency generation, microbalances, and ultrafine focusing of optical assemblies. It is also the basis of a number of scientific instrumental techniques with atomic resolution, the scanning probe microscopies such as STM, AFM, MTA, SNOM, etc., and everyday uses such as acting as the ignition source for cigarette lighters and push-start propane barbecues.

History

Discovery and Early Research: The pyroelectric effect, by which a material generates an electric potential in response to a temperature change, was studied by Carl Linnaeus and Franz Aepinus in the mid-18th century. Drawing on this knowledge, both René Just Haüy and Antoine César Becquerel posited a relationship between mechanical stress and electric charge; however, experiments by both proved inconclusive.

The first demonstration of the direct piezoelectric effect was in 1880 by the brothers Pierre Curie and Jacques Curie. They combined their knowledge of pyroelectricity with their understanding of the underlying crystal structures that gave rise to pyroelectricity to predict crystal behaviour, and demonstrated the effect using crystals of tourmaline, quartz, topaz, cane sugar, and Rochelle salt (sodium potassium tartrate tetrahydrate). Quartz and Rochelle salt exhibited the most piezoelectricity.

The Curies, however, did not predict the converse piezoelectric effect. The converse effect was mathematically deduced from fundamental thermodynamic principles by Gabriel Lippmann in 1881. The Curies immediately confirmed the existence of the converse effect, and went on to obtain quantitative proof of the complete reversibility of electro-elasto-mechanical deformations in piezoelectric crystals.

For the next few decades, piezoelectricity remained something of a laboratory curiosity. More work was done to explore and define the crystal structures that exhibited piezoelectricity. This culminated in

1910 with the publication of Woldemar Voigt's *Lehrbuch der Kristallphysik* (Textbook on Crystal Physics), which described the 20 natural crystal classes capable of piezoelectricity, and rigorously defined the piezoelectric constants using tensor analysis.

World War I and Post-war

The first practical application for piezoelectric devices was sonar, first developed during World War I. In France in 1917, Paul Langevin and his coworkers developed an ultrasonic submarine detector. The detector consisted of a transducer, made of thin quartz crystals carefully glued between two steel plates, and a hydrophone to detect the returned echo. By emitting a high-frequency chirp from the transducer, and measuring the amount of time it takes to hear an echo from the sound waves bouncing off an object, one can calculate the distance to that object.

The use of piezoelectricity in sonar, and the success of that project, created intense development interest in piezoelectric devices. Over the next few decades, new piezoelectric materials and new applications for those materials were explored and developed.

Piezoelectric devices found homes in many fields. Ceramic phonograph cartridges simplified player design, were cheap and accurate, and made record players cheaper to maintain and easier to build. The development of the ultrasonic transducer allowed for easy measurement of viscosity and elasticity in fluids and solids, resulting in huge advances in materials research. Ultrasonic time-domain reflectometers (which send an ultrasonic pulse through a material and measure reflections from discontinuities) could find flaws inside cast metal and stone objects, improving structural safety.

World War II and post-war

During World War II, independent research groups in the United States, Russia, and Japan discovered a new class of synthetic materials, called ferroelectrics, which exhibited piezoelectric constants many times higher than natural materials. This led to intense research to develop barium titanate and later lead zirconate titanate materials with specific properties for particular applications.

One significant example of the use of piezoelectric crystals was developed by Bell Telephone Laboratories. Following World War I, Frederick R. Lack, working in radio telephony in the engineering department, developed the "AT cut" crystal, a crystal that operated through a wide range of temperatures. Lack's crystal didn't need the heavy accessories previous crystal used, facilitating its use on aircraft. This development allowed Allied air forces to engage in coordinated mass attacks through the use of aviation radio.

Development of piezoelectric devices and materials in the United States was kept within the companies doing the development, mostly due to the wartime beginnings of the field, and in the interests of securing profitable patents. New materials were the first to be developed — quartz crystals were the first commercially exploited piezoelectric material, but scientists searched for higher-performance materials. Despite the advances in materials and the maturation of manufacturing processes, the United States market had not grown as quickly. Without many new applications, the growth of the United States' piezoelectric industry suffered.

In contrast, Japanese manufacturers shared their information, quickly overcoming technical and manufacturing challenges and creating new markets. Japanese efforts in materials research created piezoceramic materials competitive to the U.S. materials, but free of expensive patent restrictions. Major Japanese piezoelectric developments include new designs of piezoceramic filters for radios and televisions, piezo buzzers and audio transducers that can connect directly to electronic circuits, and the piezoelectric igniter, which generates sparks for small engine ignition systems (and gas-grill lighters) by compressing a ceramic disc. Ultrasonic transducers that transmit sound waves through air had existed for quite some time, but first saw major commercial use in early television remote controls. These transducers now are mounted on several car models as an echolocation device, helping the driver determine the distance from the rear of the car to any objects that may be in its path.

Mechanism

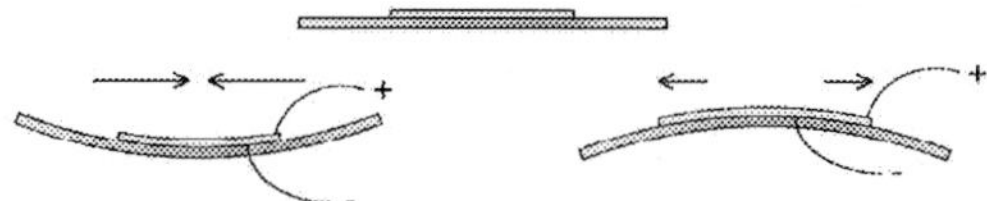

Figure: *Piezoelectric plate used to convert audio signal to sound waves*

The nature of the piezoelectric effect is closely related to the occurrence of electric dipole moments in solids. The latter may either be induced for ions on crystal lattice sites with asymmetric charge surroundings (as in $BaTiO_3$ and PZTs) or may directly be carried by molecular groups (as in cane sugar). The dipole density or polarization (dimensionality [Cm/m^3]) may easily be calculated for crystals by summing up the dipole moments per volume of the crystallographic unit cell. As every dipole is a vector, the dipole density P is a vector field. Dipoles near each other tend to be aligned in regions called Weiss domains. The domains are usually randomly oriented, but can be aligned using the process of *poling* (not the same as magnetic poling), a process by which a strong electric field is applied across the material, usually at elevated temperatures. Not all piezoelectric materials can be poled.

Of decisive importance for the piezoelectric effect is the change of polarization P when applying a mechanical stress. This might either be caused by a re-configuration of the dipole-inducing surrounding or by re-orientation of molecular dipole moments under the influence of the external stress. Piezoelectricity may then manifest in a variation of the polarization strength, its direction or both, with the details depending on 1. the orientation of P within the crystal, 2. crystal symmetry and 3. the applied mechanical stress.

The change in P appears as a variation of surface charge density upon the crystal faces, i.e. as a variation of the electrical field extending between the faces, since the units of surface charge density and polarization are the same, [C/m^2] = [Cm/m^3]. However, piezoelectricity is not caused by a change in charge density on the surface, but by dipole density in the bulk. For example, a 1 cm^3 cube of quartz with 2 kN (500 lbf) of correctly applied force can produce a voltage of 12500 V.

Piezoelectric materials also show the opposite effect, called converse piezoelectric effect, where the application of an electrical field creates mechanical deformation in the crystal.

Mathematical Description

Piezoelectricity is the combined effect of the electrical behaviour of the material:

$$D = \varepsilon E$$

where D is the electric charge density displacement (electric displacement), ε is permittivity and E is electric field strength, and

Hooke's Law:

$$S = sT$$

where S is strain, s is compliance and T is stress.

These may be combined into so-called *coupled equations*, of which the strain-charge form is:

$$\{S\} = \left[s^E\right]\{T\} + [d]\{E\}$$

$$\{D\} = [d^t]\{T\} + \left[\varepsilon^T\right]\{E\},$$

where $[d]$ is the matrix for the direct piezoelectric effect and $[d^t]$ is the matrix for the converse piezoelectric effect. The superscript E indicates a zero, or constant, electric field; the superscript T indicates a zero, or constant, stress field; and the superscript t stands for transposition of a matrix. The strain-charge for a material of the 4mm (C_{4v}) crystal class (such as a poled piezoelectric ceramic such as tetragonal PZT or $BaTiO_3$) as well as the 6mm crystal class may also be written as (ANSI IEEE 176):

$$\begin{bmatrix} S_1 \\ S_2 \\ S_3 \\ S_4 \\ S_5 \\ S_6 \end{bmatrix} = \begin{bmatrix} s_{11}^E & s_{12}^E & s_{13}^E & 0 & 0 & 0 \\ s_{21}^E & s_{22}^E & s_{23}^E & 0 & 0 & 0 \\ s_{31}^E & s_{32}^E & s_{33}^E & 0 & 0 & 0 \\ 0 & 0 & 0 & s_{44}^E & 0 & 0 \\ 0 & 0 & 0 & 0 & s_{55}^E & 0 \\ 0 & 0 & 0 & 0 & 0 & s_{66}^E = 2\left(s_{11}^E - s_{12}^E\right) \end{bmatrix} \begin{bmatrix} T_1 \\ T_2 \\ T_3 \\ T_4 \\ T_5 \\ T_6 \end{bmatrix} + \begin{bmatrix} 0 & 0 & d_{31} \\ 0 & 0 & d_{32} \\ 0 & 0 & d_{33} \\ 0 & d_{24} & 0 \\ d_{15} & 0 & 0 \\ 0 & 0 & 0 \end{bmatrix} \begin{bmatrix} E_1 \\ E_2 \\ E_3 \end{bmatrix}$$

$$\begin{bmatrix} D_1 \\ D_2 \\ D_3 \end{bmatrix} = \begin{bmatrix} 0 & 0 & 0 & 0 & d_{15} & 0 \\ 0 & 0 & 0 & d_{24} & 0 & 0 \\ d_{31} & d_{32} & d_{33} & 0 & 0 & 0 \end{bmatrix} \begin{bmatrix} T_1 \\ T_2 \\ T_3 \\ T_4 \\ T_5 \\ T_6 \end{bmatrix} + \begin{bmatrix} \varepsilon_{11} & 0 & 0 \\ 0 & \varepsilon_{22} & 0 \\ 0 & 0 & \varepsilon_{33} \end{bmatrix} \begin{bmatrix} E_1 \\ E_2 \\ E_3 \end{bmatrix}$$

where the first equation represents the relationship for the converse piezoelectric effect and the latter for the direct piezoelectric effect.

Although the above equations are the most used form in literature, some comments about the notation are necessary. Generally D and E are vectors, that is, Cartesian tensor of rank-1; and permittivity ε is Cartesian tensor of rank 2. Strain and stress are, in principle, also rank-2 tensors. But conventionally, because strain and stress are all symmetric tensors, the subscript of strain and stress can be relabelled in the following fashion: 11 → 1; 22 → 2; 33 → 3; 23 → 4; 13 → 5; 12 → 6. (Different convention may be used by different authors in literature. Say, some use 12 → 4; 23 → 5; 31 → 6 instead.)

That is why S and T appear to have the "vector form" of 6 components. Consequently, s appears to be a 6 by 6 matrix instead of rank-4 tensor. Such a re-labelled notation is often called Voigt notation.

In total, there are 4 piezoelectric coefficients, d_{ij}, e_{ij}, g_{ij}, and h_{ij} defined as follows:

$$d_{ij} = \left(\frac{\partial D_i}{\partial T_j}\right)^E = \left(\frac{\partial S_j}{\partial E_i}\right)^T$$

$$e_{ij} = \left(\frac{\partial D_i}{\partial S_j}\right)^E = -\left(\frac{\partial T_j}{\partial E_i}\right)^S$$

$$g_{ij} = -\left(\frac{\partial E_i}{\partial T_j}\right)^D = \left(\frac{\partial S_j}{\partial D_i}\right)^T$$

$$h_{ij} = -\left(\frac{\partial E_i}{\partial S_j}\right)^D = -\left(\frac{\partial T_j}{\partial D_i}\right)^S$$

where the first set of 4 terms correspond to the direct piezoelectric effect and the second set of 4 terms correspond to the converse piezoelectric effect. A formalism has been worked out for those piezoelectric crystals, for which the polarization is of the crystal-field induced type, that allows for the calculation of piezoelectrical coefficients d_{ij} from electrostatic lattice constants or higher-order Madelung constants.

Crystal Classes

Of the thirty-two crystal classes, twenty-one are non-centrosymmetric (not having a centre of symmetry), and of these, twenty exhibit direct piezoelectricity (the 21st is the cubic class 432). Ten of these represent the polar crystal classes, which show a spontaneous polarization without mechanical stress due to a non-vanishing electric dipole moment associated with their unit cell, and which exhibit pyroelectricity. If the dipole moment can be reversed by the application of an electric field, the material is said to be ferroelectric.

- Polar crystal classes: 1, 2, m, mm2, 4, 4 mm, 3, 3m, 6, 6 mm.
- Piezoelectric crystal classes: 1, 2, m, 222, mm2, 4, 4, 422, 4 mm, 42m, 3, 32, 3m, 6, 6, 622, 6 mm, 62m, 23, 43m.

For polar crystals, for which P "‘ 0 holds without applying a mechanical load, the piezoelectric effect manifests itself by changing the magnitude or the direction of P or both. For the non-polar, but piezoelectric crystals, on the other hand, a polarization P different from zero is only elicited by applying a mechanical load. For them the stress can be imagined to transform the material from a non-polar crystal class (P =0) to a polar one, having $P \neq 0$.

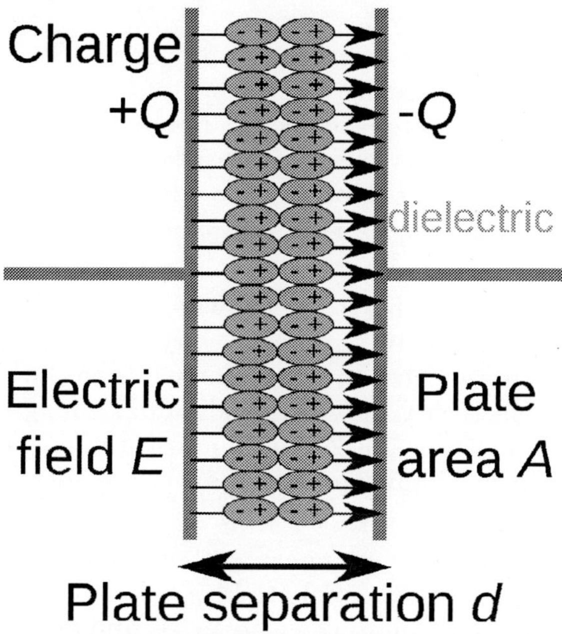

Figure: *Any spatially separated charge will result in an electric field, and therefore an electric potential. Shown here is a standard dielectric in a capacitor. In a piezoelectric device, mechanical stress, instead of an externally applied voltage, causes the charge separation in the individual atoms of the material, .*

Materials

Many materials, both natural and synthetic, exhibit piezoelectricity:

Naturally occurring crystals:

- Berlinite ($AlPO_4$), a rare phosphate mineral that is structurally identical to quartz
- Sucrose (table sugar)
- Quartz
- Rochelle salt

- Topaz
- Tourmaline-group minerals

The action of piezoelectricity in Topaz can probably be attributed to ordering of the (F,OH) in its lattice, which is otherwise centrosymmetric: Orthorhombic Bipyramidal (mmm). Topaz has anomalous optical properties which are attributed to such ordering.

Bone

Dry bone exhibits some piezoelectric properties. Studies of Fukada *et al.* showed that these are not due to the apatite crystals, which are centrosymmetric, thus non-piezoelectric, but due to collagen. Collagen exhibits the polar uniaxial orientation of molecular dipoles in its structure and can be considered as bioelectret, a sort of dielectric material exhibiting quasipermanent space charge and dipolar charge.

Potentials are thought to occur when a number of collagen molecules are stressed in the same way displacing significant numbers of the charge carriers from the inside to the surface of the specimen. Piezoelectricity of single individual collagen fibrils was measured using piezoresponse force microscopy, and it was shown that collagen fibrils behave predominantly as shear piezoelectric materials.

The piezoelectric effect is generally thought to act as a biological force sensor. This effect was exploited by research conducted at the University of Pennsylvania in the late 1970s and early 1980s, which established that sustained application of electrical potential could stimulate both resorption and growth (depending on the polarity) of bone in-vivo. Further studies in the 1990s provided the mathematical equation to confirm long bone wave propagation as to that of hexagonal (Class 6) crystals.

Other Natural Materials

Biological materials exhibiting piezoelectric properties include:

- Tendon
- Silk
- Wood due to piezoelectric texture
- Enamel
- Dentin
- DNA

- Viral proteins, including those from bacteriophage. One study has found that thin films of M13 bacteriophage can be used to construct a piezoelectric generator sufficient to operate a liquid crystal display.

Synthetic Crystals

- Gallium orthophosphate ($GaPO_4$), a quartz analogic crystal
- Langasite ($La_3Ga_5SiO_{14}$), a quartz analogic crystal

Synthetic Ceramics

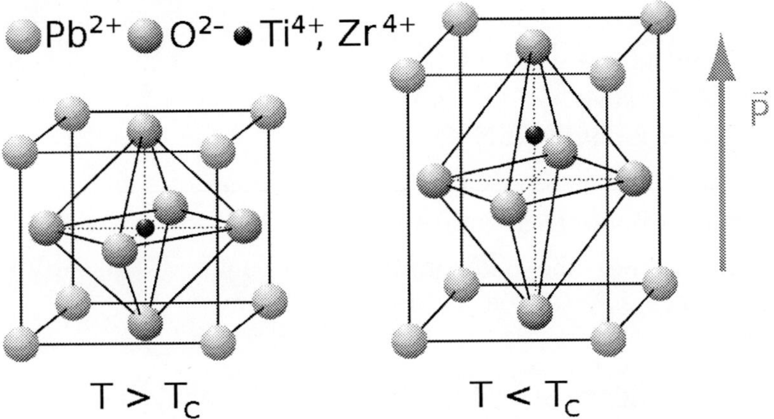

Figure: *Tetragonal unit cell of lead titanate*

The family of ceramics with perovskite or tungsten-bronze structures exhibits piezoelectricity:

- Barium titanate ($BaTiO_3$)—Barium titanate was the first piezoelectric ceramic discovered.
- Lead titanate ($PbTiO_3$)
- Lead zirconate titanate ($Pb[Zr_xTi_{1"x}]O_3$ $0 \le x \le 1$)—more commonly known as *PZT*, lead zirconate titanate is the most common piezoelectric ceramic in use today.
- Potassium niobate ($KNbO_3$)
- Lithium niobate ($LiNbO_3$)
- Lithium tantalate ($LiTaO_3$)
- Sodium tungstate (Na_2WO_3)
- Zinc oxide (ZnO)
- $Ba_2NaNb_5O_5$
- $Pb_2KNb_5O_{15}$

Lead-free Piezoceramics

More recently, there is growing concern regarding the toxicity in lead-containing devices driven by the result of restriction of hazardous substances directive regulations. To address this concern, there has been a resurgence in the compositional development of lead-free piezoelectric materials.

- Sodium potassium niobate ($(K,Na)NbO_3$). In 2004, a group of Japanese researchers led by Yasuyoshi Saito discovered a sodium potassium niobate composition with properties close to those of PZT, including a high T_C.
- Bismuth ferrite ($BiFeO_3$) is also a promising candidate for the replacement of lead-based ceramics.
- Sodium niobate $NaNbO_3$
- Bismuth titanate $Bi_4Ti_3O_{12}$
- Sodium bismuth titanate $Na_{0.5}Bi_{0.5}TiO_3$

So far, neither the environmental impact nor the stability of supplying these substances have been confirmed.

Polymers

- Polyvinylidene fluoride (PVDF): PVDF exhibits piezoelectricity several times greater than quartz. Unlike ceramics, where the crystal structure of the material creates the piezoelectric effect, in polymers the intertwined long-chain molecules attract and repel each other when an electric field is applied.

Organic Nanostructures

- A strong shear piezoelectric activity was observed in self-assembled diphenylalanine peptide nanotubes (PNTs), indicating electric polarization directed along the tube axis. Comparison with LiNbO3 and lateral signal calibration yields sufficiently high effective piezoelectric coefficient values of at least 60 pm/V (shear response for tubes of ≈200 nm in diameter). PNTs demonstrate linear deformation without irreversible degradation in a broad range of driving voltages.

Applications

Currently, industrial and manufacturing is the largest application market for piezoelectric devices, followed by the automotive industry. Strong demand also comes from medical instruments as well as information and telecommunications. The global demand for piezoelectric devices was valued at approximately US$14.8 billion in 2010. The largest material group for piezoelectric devices is piezocrystal, and piezopolymer is experiencing the fastest growth due to its low weight and small size.

Piezoelectric crystals are now used in numerous ways:

High Voltage and Power Sources

Direct piezoelectricity of some substances, like quartz, can generate potential differences of thousands of volts.

- The best-known application is the electric cigarette lighter: pressing the button causes a spring-loaded hammer to hit a piezoelectric crystal, producing a sufficiently high voltage electric current that flows across a small spark gap, thus heating and igniting the gas. The portable sparkers used to ignite gas stoves work the same way, and many types of gas burners now have built-in piezo-based ignition systems.
- A similar idea is being researched by DARPA in the United States in a project called *Energy Harvesting*, which includes an attempt to power battlefield equipment by piezoelectric generators embedded in soldiers' boots. However, these energy harvesting sources by association have an impact on the body. DARPA's effort to harness 1–2 watts from continuous shoe impact while walking were abandoned due to the impracticality and the discomfort from the additional energy expended by a person wearing the shoes. Other energy harvesting ideas include harvesting the energy from human movements in train stations or other public places and converting a dance floor to generate electricity. Vibrations from industrial machinery can also be harvested by piezoeletric materials to charge batteries for backup supplies or to power low-power microprocessors and wireless radios.
- A piezoelectric transformer is a type of AC voltage multiplier. Unlike a conventional transformer, which uses magnetic coupling between input and output, the piezoelectric transformer uses acoustic coupling. An input voltage is applied across a short length of a bar of piezoceramic material such as PZT, creating an alternating stress in the bar by the inverse piezoelectric effect and causing the whole bar to vibrate. The vibration frequency is chosen to be the resonant frequency of the block, typically in the 100 kilohertz to 1 megahertz range.

A higher output voltage is then generated across another section of the bar by the piezoelectric effect. Step-up ratios of more than 1000:1 have been demonstrated. An extra feature of this transformer is that, by operating it above its resonant frequency, it can be made to appear as an inductive load, which is useful in circuits that require a controlled soft start. These devices can be used in DC-AC inverters to drive cold cathode fluorescent lamps. Piezo transformers are some of the most compact high voltage sources.

Sensors

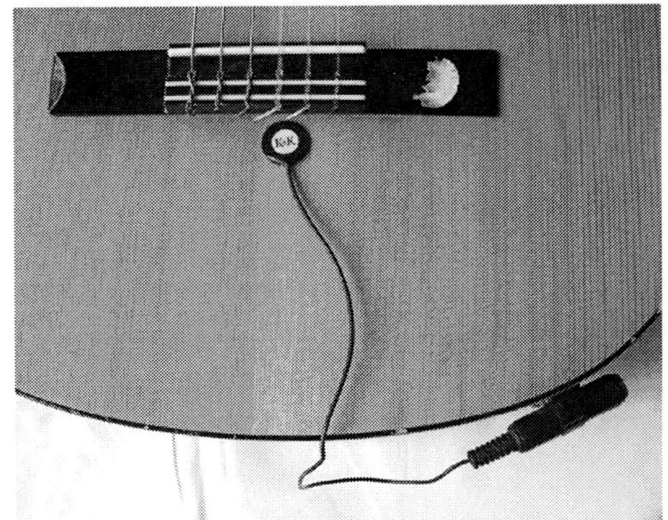

Figure: *Piezoelectric disk used as a guitar pickup*

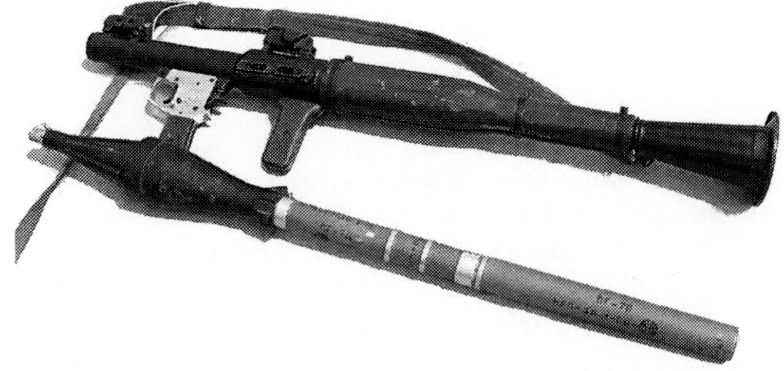

Figure: *Many rocket-propelled grenades used a piezoelectric fuze. For example: RPG-7*

The principle of operation of a piezoelectric sensor is that a physical dimension, transformed into a force, acts on two opposing faces of the sensing element. Depending on the design of a sensor, different "modes" to load the piezoelectric element can be used: longitudinal, transversal and shear. Detection of pressure variations in the form of sound is the most common sensor application, e.g. piezoelectric microphones (sound waves bend the piezoelectric material, creating a changing voltage) and piezoelectric pickups for acoustic-electric guitars. A piezo sensor attached to the body of an instrument is known as a contact microphone.

Piezoelectric sensors especially are used with high frequency sound in ultrasonic transducers for medical imaging and also industrial nondestructive testing (NDT).

For many sensing techniques, the sensor can act as both a sensor and an actuator – often the term *transducer* is preferred when the device acts in this dual capacity, but most piezo devices have this property of reversibility whether it is used or not. Ultrasonic transducers, for example, can inject ultrasound waves into the body, receive the returned wave, and convert it to an electrical signal (a voltage). Most medical ultrasound transducers are piezoelectric.

In addition to those mentioned above, various sensor applications include:

- Piezoelectric elements are also used in the detection and generation of sonar waves.
- Power monitoring in high power applications (e.g. medical treatment, sonochemistry and industrial processing).
- Piezoelectric microbalances are used as very sensitive chemical and biological sensors.
- Piezos are sometimes used in strain gauges.
- Piezoelectric transducers are used in electronic drum pads to detect the impact of the drummer's sticks, and to detect muscle movements in medical acceleromyography.
- Automotive engine management systems use piezoelectric transducers to detect Engine knock (Knock Sensor, KS), also known as detonation, at certain hertz frequencies. A piezoelectric transducer is also used in fuel injection systems to measure manifold absolute pressure (MAP sensor) to determine engine load, and ultimately the fuel injectors milliseconds of on time.

- Ultrasonic piezo sensors are used in the detection of acoustic emissions in acoustic emission testing.

Actuators

Figure: *Metal disk with piezoelectric disk attached, used in a buzzer*

As very high electric fields correspond to only tiny changes in the width of the crystal, this width can be changed with better-than-μm precision, making piezo crystals the most important tool for positioning objects with extreme accuracy — thus their use in actuators. Multilayer ceramics, using layers thinner than 100 μm, allow reaching high electric fields with voltage lower than 150 V. These ceramics are used within two kinds of actuators: direct piezo actuators and Amplified piezoelectric actuators. While direct actuator's stroke is generally lower than 100 μm, amplified piezo actuators can reach millimeter strokes.

- Loudspeakers: Voltage is converted to mechanical movement of a piezoelectric polymer film.
- Piezoelectric motors: Piezoelectric elements apply a directional force to an axle, causing it to rotate. Due to the extremely small distances involved, the piezo motor is viewed as a high-precision replacement for the stepper motor.
- Piezoelectric elements can be used in laser mirror alignment, where their ability to move a large mass (the mirror mount) over microscopic distances is exploited to electronically align

some laser mirrors. By precisely controlling the distance between mirrors, the laser electronics can accurately maintain optical conditions inside the laser cavity to optimize the beam output.

- A related application is the acousto-optic modulator, a device that scatters light off of soundwaves in a crystal, generated by piezoelectric elements. This is useful for fine-tuning a laser's frequency.
- Atomic force microscopes and scanning tunnelling microscopes employ converse piezoelectricity to keep the sensing needle close to the probe.
- Inkjet printers: On many inkjet printers, piezoelectric crystals are used to drive the ejection of ink from the inkjet print head towards the paper.
- Diesel engines: High-performance common rail diesel engines use piezoelectric fuel injectors, first developed by Robert Bosch GmbH, instead of the more common solenoid valve devices.
- Active vibration control using amplified actuators.
- X-ray shutters.
- XY stages for micro scanning used in infrared cameras.
- Moving the patient precisely inside active CT and MRI scanners where the strong radiation or magnetism precludes electric motors.
- Crystal earpieces are sometimes used in old or low power radios.

Frequency Standard

The piezoelectrical properties of quartz are useful as standard of frequency.

- Quartz clocks employ a crystal oscillator made from a quartz crystal that uses a combination of both direct and converse piezoelectricity to generate a regularly timed series of electrical pulses that is used to mark time. The quartz crystal (like any elastic material) has a precisely defined natural frequency (caused by its shape and size) at which it prefers to oscillate, and this is used to stabilize the frequency of a periodic voltage applied to the crystal.
- The same principle is critical in all radio transmitters and receivers, and in computers where it creates a clock pulse. Both of these usually use a frequency multiplier to reach gigahertz ranges.

Piezoelectric Motors

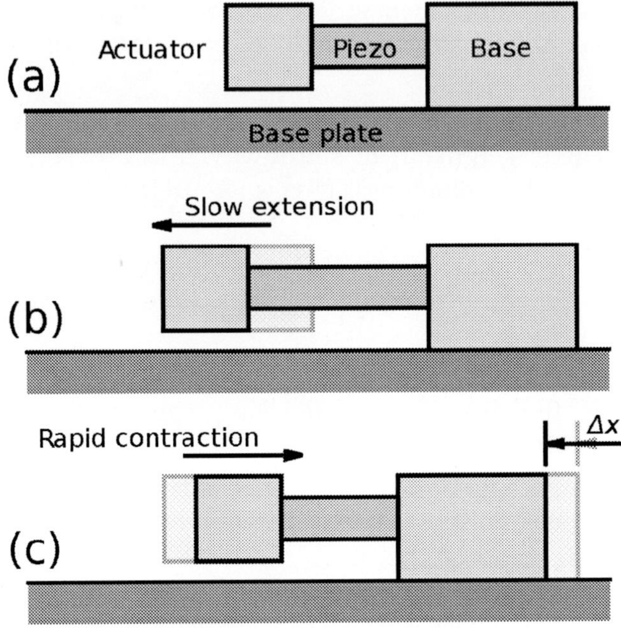

Figure: *A slip-stick actuator.*

Types of piezoelectric motor include:

- The travelling-wave motor used for auto-focus in reflex cameras
- Inchworm motors for linear motion
- Rectangular four-quadrant motors with high power density (2.5 watt/cm^3) and speed ranging from 10 nm/s to 800 mm/s.
- Stepping piezo motor, using stick-slip effect.

All these motors, except the stepping stick-slip motor work on the same principle. Driven by dual orthogonal vibration modes with a phase difference of 90°, the contact point between two surfaces vibrates in an elliptical path, producing a frictional force between the surfaces. Usually, one surface is fixed causing the other to move. In most piezoelectric motors the piezoelectric crystal is excited by a sine wave signal at the resonant frequency of the motor. Using the resonance effect, a much lower voltage can be used to produce a high vibration amplitude.

Stick-slip motor works using the inertia of a mass and the friction of a clamp. Such motors can be very small. Some are used for camera sensor displacement, thus allowing an anti-shake function.

Reduction of Vibrations and Noise

Different teams of researchers have been investigating ways to reduce vibrations in materials by attaching piezo elements to the material. When the material is bent by a vibration in one direction, the vibration-reduction system responds to the bend and sends electric power to the piezo element to bend in the other direction. Future applications of this technology are expected in cars and houses to reduce noise.

In a demonstration at the Material Vision Fair in Frankfurt in November 2005, a team from TU Darmstadt in Germany showed several panels that were hit with a rubber mallet, and the panel with the piezo element immediately stopped swinging.

Piezoelectric ceramic fibre technology is being used as an electronic damping system on some HEAD tennis rackets.

Infertility Treatment

In people with previous total fertilization failure, piezoelectric activation of oocytes together with intracytoplasmic sperm injection (ICSI) seems to improve fertilization outcomes.

Surgery

A recent application of piezoelectric ultrasound sources is piezoelectric surgery, also known as piezosurgery. Piezosurgery is a minimally invasive technique that aims to cut a target tissue with little damage to neighbouring tissues. For example, Hoigne *et al.* reported its use in hand surgery for the cutting of bone, using frequencies in the range 25–29 kHz, causing microvibrations of 60–210 μm. It has the ability to cut mineralized tissue without cutting neurovascular tissue and other soft tissue, thereby maintaining a blood-free operating area, better visibility and greater precision.

Chapter 6

Mechanical Properties of Materials

Concepts of Stress and Strain

Strength of Materials

Mechanics of materials, also called strength of materials, is a subject which deals with the behaviour of objects withstanding stresses and strains. This theory was established on the basis of mathematical modelling in first and second principal stress, specifically because types of stress state in construction parts such beam and shell are possible to approximate as one or two dimensional one. An important founding pioneer in mechanics of materials was Stephen Timoshenko.

The study of strength of materials often refers to various methods of calculating stresses in structural members, such as beams, columns and shafts. The methods employed to predict the response of a structure under loading and its susceptibility to various failure modes may take into account various properties of the materials other than material yield strength and ultimate strength; for example, failure by buckling is dependent on material stiffness and thus Young's Modulus.

Definition

In materials science, the strength of a material is its ability to withstand an applied stress without failure. The field of strength of materials deals with loads, deformations and the forces acting on a material. A load applied to a mechanical member will induce internal forces within the member called stresses. The stresses acting on the material cause deformation of the material. Deformation of the material is called strain, while the intensity of the internal forces is called stress. The applied stress may be tensile, compressive, or shear. The strength of any material relies on three different types of analytical method: strength, stiffness and stability, where strength refers to the load carrying capacity, stiffness refers to the deformation or elongation, and stability refers to the ability to maintain its initial configuration. Material yield strength refers to the point on the engineering stress-strain curve (as opposed to true stress-strain curve) beyond which the material experiences deformations that will not be completely reversed upon removal of the loading. The ultimate strength refers to the point on the engineering stress-strain curve corresponding to the stress that produces fracture.

Types of Loadings

- *Transverse loading* - Forces applied perpendicular to the longitudinal axis of a member. Transverse loading causes the member to bend and deflect from its original position, with internal tensile and compressive strains accompanying the change in curvature of the member. Transverse loading also induces shear forces that cause shear deformation of the material and increase the transverse deflection of the member.
- *Axial loading* - The applied forces are collinear with the longitudinal axis of the member. The forces cause the member to either stretch or shorten.
- *Torsional loading* - Twisting action caused by a pair of externally applied equal and oppositely directed force couples acting on parallel planes or by a single external couple applied to a member that has one end fixed against rotation.

Stress (Mechanics)

In continuum mechanics, stress is a physical quantity that expresses the internal forces that neighbouring particles of a continuous material exert on each other. For example, when a solid vertical bar is supporting a weight, each particle in the bar pulls on the particles immediately above and below it. When a liquid is under pressure, each particle gets pushed inwards by all the surrounding particles, and, in reaction, pushes them outwards. These macroscopic forces are actually the average of a very large number of intermolecular forces and collisions between the molecules in those particles.

Stress inside a body may arise by various mechanisms, such as reaction to external forces applied to the bulk material (like gravity) or to its surface (like contact forces, external pressure, or friction). Any strain (deformation) of a solid material generates an internal elastic stress, analogous to the reaction force of a spring, that tends to restore the material to its original undeformed state. In liquids and gases, only deformations that change the volume generate persistent elastic stress. However, if the deformation is gradually changing with time, even in fluids there will usually be some viscous stress, opposing that change. Elastic and viscous stresses are usually combined under the name mechanical stress.

Significant stress may exist even when deformation is negligible (a common assumption when modelling the flow of water) or non-existent. Stress may exist in the absence of external forces; such built-in stress is important, for example, in prestressed concrete and tempered glass. Stress may also be imposed on a material without the application of net forces, for example by changes in temperature or chemical composition, or by external electromagnetic fields (as in piezoelectric and magnetostrictive materials).

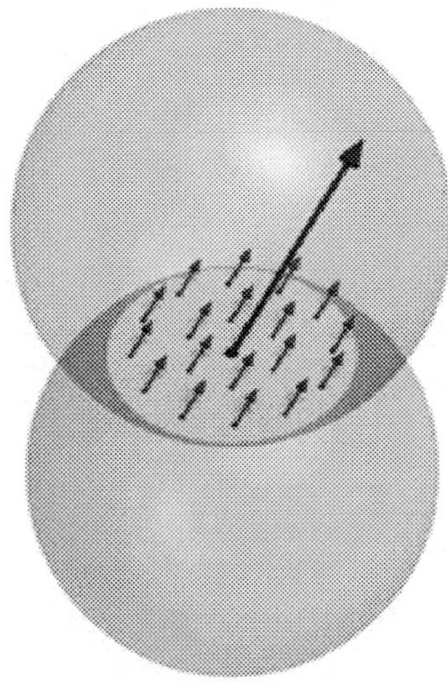

Figure: *The stress across a surface element (yellow disk) is the force that the material on one side (top ball) exerts on the material on the other side (bottom ball), divided by the area of the surface.*

Quantitatively, the stress is expressed by the Cauchy traction vector *T* defined as the traction force *F* between adjacent parts of the material across an imaginary separating surface *S*, divided by the area of *S*. In a fluid at rest the force is perpendicular to the surface, and is the familiar pressure. In a solid, or in a flow of viscous liquid, the force *F* may not be perpendicular to *S*; hence the stress across a surface must be regarded a vector quantity, not a scalar. Moreover, the direction and magnitude generally depend on the orientation of *S*. Thus the stress state of the material must be described by tensor, called the (Cauchy) stress tensor; which is a linear function that relates the normal vector *n* of a surface *S* to the stress *T* across *S*. With respect to any chosen coordinate system, the Cauchy stress tensor can be represented as a symmetric matrix of 3x3 real numbers. Even within a homogeneous body, the stress tensor may vary from place to place, and may change over time; therefore, the stress within a material is, in general, a time-varying tensor field.

The relation between mechanical stress, deformation, and the rate of change of deformation can be quite complicated, although a linear approximation may be adequate in practice if the quantities are small enough. Stress that exceeds certain strength limits of the material will result in permanent deformation (such as plastic flow, fracture, cavitation) or even change its crystal structure and chemical composition.

In some branches of engineering, the term stress is occasionally used in a looser sense as a synonym of "internal force". For example, in the analysis of trusses, it may refer to the total traction or compression force acting on a beam, rather than the force divided by the area of its cross-section.

History

Since ancient times humans have been consciously aware of stress inside materials. Until the 17th century the understanding of stress was largely intuitive and empirical; and yet it resulted in some surprisingly sophisticated technology, like the composite bow and glass blowing.

Over several millennia, architects and builders, in particular, learned how to put together carefully shaped wood beams and stone blocks to withstand, transmit, and distribute stress in the most effective manner, with ingenious devices such as the capitals, arches, cupolas, trusses and the flying buttresses of Gothic cathedrals.

Ancient and medieval architects did develop some geometrical methods and simple formulas to compute the proper sizes of pillars and beams, but the scientific understanding of stress became possible only after the necessary tools were invented in the 17th and 18th centuries: Galileo's rigorous experimental method, Descartes's coordinates and analytic geometry, and Newton's laws of motion and equilibrium and calculus of infinitesimals. With those tools, Cauchy was able to give the first rigorous and general mathematical model for stress in a homogeneous medium. Cauchy observed that the force across an imaginary surface was a linear function of its normal vector; and, moreover, that it must be a symmetric function (with zero total momentum).

The understanding of stress in liquids started with Newton himself, who provided a differential formula for friction forces (shear stress) in laminar parallel flow.

Overview

Definition: Stress is defined as the average force per unit area that some particle of a body exerts on an adjacent particle, across an imaginary surface that separates them. Being derived from a fundamental physical quantity (force) and a purely geometrical quantity (area), stress is also a fundamental quantity, like velocity, torque or energy, that can be quantified and analyzed without explicit consideration of the nature of the material or of its physical causes. Following the basic premises of continuum mechanics, stress is a macroscopic concept. Namely, the particles considered in its definition and analysis should be just small enough to be treated as homogeneous in composition and state, but still large enough to ignore quantum effects and the detailed motions of molecules. Thus, the force between two particles is actually the average of a very large number of atomic forces between their molecules; and physical quantities like mass, velocity, and forces that act through the bulk of three-dimensional bodies, like gravity, are assumed to be smoothly distributed over them. Depending on the context, one may also assume that the particles are large enough to allow the averaging out of other microscopic features, like the grains of a metal rod or the fibres of a piece of wood.

Normal and Shear Stress

In general, the stress T that a particle P applies on another particle Q across a surface S can have any direction relative to S. The vector T may be regarded as the sum of two components: the normal stress (Compression or Tension) perpendicular to the surface, and the shear stress that is parallel to it. If the normal unit vector n of the surface (pointing from Q towards P) is assumed fixed, the normal component can be expressed by a single number, the dot product $T \cdot n$. This number will be positive if P is "pulling" on Q (tensile stress), and negative if P is "pushing" against Q (compressive stress) The shear component is then the vector T - (T·n)n.

Units

The dimension of stress is that of pressure, and therefore its coordinates are commonly measured in the same units as pressure: namely, pascals (Pa, that is, newtons per square metre) in the International System, or pounds per square inch (psi) in the Imperial system.

Causes and Effects

Stress in a material body may be due to multiple physical causes, including external influences and internal physical processes. Some of these agents (like gravity, changes in temperature and phase, and electromagnetic fields) act on the bulk of the material, varying continuously with position and time. Other agents (like external loads and friction, ambient pressure, and contact forces) may create stresses and forces that are concentrated on certain surfaces, lines, or points; and possibly also on very short time intervals (as in the impulses due to collisions). In general, the stress distribution in the body is expressed as a piecewise continuous function of space and time.

Conversely, stress is usually correlated with various effects on the material, possibly including changes in physical properties like birefringence, polarization, and permeability. The imposition of stress by an external agent usually creates some strain (deformation) in the material, even if it is too small to be detected. In a solid material, such strain will in turn generate an internal elastic stress, analogous to the reaction force of a stretched spring, tending to restore the material to its original undeformed state. Fluid materials (liquids, gases and plasmas) by definition can only oppose deformations that would change their volume. However, if the deformation is changing with time, even in fluids there will usually be some viscous stress, opposing that change.

The relation between stress and its effects and causes, including deformation and rate of change of deformation, can be quite complicated (although a linear approximation may be adequate in practice if the quantities are small enough). Stress that exceeds certain strength limits of the material will result in permanent deformation (such as plastic flow, fracture, cavitation) or even change its crystal structure and chemical composition.

Simple Stresses

In some situations, the stress within a body may adequately be described by a single number, or by a single vector (a number and a direction). Three such simple stress situations, that are often encountered in engineering design, are the *uniaxial normal stress*, the *simple shear stress*, and the *isotropic normal stress*.

Uniaxial Normal Stress

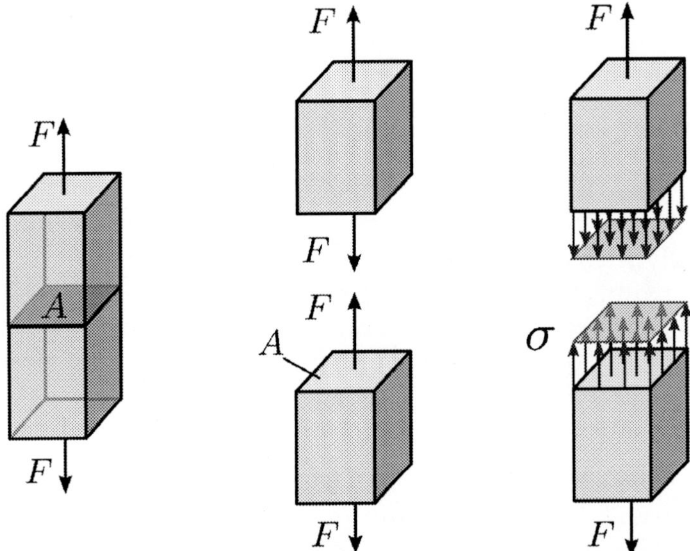

Figure: *Idealized stress in a straight bar with uniform cross-section.*

A common situation with a simple stress pattern is when a straight rod, with uniform material and cross section, is subjected to tension by opposite forces of magnitude F along its axis. If the system is in equilibrium and not changing with time, and the weight of the bar can be neglected, then through each transversal section of the bar the top part must pull on the bottom part with the same force F Therefore the stress throughout the bar, across any *horizontal* surface, can be described by the number $\sigma = F/A$, where A is the area of the cross-section. On the other hand, if one imagines the bar being cut along its length, parallel to the axis, there will be no force (hence no stress) between the two halves across the cut.

This type of stress may be called (simple) normal stress or uniaxial stress; specifically, (uniaxial, simple, etc.) tensile stress. If the load is compression on the bar, rather than stretching it, the analysis is the same except that the force F and the stress change sign, and the stress is called compressive stress.

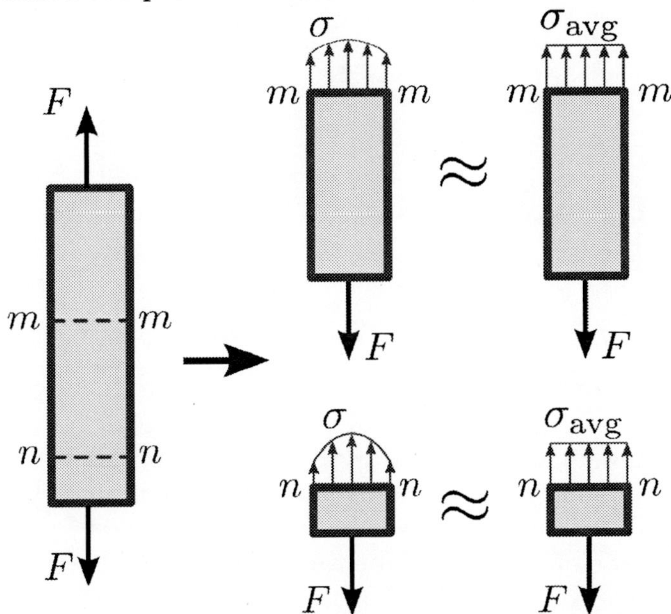

***Figure:** The ratio $\sigma = F / A$ may be only an average stress. The stress may be unevenly distributed over the cross section (m–m), especially near the attachment points (n–n).*

This analysis assumes the stress is evenly distributed over the entire cross-section. In practice, depending on how the bar is attached at the ends and how it was manufactured, this assumption may not be valid. In that case, the value $\sigma = F/A$ will be only the average stress, called engineering stress or nominal stress. However, if the bar's length L is many times its diameter D, and it has no gross defects or built-in stress, then the stress can be assumed to be uniformly distributed over any cross-section that is more than a few times D from both ends. (This observation is known as the Saint-Venant's principle).

Normal stress occurs in many other situations besides axial tension and compression. If an elastic bar with uniform and symmetric cross-section is bent in one of its planes of symmetry, the resulting bending stress will still be normal (perpendicular to the cross-section), but will vary over the cross section: the outer part will be under tensile stress, while the inner part will be compressed. Another variant of normal stress is the hoop stress that occurs on the walls of a cylindrical pipe or vessel filled with pressurized fluid.

Simple Shear Stress

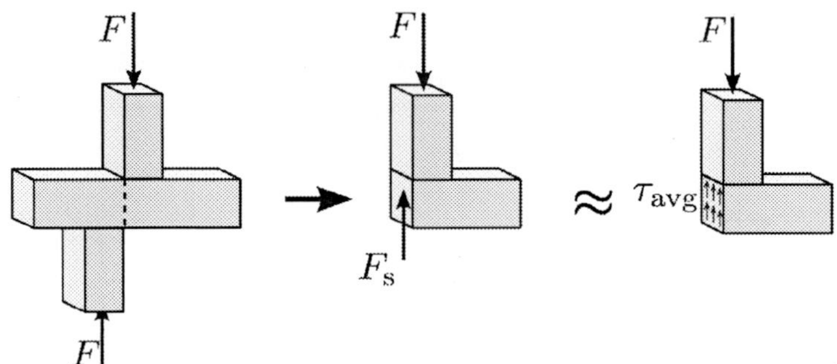

Figure: *Shear stress in a horizontal bar loaded by two offset blocks.*

Another simple type of stress occurs when a uniformly thick layer of elastic material like glue or rubber is firmly attached to two stiff bodies that are pulled in opposite directions by forces parallel to the layer; or a section of a soft metal bar that is being cut by the jaws of a scissors-like tool. Let F be the magnitude of those forces, and M be the midplane of that layer. Just as in the normal stress case, the part of the layer on one side of M must pull the other part with the same force F. Assuming that the direction of the forces is known, the stress across M can be expressed by the single number $\tau = F/A$, where F is the magnitude of those forces and A is the area of the layer.

However, unlike normal stress, this simple shear stress is directed parallel to the cross-section considered, rather than perpendicular to it. For any plane S that is perpendicular to the layer, the net internal force across S, and hence the stress, will be zero.

As in the case of an axially loaded bar, in practice the shear stress may not be uniformly distributed over the layer; so, as before, the ratio F/A will only be an average ("nominal", "engineering") stress. However, that average is often sufficient for practical purposes. Shear stress is observed also when a cyindrical bar such as a shaft is subjected to opposite torques at its ends. In that case, the shear stress on each cross-section is parallel to the cross-section, but oriented tangentially relative to the axis, and increases with distance from the axis. Significant shear stress occurs in the middle plate (the "web") of I-beams under bending loads, due to the web constraining the end plates ("flanges").

Isotropic Stress

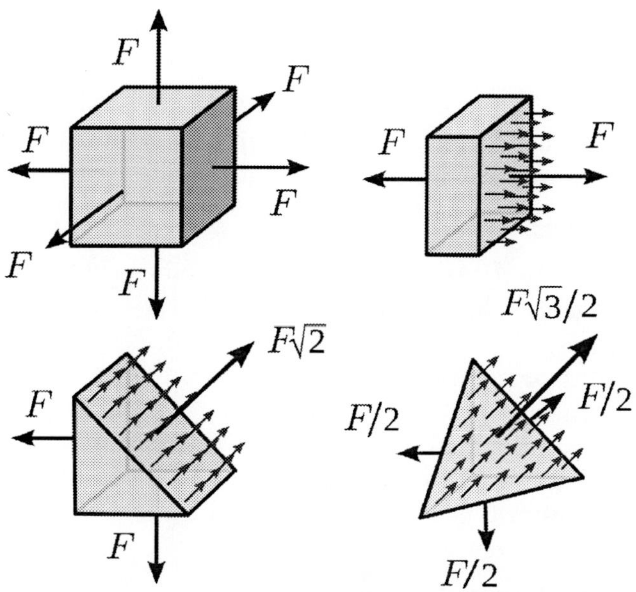

Figure: *Isotropic tensile stress. Top left: Each face of a cube of homogeneous material is pulled by a force with magnitude* F*, applied evenly over the entire face whose area is* A*. The force across any section* S *of the cube must balance the forces applied below the section. In the three sections shown, the forces are* F *(top right),* F $\sqrt{2}$ *(bottom left), and* F $\sqrt{3}/2$ *(bottom right); and the area of* S *is* A*,* A $\sqrt{2}$ *and* A $\sqrt{3}/2$*, respectively. So the stress across* S *is* F/A *in all three cases.*

Another simple type of stress occurs when the material body is under equal compression or tension in all directions. This is the case, for example, in a portion of liquid or gas at rest, whether enclosed in some container or as part of a larger mass of fluid; or inside a cube of elastic material that is being pressed or pulled on all six faces by equal perpendicular forces — provided, in both cases, that the material is homogeneous, without built-in stress, and that the effect of gravity and other external forces can be neglected.

In these situations, the stress across any imaginary internal surface turns out to be equal in magnitude and always directed perpendicularly to the surface independently of the surface's orientation. This type of stress may be called isotropic normal or just isotropic; if it is compressive, it is called hydrostatic pressure or just pressure. Gases by definition cannot withstand tensile stresses, but liquids may withstand very small amounts of isotropic tensile stress.

Cylinder Stresses

Parts with rotational symmetry, such as wheels, axles, pipes, and pillars, are very common in engineering. Often the stress patterns that occur in such parts have rotational or even cylindrical symmetry. The analysis of such cylinder stresses can take advantage of the symmetry to reduce the dimension of the domain and/or of the stress tensor.

General Stress

Often, mechanical bodies experience more than one type of stress at the same time; this is called combined stress. In normal and shear stress, the magnitude of the stress is maximum for surfaces that are perpendicular to a certain direction *d*, and zero across any surfaces that are parallel to *d*. When the stress is zero only across surfaces that are perpendicular to one particular direction, the stress is called biaxial, and can be viewed as the sum of two normal or shear stresses. In the most general case, called triaxial stress, the stress is nonzero across every surface element.

The Cauchy Stress Tensor

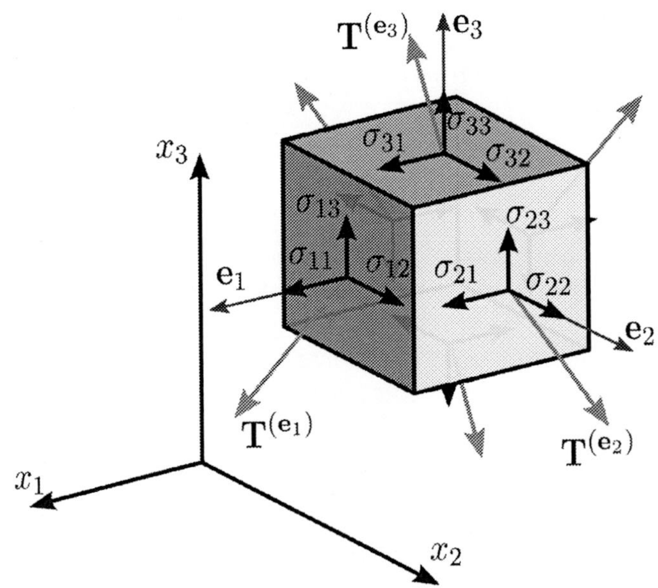

Figure: *Components of stress in three dimensions*

Combined stresses cannot be described by a single vector. Even if the material is stressed in the same way throughout the volume of the body, the stress across any imaginary surface will depend on the orientation of that surface, in a non-trivial way.

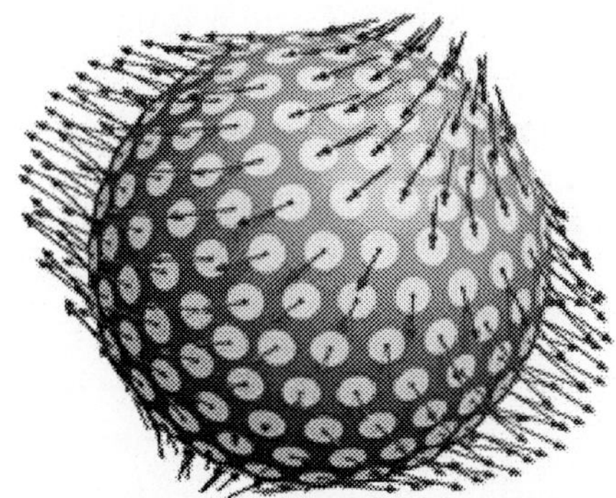

Figure: *Illustration of typical stresses (arrows) across various surface elements on the boundary of a particle (sphere), in a homogeneous material under uniform (but not isotropic) triaxial stress. The normal stresses on the principal axes are +5, +2, and –3 units.*

However, Cauchy observed that the stress vector T across a surface will always be a linear function of the surface's normal vector n, the unit-length vector that is perpendicular to it. That is, $T = \sigma(n)$, where the function σ satisfies

$$\sigma(\alpha u + \beta v) = \alpha\sigma(u) + \beta\sigma(v)$$

for any vectors u, v and any real numbers α, β. The function σ, now called the (Cauchy) stress tensor, completely describes the stress state of a uniformly stressed body. (Today, any linear connection between two physical vector quantities is called a tensor, reflecting Cauchy's original use to describe the "tensions" (stresses) in a material.) In tensor calculus,σ is classified as second-order tensor of type (0,2).

Like any linear map between vectors, the stress tensor can be represented in any chosen Cartesian coordinate system by a 3×3 matrix of real numbers. Depending on whether the coordinates are numbered x_1, x_2, x_3 or named x, y, z, the matrix may be written as

$$\begin{bmatrix} \sigma_{11} & \sigma_{12} & \sigma_{13} \\ \sigma_{21} & \sigma_{22} & \sigma_{23} \\ \sigma_{31} & \sigma_{32} & \sigma_{33} \end{bmatrix} \text{or} \begin{bmatrix} \sigma_{xx} & \sigma_{xy} & \sigma_{xz} \\ \sigma_{yx} & \sigma_{yy} & \sigma_{yz} \\ \sigma_{zx} & \sigma_{zy} & \sigma_{zz} \end{bmatrix}$$

The stress vector $T = \sigma(n)$ across a surface with normal vector with coordinates n_1, n_2, n_3 is then a matrix product $T = \sigma n$, that is

$$\begin{bmatrix} T_1 \\ T_2 \\ T_3 \end{bmatrix} = \begin{bmatrix} \sigma_{11} & \sigma_{21} & \sigma_{31} \\ \sigma_{12} & \sigma_{22} & \sigma_{32} \\ \sigma_{13} & \sigma_{23} & \sigma_{33} \end{bmatrix} \begin{bmatrix} n_1 \\ n_2 \\ n_3 \end{bmatrix}$$

The linear relation between T and n follows from the fundamental laws of conservation of linear momentum and static equilibrium of forces, and is therefore mathematically exact, for any material and any stress situation.

The components of the Cauchy stress tensor at every point in a material satisfy the equilibrium equations (Cauchy's equations of motion for zero acceleration). Moreover, the principle of conservation of angular momentum implies that the stress tensor is symmetric, that is $\sigma_{12} = \sigma_{21}, \sigma_{13} = \sigma_{31}$, and $\sigma_{23} = \sigma_{32}$. Therefore, the stress state of the medium at any point and instant can be specified by only six independent parameters, rather than nine. These may be written

$$\begin{bmatrix} \sigma_x & \tau_{xy} & \tau_{xz} \\ \tau_{xy} & \sigma_y & \tau_{yz} \\ \tau_{xz} & \tau_{yz} & \sigma_z \end{bmatrix}$$

where the elements $\sigma_x, \sigma_y, \sigma_z$ are called the orthogonal normal stresses (relative to the chosen coordinate system), $\tau_{xy}, \tau_{xz}, \tau_{yz}$ and the orthogonal shear stresses.

Change of Coordinates

The Cauchy stress tensor obeys the tensor transformation law under a change in the system of coordinates. A graphical representation of this transformation law is the Mohr's circle of stress distribution.

As a symmetric 3×3 real matrix, the stress tensor σ has three mutually orthogonal unit-length eigenvectors e_1, e_2, e_3 and three real eigenvalues $\lambda_1, \lambda_2, \lambda_3$, such that $\sigma e_i = \lambda_i e_i$.

Therefore, in a coordinate system with axes e_1, e_2, e_3, the stress tensor is a diagonal matrix, and has only the three normal components $\lambda_1, \lambda_2, \lambda_3$ the principal stresses. If the three eigenvalues are equal, the stress is an isotropic compression or tension, always perpendicular to any surface; there is no shear stress, and the tensor is a diagonal matrix in any coordinate frame.

Stress as a Tensor Field

In general, stress is not uniformly distributed over a material body, and may vary with time. Therefore the stress tensor must be defined for each point and each moment, by considering an infinitesimal particle of the medium surrounding that point, and taking the average stresses in that particle as being the stresses at the point.

Stress in Thin Plates

Man-made objects are often made from stock plates of various materials by operations that do not change their essentially two-dimensional character, like cutting, drilling, gentle bending and welding along the edges. The description of stress in such bodies can be simplified by modelling those parts as two-dimensional surfaces rather than three-dimensional bodies.

In that view, one redefines a "particle" as being an infinitesimal patch of the plate's surface, so that the boundary between adjacent particles becomes an infinitesimal line element; both are implicitly extended in the third dimension, straight through the plate. "Stress" is then redefined as being a measure of the internal forces between two adjacent "particles" across their common line element, divided by the length of that line. Some components of the stress tensor can be ignored, but since particles are not infinitesimal in the third dimension one can no longer ignore the torque that a particle applies on its neighbours. That torque is modelled as a bending stress that tends to change the curvature of the plate. However, these simplifications may not hold at welds, at sharp bends and creases (where the radius of curvature is comparable to the thickness of the plate).

Stress in thin Beams

The analysis of stress can be considerably simplified also for thin bars, beams or wires of uniform (or smoothly varying) composition and cross-section that are subjected to moderate bending and twisting. For those bodies may consider only cross-sections that are perpendicular to the bar's axis, and redefine a "particle" as being a piece of wire with infinitesimal length between two such cross sections. The ordinary stress is then reduced to a scalar (tension or compression of the bar), but one must take into account also a bending stress (that tries to change the bar's curvature, in some direction perpendicular to the axis) and a torsional stress (that tries to twist or un-twist it about its axis).

Other Descriptions of Stress

The Cauchy stress tensor is used for stress analysis of material bodies experiencing small deformations where the differences in stress distribution in most cases can be neglected. For large deformations, also called finite deformations, other measures of stress, such as the first and second Piola–Kirchhoff stress tensors, the Biot stress tensor, and the Kirchhoff stress tensor, are required.

Solids, liquids, and gases have stress fields. Static fluids support normal stress but will flow under shear stress. Moving viscous fluids can support shear stress (dynamic pressure).

Solids can support both shear and normal stress, with ductile materials failing under shear and brittle materials failing under normal stress. All materials have temperature dependent variations in stress-related properties, and non-Newtonian materials have rate-dependent variations.

Stress Analysis

Stress analysis is a branch of applied physics that covers the determination of the internal distribution of stresses in solid objects. It is an essential tool in engineering for the study and design of structures such as tunnels, dams, mechanical parts, and structural frames, under prescribed or expected loads. It is also important in many other disciplines; for example, in geology, to study phenomena like plate tectonics, vulcanism and avalanches; and in biology, to understand the anatomy of living beings.

Goals and Assumptions

Stress analysis is generally concerned with objects and structures that can be assumed to be in macroscopic static equilibrium. By Newton's laws of motion, any external forces are being applied to such a system must be balanced by internal reaction forces, which are almost always surface contact forces between adjacent particles — that is, as stress. Since every particle needs to be in equilibrium, this reaction stress will generally propagate from particle, creating a stress distribution throughout the body.

The typical problem in stress analysis is to determine these internal stresses, given the external forces that are acting on the system. The latter may be body forces (such as gravity or magnetic attraction), that act throughout the volume of a material; or concentrated loads (such as friction between an axle and a bearing, or the weight of a train wheel on a rail), that are imagined to act over a two-dimensional area, or along a line, or at single point.

In stress analysis one normally disregards the physical causes of the forces or the precise nature of the materials. Instead, one assumes that the stresses are related to deformation (and, in non-static problems, to the rate of deformation) of the material by known constitutive equations.

Methods

Stress analysis may be carried out experimentally, by applying loads to the actual artifact or to scale model, and measuring the resulting stresses, by any of several available methods. This approach is often used for safety certification and monitoring. However, most stress analysis is done by mathematical methods, especially during design.

The basic stress analysis problem can be formulated by Euler's equations of motion for continuous bodies (which are consequences of Newton's laws for conservation of linear momentum and angular momentum) and the Euler-Cauchy stress principle, together with the appropriate constitutive equations. Thus one obtains a system of partial differential equations involving the stress tensor field and the strain tensor field, as unknown functions to be determined. The external body forces appear as the independent ("right-hand side") term in the differential equations, while the concentrated forces appear as boundary conditions. The basic stress analysis problem is therefore a boundary-value problem.

Stress analysis for elastic structures is based on the theory of elasticity and infinitesimal strain theory. When the applied loads cause permanent deformation, one must use more complicated constitutive equations, that can account for the physical processes involved (plastic flow, fracture, phase change, etc.).

However, engineered structures are usually designed so that the maximum expected stresses are well within the range of linear elasticity (the generalization of Hooke's law for continuous media); that is, the deformations caused by internal stresses are linearly related to them. In this case the differential equations that define the stress tensor are linear, and the problem becomes much easier. For one thing, the stress at any point will be a linear function of the loads, too. For small enough stresses, even non-linear systems can usually be assumed to be linear.

Stress analysis is simplified when the physical dimensions and the distribution of loads allow the structure to be treated as one- or two-dimensional. In the analysis of trusses, for example, the stress field may be assumed to be uniform and uniaxial over each member. Then the differential equations reduce to a finite set of equations (usually linear) with finitely many unknowns. In other contexts one may be able to reduce the three-dimensional problem to a two-dimensional one, and/or replace the general stress and strain tensors by simpler models like uniaxial tension/compression, simple shear, etc.

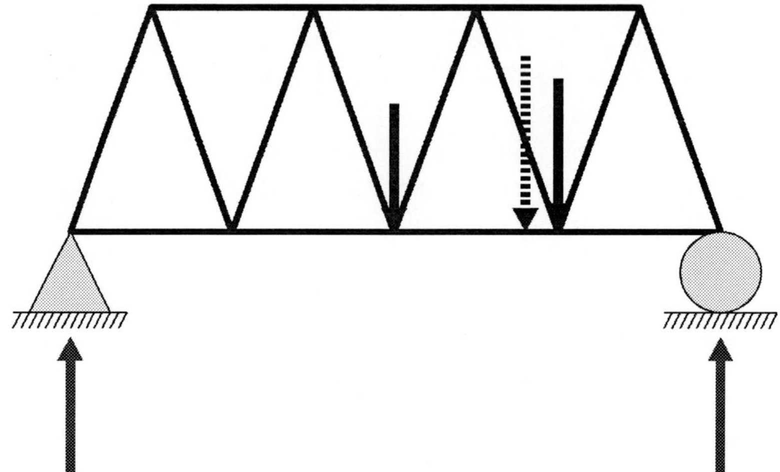

Figure: *Simplified model of a truss for stress analysis, assuming unidimensional elements under uniform axial tension or compression.*

Still, for two- or three-dimensional cases one must solve a partial differential equation problem. Anlytical or closed-form solutions to the differential equations can be obtained when the geometry, constitutive relations, and boundary conditions are simple enough. Otherwise one must generally resort to numerical approximations such as the finite element method, the finite difference method, and the boundary element method.

Theoretical Background

The mathematical description of stress is founded on Euler's laws for the motion of continuous bodies. They can be derived from Newton's laws, but may also be taken as axioms describing the motions of such bodies.

Alternative Measures of Stress

Other useful stress measures include the first and second Piola–Kirchhoff stress tensors, the Biot stress tensor, and the Kirchhoff stress tensor.

Piola–Kirchhoff Stress Tensor

In the case of finite deformations, the *Piola–Kirchhoff stress tensors* express the stress relative to the reference configuration. This is in contrast to the Cauchy stress tensor which expresses the stress relative to the present configuration. For infinitesimal deformations or rotations, the Cauchy and Piola–Kirchhoff tensors are identical.

Whereas the Cauchy stress tensor, σ relates stresses in the current configuration, the deformation gradient and strain tensors are described by relating the motion to the reference configuration; thus not all tensors describing the state of the material are in either the reference or current configuration.

Describing the stress, strain and deformation either in the reference or current configuration would make it easier to define constitutive models (for example, the Cauchy Stress tensor is variant to a pure rotation, while the deformation strain tensor is invariant; thus creating problems in defining a constitutive model that relates a varying tensor, in terms of an invariant one during pure rotation; as by definition constitutive models have to be invariant to pure rotations). The 1st Piola–Kirchhoff stress tensor, P is one possible solution to this problem. It defines a family of tensors, which describe the configuration of the body in either the current or the reference state.

The 1st Piola–Kirchhoff stress tensor, P relates forces in the *present* configuration with areas in the *reference* ("material") configuration.

$$\mathrm{P} = J\sigma\mathrm{F}^{-T}$$

where F is the deformation gradient and $J = \det \mathrm{F}$ is the Jacobian determinant.

In terms of components with respect to an orthonormal basis, the first Piola–Kirchhoff stress is given by

$$P_{iL} = J\,\sigma_{ik}\,F_{Lk}^{-1} = J\,\sigma_{ik}\,\frac{\partial X_L}{\partial x_k}$$

Because it relates different coordinate systems, the 1st Piola–Kirchhoff stress is a two-point tensor. In general, it is not symmetric. The 1st Piola–Kirchhoff stress is the 3D generalization of the 1D concept of engineering stress.

If the material rotates without a change in stress state (rigid rotation), the components of the 1st Piola–Kirchhoff stress tensor will vary with material orientation.

The 1st Piola–Kirchhoff stress is energy conjugate to the deformation gradient. 2nd Piola–Kirchhoff stress tensor

Whereas the 1st Piola–Kirchhoff stress relates forces in the current configuration to areas in the reference configuration, the 2nd Piola–Kirchhoff stress tensor S relates forces in the reference configuration to areas in the reference configuration. The force in the reference configuration is obtained via a mapping that preserves the relative relationship between the force direction and the area normal in the reference configuration.

$$S = J F^{-1} \cdot \sigma \cdot F^{-T} .$$

In index notation with respect to an orthonormal basis,

$$S_{IL} = J F_{Ik}^{-1} F_{Lm}^{-1} \sigma_{km} = J \frac{\partial X_I}{\partial x_k} \frac{\partial X_L}{\partial x_m} \sigma_{km}$$

This tensor, a one-point tensor, is symmetric.

If the material rotates without a change in stress state (rigid rotation), the components of the 2nd Piola–Kirchhoff stress tensor remain constant, irrespective of material orientation.

The 2nd Piola–Kirchhoff stress tensor is energy conjugate to the Green–Lagrange finite strain tensor.

Stress-Strain Diagrams

Stress–strain Curve

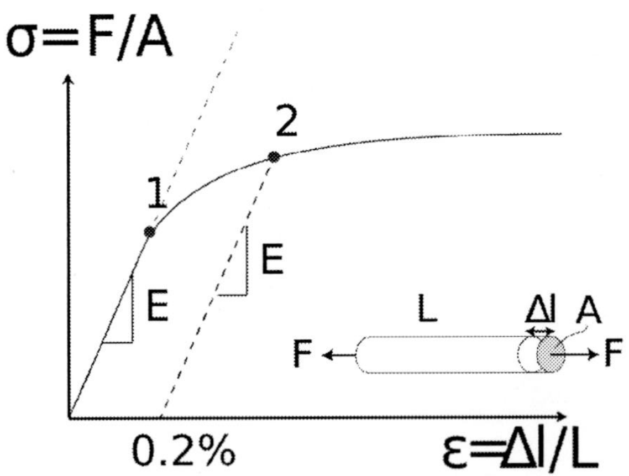

Figure: *Stress–strain curve showing typical yield behaviour for nonferrous alloys. Stress (σ) is shown as a function of strain (ϵ)*

1: Elastic (proportionality) limit

2: Offset yield strength (0.2% proof strength)

The relationship between the stress and strain that a particular material displays is known as that material's Stress-Strain curve. It is unique for each material and is found by recording the amount of deformation (strain) at distinct intervals of tensile or compressive loading (stress). These curves reveal many of the properties of a material (including data to establish the Modulus of Elasticity, E).

Stress–strain curves of various materials vary widely, and different tensile tests conducted on the same material yield different results, depending upon the temperature of the specimen and the speed of the loading. It is possible, however, to distinguish some common characteristics among the stress-strain curves of various groups of materials and, on this basis, to divide materials into two broad categories; namely, the ductile materials and the brittle materials.

Consider a bar of cross sectional area A being subjected to equal and opposite forces F pulling at the ends. If this were a rope, we would say that it is experiencing a tension force. Taking this concept over, we say that the bar is under tension, and is experiencing a stress that we define to be the ratio of the force to the cross sectional area:

Stress = F/A

This stress is called the tensile stress because every part of the object is subjected to a tension. The SI unit of stress is the Newton per square metre, which is called the Pascal.

1 Pascal = 1 Pa = 1 N/m^2

Shear Stress and Strain

Now consider a force that is applied tangentially to an object

The ratio of the shearing force to the area A is called the shear stress

If the object is twisted through an angle q, then the strain is

Shear Strain = tan q

Finally, we can define the shear modulus, MS, as

The shear modulus is also known as the torsion modulus

Ductile Materials

1. Ultimate Strength
2. Yield Strength
3. Rupture
4. Strain hardening region

5. Necking region.

A: Apparent stress (F/A_0)

B: Actual stress (F/A)

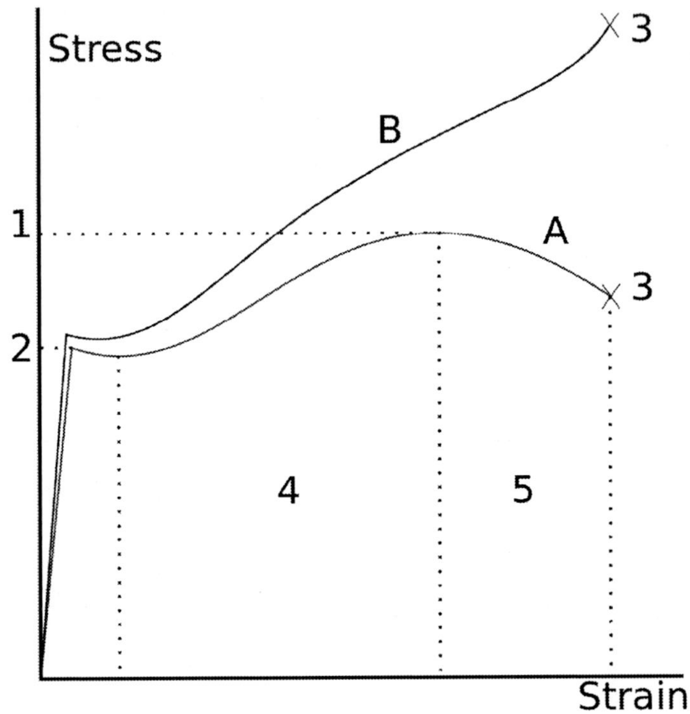

Figure: *A stress–strain curve typical of structural steel*

Ductile materials, which includes structural steel, as well as many alloys of other metals, are characterized by their ability to yield at normal temperatures.

Low carbon steel generally exhibits a very linear stress–strain relationship up to a well defined yield point. The linear portion of the curve is the elastic region and the slope is the modulus of elasticity or Young's Modulus.

After the yield point, the curve typically decreases slightly because of dislocations escaping from Cottrell atmospheres. As deformation continues, the stress increases on account of strain hardening until it reaches the ultimate strength. Until this point, the cross-sectional area decreases uniformly because of Poisson contractions. The actual rupture point is in the same vertical line as the visual rupture point.

However, beyond this point a *neck* forms where the local cross-sectional area decreases more quickly than the rest of the sample resulting in an increase in the true stress. On an engineering stress–strain curve this is seen as a decrease in the *apparent stress*. However if the curve is plotted in terms of *true stress* and *true strain* the stress will continue to rise until failure. Eventually the neck becomes unstable and the specimen ruptures (fractures).

Less ductile materials such as aluminum and medium to high carbon steels do not have a well-defined yield point.There are generally two types of yield points,upper and lower yield point. For these materials the yield strength is typically determined by the "offset yield method", by which a line is drawn parallel to the linear elastic portion of the curve and intersecting the abscissa at some arbitrary value (generally from 0.1% to 0.2%). The intersection of this line and the stress–strain curve is reported as the yield point.

The elastic region is the portion of the curve where the material will return to its original shape if the load is removed. The plastic region is the portion where some permanent deformation will occur, even if the load is removed. Failure point is when the object ruptures.

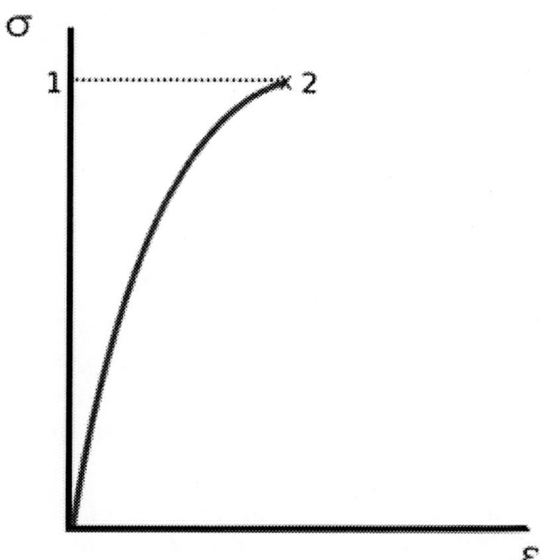

Figure: *Stress–strain curve for brittle materials.*

Brittle materials, which includes cast iron, glass, and stone, are characterized by the fact that rupture occurs without any noticeable prior change in the rate of elongation.

Brittle materials such as concrete or carbon fibre do not have a yield point, and do not strain-harden. Therefore the ultimate strength and breaking strength are the same. A typical stress-strain curve. Typical brittle materials like glass do not show any plastic deformation but fail while the deformation is elastic. One of the characteristics of a brittle failure is that the two broken parts can be reassembled to produce the same shape as the original component as there will not be a neck formation like in the case of ductile materials.

A typical stress strain curve for a brittle material will be linear. Testing of several identical specimen, cast iron, or soil, tensile strength is negligible compared to the compressive strength and it is assumed zero for many engineering applications.

Glass fibres have a tensile strength stronger than steel, but bulk glass usually does not. This is because of the stress intensity factor associated with defects in the material. As the size of the sample gets larger, the size of defects also grows. In general, the tensile strength of a rope is always less than sum of the tensile strength of its individual fibres.

Elastic Deformation

In materials science, deformation is a change in the shape or size of an object due to an applied force (the deformation energy in this case is transferred through work) or a change in temperature (the deformation energy in this case is transferred through heat). The first case can be a result of tensile (pulling) forces, compressive (pushing) forces, shear, bending or torsion (twisting).

In the second case, the most significant factor, which is determined by the temperature, is the mobility of the structural defects such as grain boundaries, point vacancies, line and screw dislocations, stacking faults and twins in both crystalline and non-crystalline solids. The movement or displacement of such mobile defects is thermally activated, and thus limited by the rate of atomic diffusion. Deformation is often described as strain.

As deformation occurs, internal inter-molecular forces arise that oppose the applied force. If the applied force is not too large these forces may be sufficient to completely resist the applied force, allowing the object to assume a new equilibrium state and to return to its original state when the load is removed. A larger applied force may lead to a permanent deformation of the object or even to its structural failure.

In the figure it can be seen that the compressive loading (indicated by the arrow) has caused deformation in the cylinder so that the original shape (dashed lines) has changed (deformed) into one with bulging sides. The sides bulge because the material, although strong enough to not crack or otherwise fail, is not strong enough to support the load without change, thus the material is forced out laterally. Internal forces (in this case at right angles to the deformation) resist the applied load.

Elasticity (Physics)

In physics, elasticity is a physical property of materials which return to their original shape after they are deformed.

Overview

When an elastic material is deformed due to an external force, it experiences internal forces that oppose the deformation and restore it to its original state if the external force is no longer applied. There are various elastic moduli, such as Young's modulus, the shear modulus, and the bulk modulus, all of which are measures of the inherent stiffness of a material as a resistance to deformation under an applied load. The various moduli apply to different kinds of deformation. For instance, Young's modulus applies to uniform extension, whereas the shear modulus applies to shearing.

The elasticity of materials is described by a stress-strain curve, which shows the relation between stress (the average restorative internal force per unit area) and strain (the relative deformation). For most metals or crystaline materials, the curve is linear for small deformations, and so the stress-strain relationship can adequately be described by Hooke's law and higher-order terms can be ignored.

However, for larger stresses beyond the elastic limit, the relation is no longer linear. For even higher stresses, materials exhibit plastic behaviour, that is, they deform irreversibly and do not return to their original shape after stress is no longer applied. For rubber-like materials such as elastomers, the gradient of the stress-strain curve increases with stress, meaning that rubbers progressively become more difficult to stretch, while for most metals, the gradient decreases at very high stresses, meaning that they progressively become easier to stretch. Elasticity is not exhibited only by solids; non-Newtonian fluids, such as viscoelastic fluids, will also exhibit elasticity in certain conditions. In response to a small, rapidly applied and removed strain, these fluids may deform and then return to their original shape. Under larger strains, or strains applied for longer periods of time, these fluids may start to flow like a viscous liquid.

Hooke's Law

As noted above, for small deformations, most elastic materials such as springs exhibit linear elasticity. This idea was first formulated by Robert Hooke in 1675 as a Latin anagram, "ceiiinosssttuv". He published the answer in 1678: "*Ut tensio, sic vis*" meaning "*As the extension, so the force*", a linear relationship commonly referred to as Hooke's law. This law can be stated as a relationship between force F and displacement x,

$$F = -kx,$$

where k is a constant known as the *rate* or *spring constant.* It can also be stated as a relationship between stress σ and strain ε:

$$\sigma = E\varepsilon,$$

where E Is known as the elastic modulus or Young's modulus.

Although the general proportionality constant between stress and strain in three dimensions is a 4th order tensor, systems that exhibit symmetry, such as a one-dimensional rod, can often be reduced to applications of Hooke's law.

Factors Affecting Elasticity

For isotropic materials, the presence of fractures affects the Young and the shear modulus perpendicular to the planes of the cracks, which decrease (Young's modulus faster than the shear modulus) as the fracture density increases, indicating that the presence of cracks makes bodies brittler. Microscopically, the stress-strain relationship of materials is in general governed by the Helmholtz free energy, a thermodynamic quantity.

Molecules settle in the configuration which minimizes the free energy, subject to constraints derived from their structure, and, depending on whether the energy or the entropy term dominates the free energy, materials can broadly be classified as *energy-elastic* and *entropy-elastic.* As such, microscopic factors affecting the free energy, such as the equilibriumdistance between molecules, can affect the elasticity of materials: for instance, in inorganic materials, as the equilibrium distance between molecules at 0 K increases, the bulk modulus decreases. The effect of temperature on elasticity is difficult to isolate, because there are numerous factors affecting it. For instance, the bulk modulus of a material is dependent on the form of its lattice, its behaviour under expansion, as well as the vibrations of the molecules, all of which are dependent on temperature.

Plasticity (Physics)

In physics and materials science, plasticity describes the deformation of a material undergoing non-reversible changes of shape in response to applied forces. For example, a solid piece of metal being bent or pounded into a new shape displays plasticity as permanent changes occur within the material itself. In engineering, the transition from elastic behaviour to plastic behaviour is called yield.

Plastic deformation is observed in most materials including metals, soils, rocks, concrete, foams, bone and skin. However, the physical mechanisms that cause plastic deformation can vary widely. At the crystal scale, plasticity in metals is usually a consequence of dislocations. In most crystalline materials such defects are relatively rare. But there are also materials where defects are numerous and are part of the very crystal structure, in such cases plastic crystallinity can result. In brittle materials such as rock, concrete, and bone, plasticity is caused predominantly by slip at microcracks.

For many ductile metals, tensile loading applied to a sample will cause it to behave in an elastic manner. Each increment of load is accompanied by a proportional increment in extension, and when the load is removed, the piece returns exactly to its original size. However, once the load exceeds some threshold (the yield strength), the extension increases more rapidly than in the elastic region, and when the load is removed, some amount of the extension remains.

However, elastic deformation is an approximation and its quality depends on the considered time frame and loading speed. If the deformation behaviour includes elastic deformation as indicated in the adjacent graph it is also often referred to as elastic-plastic or elasto-plastic deformation. Perfect plasticity is a property of materials to undergo irreversible deformation without any increase in stresses or loads. Plastic materials with hardening necessitate increasingly higher stresses to result in further plastic deformation. Generally plastic deformation is also dependent on the deformation speed, i.e. usually higher stresses have to be applied to increase the rate of deformation and such materials are said to deform visco-plastically.

Contributing Properties

The plasticity of a material is directly proportional to the ductility and malleability of the material.

Physical Mechanisms

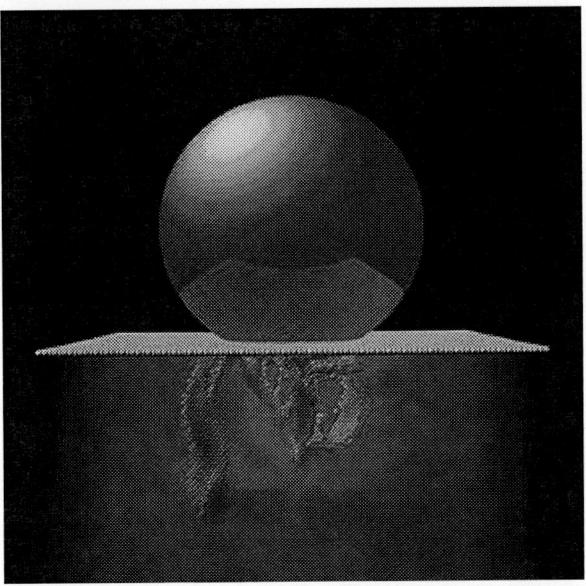

Figure: *Plasticity under a spherical Nanoindenter in (111) Copper. All particles in ideal lattice positions are omitted and the colour code refers to the von Mises stress field.*

Plasticity in Metals

Plasticity in a crystal of pure metal is primarily caused by two modes of deformation in the crystal lattice, slip and twinning. Slip is a shear deformation which moves the atoms through many interatomic distances relative to their initial positions. Twinning is the plastic deformation which takes place along two planes due to set of forces applied on a given metal piece. Most metals show more plasticity when hot than when cold. Lead shows sufficient plasticity at room temperature. But cast iron does not possess sufficient plasticity for any forging operation even when hot. This property is of importance in forming, shaping and extruding operations on metals. But most metals are rendered plastic by heating and hence shaped hot.

Slip Systems

Crystalline materials contain uniform planes of atoms organized with long-range order. Planes may slip past each other along their close-packed directions, as is shown on the slip systems wiki page. The result is a permanent change of shape within the crystal and plastic deformation. The presence of dislocations increases the likelihood of planes slipping.

Reversible Plasticity

On the nano scale the primary plastic deformation in simple fcc metals is reversible, as long as there is no material transport in form of cross-glide.

Shear Banding

The presence of other defects within a crystal may entangle dislocations or otherwise prevent them from gliding. When this happens, plasticity is localized to particular regions in the material. For crystals, these regions of localized plasticity are called shear bands.

Plasticity in Amorphous Materials

Crazing: In amorphous materials, the discussion of "dislocations" is inapplicable, since the entire material lacks long range order. These materials can still undergo plastic deformation. Since amorphous materials, like polymers, are not well-ordered, they contain a large amount of free volume, or wasted space. Pulling these materials in tension opens up these regions and can give materials a hazy appearance. This haziness is the result of *crazing*, where fibrils are formed within the material in regions of high hydrostatic stress. The material may go from an ordered appearance to a "crazy" pattern of strain and stretch marks.

Plasticity in Martensitic Materials

Some materials, especially those prone to Martensitic transformations, deform in ways that are not well described by the classic theories of plasticity and elasticity. One of the best-known examples of this is nitinol, which exhibits pseudoelasticity: deformations which are reversible in the context of mechanical design, but irreversible in terms of thermodynamics.

Plasticity in Cellular Materials

These materials plastically deform when the bending moment exceeds the fully plastic moment. This applies to open cell foams where the bending moment is exerted on the cell walls. The foams can be made of any material with a plastic yield point which includes rigid polymers and metals. This method of modelling the foam as beams is only valid if the ratio of the density of the foam to the density of the matter is less than 0.3. This is because beams yield axially instead of bending. In closed cell foams, the yield strength is increased if the material is under tension because of the membrane that spans the face of the cells.

Plasticity in Soils and Sand

Soils, particularly clays, display a significant amount of inelasticity under load. The causes of plasticity in soils can be quite complex and are strongly dependent on the microstructure, chemical composition, and water content. Plastic behaviour in soils is caused primarily by the rearrangement of clusters of adjacent grains.

Plasticity in Rocks and Concrete

Inelastic deformations of rocks and concrete are primarily caused by the formation of microcracks and sliding motions relative to these cracks. At high temperatures and pressures, plastic behaviour can also be affected by the motion of dislocations in individual grains in the microstructure.

Mathematical Descriptions of Plasticity

Deformation theory:

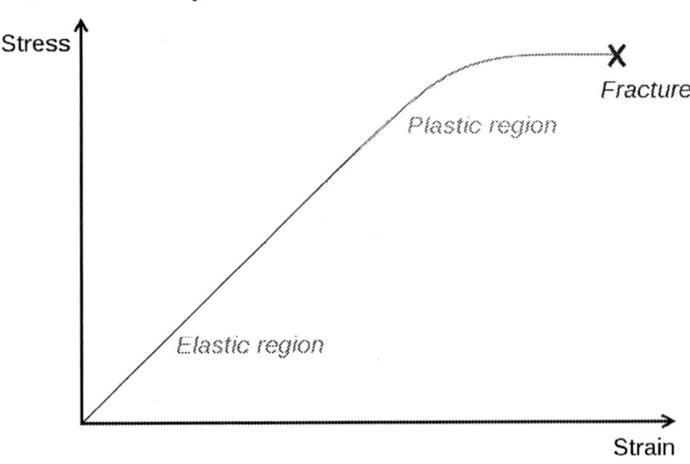

Figure: *An idealized uniaxial stress-strain curve showing elastic and plastic deformation regimes for the deformation theory of plasticity*

There are several mathematical descriptions of plasticity. One is deformation theory where the Cauchy stress tensor (of order d in d dimensions) is a function of the strain tensor. Although this description is accurate when a small part of matter is subjected to increasing loading (such as strain loading), this theory cannot account for irreversibility.

Ductile materials can sustain large plastic deformations without fracture. However, even ductile metals will fracture when the strain becomes large enough - this is as a result of work hardening of the material, which causes it to become brittle. Heat treatment such as annealing can restore the ductility of a worked piece, so that shaping can continue.

Flow Plasticity Theory

In 1934, Egon Orowan, Michael Polanyi and Geoffrey Ingram Taylor, roughly simultaneously, realized that the plastic deformation of ductile materials could be explained in terms of the theory of dislocations. The more correct mathematical theory of plasticity, flow plasticity theory, uses a set of non-linear, non-integrable equations to describe the set of changes on strain and stress with respect to a previous state and a small increase of deformation.

Yield Criteria

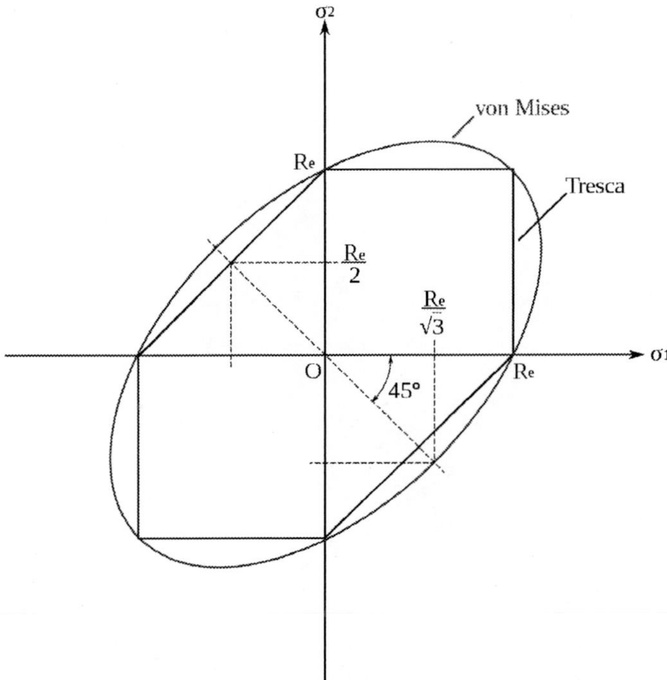

Figure: *Comparison of Tresca criterion to Von Mises criterion*

If the stress exceeds a critical value, as was mentioned above, the material will undergo plastic, or irreversible, deformation. This critical stress can be tensile or compressive. The Tresca and the von Mises criteria are commonly used to determine whether a material has yielded. However, these criteria have proved inadequate for a large range of materials and several other yield criteria are in widespread use.

Tresca Criterion

This criterion is based on the notion that when a material fails, it does so in shear, which is a relatively good assumption when considering metals. Given the principal stress state, we can use Mohr's circle to solve for the maximum shear stresses our material will experience and conclude that the material will fail if:

$$\sigma_1 - \sigma_3 \geq \sigma_0$$

Where σ_1 is the maximum normal stress, σ_3 is the minimum normal stress, and σ_0 is the stress under which the material fails in uniaxial loading. A yield surface may be constructed, which provides a visual representation of this concept. Inside of the yield surface, deformation is elastic. On the surface, deformation is plastic. It is impossible for a material to have stress states outside its yield surface.

Huber-von Mises Criterion

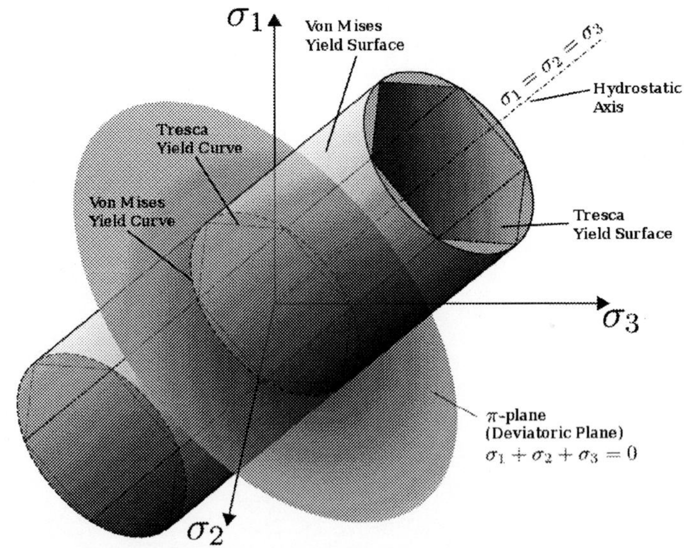

Figure: *The von Mises yield surfaces in principal stress coordinates circumscribes a cylinder around the hydrostatic axis. Also shown is Tresca's hexagonal yield surface.*

This criterion is based on the Tresca criterion but takes into account the assumption that hydrostatic stresses do not contribute to material failure. M.T. Huber was the first (1904, Lwσw) who proposed the criterion of shear energy. Von Mises solves for an effective stress under uniaxial loading, subtracting out hydrostatic stresses, and claims that all effective stresses greater than that which causes material failure in uniaxial loading will result in plastic deformation.

$$\sigma^2_{\text{effective}} = \tfrac{1}{2}\left[(\sigma_{11}-\sigma_{22})^2+(\sigma_{22}-\sigma_{33})^2+(\sigma_{11}-\sigma_{33})^2\right] + 6\left(\sigma^2_{12}+\sigma^2_{13}+\sigma^2_{23}\right)$$

Again, a visual representation of the yield surface may be constructed using the above equation, which takes the shape of an ellipse. Inside the surface, materials undergo elastic deformation. Reaching the surface means the material undergoes plastic deformations. It is physically impossible for a material to go beyond its yield surface.

Strain Rate Effects and Impact Behaviour

Strain Rate

Strain rate is the rate of change in strain (deformation) of a material with respect to time.

The strain rate at some point within the material measures the rate at which the distances of adjacent parcels of the material change with time in the neighbourhood of that point. It comprises both the rate at which the material is expanding or shrinking (expansion rate), and also the rate at which it is being deformed by progressive shearing without changing its volume (shear rate). It is zero if these distances do not change, as happens when all particles in some region are moving with the same velocity (same speed and direction) and/or rotating with the same angular velocity, as if that part of the medium were a rigid body.

The strain rate is a concept of materials science and continuum mechanics, that plays an essential role in the physics of fluids and deformable solids. In an isotropic Newtonian fluid, in particular, the viscous stress is a linear function of the rate of strain, defined by two coefficients, one relating to the expansion rate (the bulk viscosity coefficient) and one relating to the shear rate (the “ordinary” viscosity coefficient).

Definition

In physics the strain rate is generally defined as the derivative of the strain with respect to time. Its precise definition depends on how strain is measured.

Simple Deformations

In simple contexts, a single number may suffice to describe the strain, and therefore the strain rate. For example, when a long and uniform rubber band is gradually stretched by pulling at the ends, the strain can be defined as the ratio ϵ between the amount of stretching and the original length of the band:

$$\epsilon(t) = \frac{L(t) - L_0}{L_0}$$

where L_0 is the original length and $L(t)$ its length at each time t. Then the strain rate will be

$$\dot{\epsilon}(t) = \frac{d\epsilon}{dt} = \frac{d}{dt}\left(\frac{L(t) - L_0}{L_0}\right) = \frac{1}{L_0}\frac{dL}{dt}(t) = \frac{v(t)}{L_0}$$

where $v(t)$ is the speed at which the ends are moving away from each other.

The strain rate can also be expressed by a single number when the material is being subjected to parallel shear without change of volume; namely, when the deformation can be described as a set of infinitesimally thin parallel layers sliding against each other as if they were rigid sheets, in the same direction, without changing their spacing. This description fits the laminar flow of a fluid between two solid plates that slide parallel to each other (a Couette flow) or inside a circular pipe of constant cross-section (a Poiseuille flow). In those cases, the state of the material at some time t can be described by the displacement $X(y,t)$ of each layer, since an arbitrary starting time, as a function of its distance y from the fixed wall. Then the strain in each layer can be expressed as the limit of the ratio between the current relative displacement $X(y+d,t) - X(y,t)$ of a nearby layer, divided by the spacing d between the layers:

$$\epsilon(y,t) = \lim_{d \to 0} \frac{X(y+d,t) - X(y,t)}{d} = \frac{\partial X}{\partial y}(y,t)$$

Therefore the strain rate is

$$\dot{\epsilon}(y,t) = \left(\frac{\partial}{\partial t}\frac{\partial X}{\partial y}\right)(y,t) == \left(\frac{\partial}{\partial y}\frac{\partial X}{\partial t}\right)(y,t) = \frac{\partial V}{\partial y}(y,t)$$

where $V(y,t)$ is the current linear speed of the material at distance y from the wall.

Hardness of Materials

Hardness is a measure of how resistant solid matter is to various kinds of permanent shape change when a force is applied. Macroscopic hardness is generally characterized by strong intermolecular bonds, but the behaviour of solid materials under force is complex; therefore, there are different measurements of hardness: *scratch hardness*, *indentation hardness*, and *rebound hardness*.

Hardness is dependent on ductility, elastic stiffness, plasticity, strain, strength, toughness, viscoelasticity, and viscosity.

Common examples of hard matter are ceramics, concrete, certain metals, and superhard materials, which can be contrasted with soft matter.

Measuring Hardness

There are three main types of hardness measurements: *scratch*, *indentation*, and *rebound*. Within each of these classes of measurement there are individual measurement scales. For practical reasons conversion tables are used to convert between one scale and another.

Scratch Hardness

Scratch hardness is the measure of how resistant a sample is to fracture or permanent plastic deformation due to friction from a sharp object. The principle is that an object made of a harder material will scratch an object made of a softer material. When testing coatings, scratch hardness refers to the force necessary to cut through the film to the substrate. The most common test is Mohs scale, which is used in mineralogy. One tool to make this measurement is the sclerometer.

Another tool used to make these tests is the pocket hardness tester. This tool consists of a scale arm with graduated markings attached to a four wheeled carriage.

A scratch tool with a sharp rim is mounted at a predetermined angle to the testing surface. In order to use it a weight of known mass is added to the scale arm at one of the graduated markings, the tool is then drawn across the test surface. The use of the weight and markings allows a known pressure to be applied without the need for complicated machinery.

Indentation Hardness

Indentation hardness measures the resistance of a sample to material deformation due to a constant compression load from a sharp object; they are primarily used in engineering and metallurgy fields. The tests work on the basic premise of measuring the critical dimensions of an indentation left by a specifically dimensioned and loaded indenter.

Common indentation hardness scales are Rockwell, Vickers, Shore, and Brinell.

Rebound Hardness

Rebound hardness, also known as *dynamic hardness*, measures the height of the "bounce" of a diamond-tipped hammer dropped from a fixed height onto a material. This type of hardness is related to elasticity. The device used to take this measurement is known as a scleroscope.

Two scales that measures rebound hardness are the Leeb rebound hardness test and Bennett hardness scale.

Hardening

There are five hardening processes: Hall-Petch strengthening, work hardening, solid solution strengthening, precipitation hardening, and martensitic transformation.

Physics

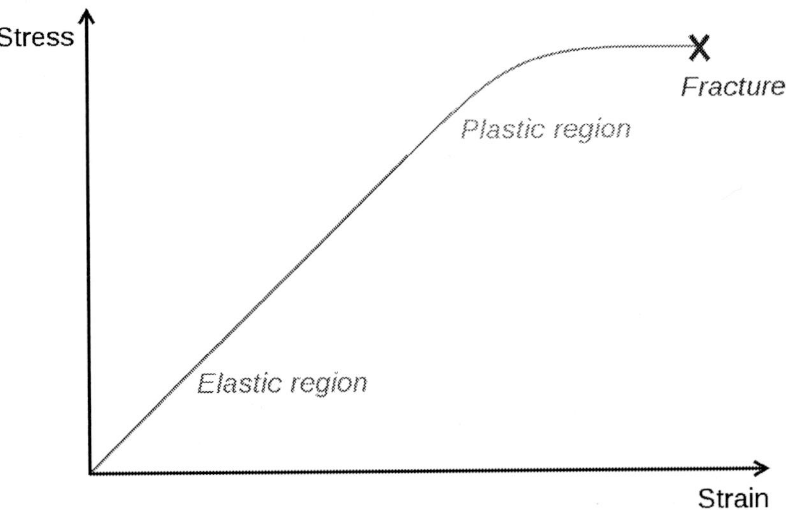

Figure: *Diagram of a stress-strain curve, showing the relationship between stress (force applied per unit area) and strain or deformation of a ductile metal.*

In solid mechanics, solids generally have three responses to force, depending on the amount of force and the type of material:

- They exhibit elasticity—the ability to temporarily change shape, but return to the original shape when the pressure is removed. "Hardness" in the elastic range—a small temporary change in shape for a given force—is known as stiffness in the case of a given object, or a high elastic modulus in the case of a material.
- They exhibit plasticity—the ability to permanently change shape in response to the force, but remain in one piece. The yield strength is the point at which elastic deformation gives way to plastic deformation. Deformation in the plastic range is non-linear, and is described by the stress-strain curve. This response produces the observed properties of scratch and indentation hardness, as described and measured in materials science. Some materials exhibit both elasticity and viscosity when undergoing plastic deformation; this is called viscoelasticity.
- They fracture—split into two or more pieces.

Strength is a measure of the extent of a material's elastic range, or elastic and plastic ranges together. This is quantified as compressive strength, shear strength, tensile strength depending on the direction of the forces involved. Ultimate strength is an engineering measure of the maximum load a part of a specific material and geometry can withstand.

Brittleness, in technical usage, is the tendency of a material to fracture with very little or no detectable deformation beforehand. Thus in technical terms, a material can be both brittle and strong. In everyday usage "brittleness" usually refers to the tendency to fracture under a small amount of force, which exhibits both brittleness and a lack of strength (in the technical sense). For perfectly brittle materials, yield strength and ultimate strength are the same, because they do not experience detectable plastic deformation. The opposite of brittleness is ductility.

The toughness of a material is the maximum amount of energy it can absorb before fracturing, which is different from the amount of force that can be applied. Toughness tends to be small for brittle materials, because elastic and plastic deformations allow materials to absorb large amounts of energy. Hardness increases with decreasing particle size. This is known as the Hall-Petch relationship. However, below a critical grain-size, hardness decreases with decreasing grain size. This is known as the inverse Hall-Petch effect. Hardness of a material to deformation is dependent on its microdurability or small-scale shear modulus in any direction, not to any rigidity or stiffness properties such as its bulk modulus or Young's modulus. Stiffness is often confused for hardness. Some materials are stiffer than diamond (e.g. osmium) but are not harder, and are prone to spalling and flaking in squamose or acicular habits.

Mechanisms and Theory

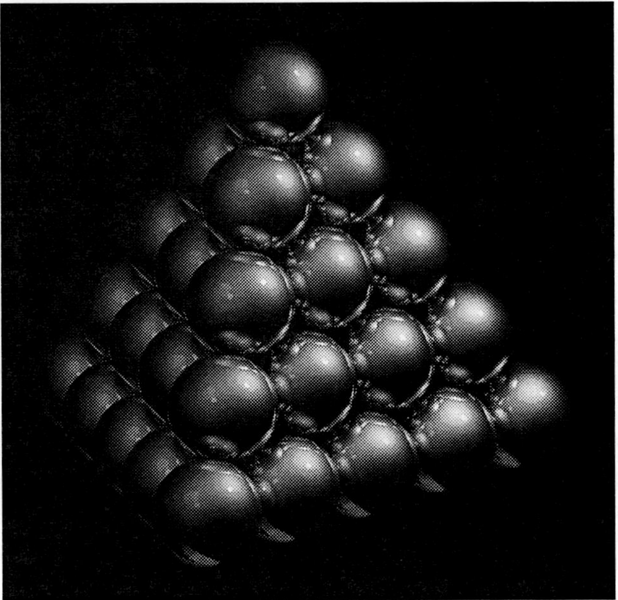

Figure: *A representation of the crystal lattice showing the planes of atoms.*

The key to understanding the mechanism behind hardness is understanding the metallic microstructure, or the structure and arrangement of the atoms at the atomic level. In fact, most important metallic properties critical to the manufacturing of today's goods are determined by the microstructure of a material. At the atomic level, the atoms in a metal are arranged in an orderly three-dimensional array called a crystal lattice. In reality, however, a given specimen of a metal likely never contains a consistent single crystal lattice. A given sample of metal will contain many grains, with each grain having a fairly consistent array pattern. At an even smaller scale, each grain contains irregularities.

There are two types of irregularities at the grain level of the microstructure that are responsible for the hardness of the material.

These irregularities are point defects and line defects. A point defect is an irregularity located at a single lattice site inside of the overall three-dimensional lattice of the grain. There are three main point defects. If there is an atom missing from the array, a vacancy defect is formed. If there is a different type of atom at the lattice site that should normally be occupied by a metal atom, a substitutional defect is formed. If there exists an atom in a site where there should normally not be, an interstitial defect is formed. This is possible because space exists between atoms in a crystal lattice. While point defects are irregularities at a single site in the crystal lattice, line defects are irregularities on a plane of atoms. Dislocations are a type of line defect involving the misalignment of these planes. In the case of an edge dislocation, a half plane of atoms is wedged between two planes of atoms. In the case of a screw dislocation two planes of atoms are offset with a helical array running between them.

Dislocations provide a mechanism for planes of atoms to slip and thus a method for plastic or permanent deformation. Planes of atoms can flip from one side of the dislocation to the other effectively allowing the dislocation to traverse through the material and the material to deform permanently. The movement allowed by these dislocations causes a decrease in the material's hardness. The way to inhibit the movement of planes of atoms, and thus make them harder, involves the interaction of dislocations with each other and interstitial atoms. When a dislocation intersects with a second dislocation, it can no longer traverse through the crystal lattice. The intersection of dislocations creates an anchor point and does not allow the planes of atoms to continue to slip over one another A dislocation can also be anchored by the interaction with interstitial atoms. If a dislocation comes in contact with two or more interstitial atoms, the slip of the planes will again be disrupted. The interstitial atoms create anchor points, or pinning points, in the same manner as intersecting dislocations.

By varying the presence of interstitial atoms and the density of dislocations, a particular metal's hardness can be controlled. Although seemingly counter-intuitive, as the density of dislocations increases, there are more intersections created and consequently more anchor points. Similarly, as more interstitial atoms are added, more pinning points that impede the movements of dislocations are formed. As a result, the more anchor points added, the harder the material will become.

Chapter 7

Magnetic Materials

A magnet is a material or object that produces a magnetic field. This magnetic field is invisible but is responsible for the most notable property of a magnet: a force that pulls on other ferromagnetic materials, such as iron, and attracts or repels other magnets.

A permanent magnet is an object made from a material that is magnetized and creates its own persistent magnetic field. An everyday example is a refrigerator magnet used to hold notes on a refrigerator door. Materials that can be magnetized, which are also the ones that are strongly attracted to a magnet, are called ferromagnetic (or ferrimagnetic). These include iron, nickel, cobalt, some alloys of rare earth metals, and some naturally occurring minerals such as lodestone. Although ferromagnetic (and ferrimagnetic) materials are the only ones attracted to a magnet strongly enough to be commonly considered magnetic, all other substances respond weakly to a magnetic field, by one of several other types of magnetism.

Ferromagnetic materials can be divided into magnetically "soft" materials like annealed iron, which can be magnetized but do not tend to stay magnetized, and magnetically "hard" materials, which do. Permanent magnets are made from "hard" ferromagnetic materials such as alnico and ferrite that are subjected to special processing in a powerful magnetic field during manufacture, to align their internal microcrystalline structure, making them very hard to demagnetize. To demagnetize a saturated magnet, a certain magnetic field must be applied, and this threshold depends on coercivity of the respective material. "Hard" materials have high coercivity, whereas "soft" materials have low coercivity.

An electromagnet is made from a coil of wire that acts as a magnet when an electric current passes through it but stops being a magnet when the current stops. Often, the coil is wrapped around a core of "soft" ferromagnetic material such as steel, which greatly enhances the magnetic field produced by the coil.

The overall strength of a magnet is measured by its magnetic moment or, alternatively, the total magnetic flux it produces. The local strength of magnetism in a material is measured by its magnetization.

Discovery and Development

Ancient people learned about magnetism from lodestones, naturally magnetized pieces of iron ore. They are naturally created magnets, which attract pieces of iron. The word *magnet* in Greek meant "stone from Magnesia", a part of ancient Greece where lodestones were found. Lodestones suspended so they could turn were the first magnetic compasses. The earliest known surviving descriptions of magnets and their properties are from Greece, India, and China around 2500 years ago. The properties of lodestones and their affinity for iron were written of by Pliny the Elder in his encyclopedia *Naturalis Historia.*

By the 12th to 13th centuries AD, magnetic compasses were used in navigation in China, Europe, and elsewhere.

Background on the Physics of Magnetism and Magnets

Magnetic Moment

A magnet's magnetic moment (also called magnetic dipole moment and usually denoted μ) is a vector that characterizes the magnet's overall magnetic properties. For a bar magnet, the direction of the magnetic moment points from the magnet's south pole to its north pole, and the magnitude relates to how strong and how far apart these poles are. In SI units, the magnetic moment is specified in terms of $A \cdot m^2$.

A magnet both produces its own magnetic field and responds to magnetic fields. The strength of the magnetic field it produces is at any given point proportional to the magnitude of its magnetic moment.

In addition, when the magnet is put into an external magnetic field, produced by a different source, it is subject to a torque tending to orient the magnetic moment parallel to the field. The amount of this torque is proportional both to the magnetic moment and the external field. A magnet may also be subject to a force driving it in one direction or another, according to the positions and orientations of the magnet and source. If the field is uniform in space, the magnet is subject to no net force, although it is subject to a torque.

A wire in the shape of a circle with area A and carrying current I is a magnet, with a magnetic moment of magnitude equal to IA.

Magnetization

The magnetization of a magnetized material is the local value of its magnetic moment per unit volume, usually denoted M, with units A/m. It is a vector field, rather than just a vector (like the magnetic moment), because different areas in a magnet can be magnetized with different directions and strengths. A good bar magnet may have a magnetic moment of magnitude 0.1 $A \cdot m^2$ and a volume of 1 cm^3, or 1×10^{-6} m^3, and therefore an average magnetization magnitude is 100,000 A/m. Iron can have a magnetization of around a million amperes per metre. Such a large value explains why iron magnets are so effective at producing magnetic fields.

Modelling Magnets

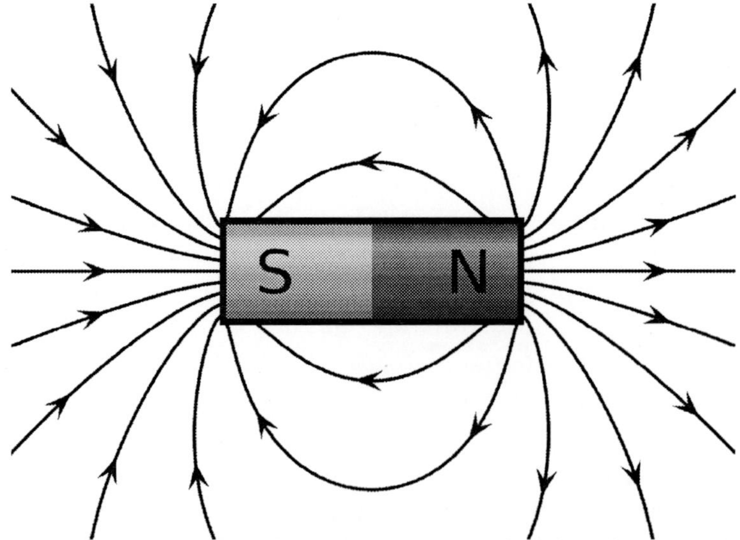

Figure: *Field of a cylindrical bar magnet calculated with Ampère's model*

Two different models exist for magnets: magnetic poles and atomic currents.

Although for many purposes it is convenient to think of a magnet as having distinct north and south magnetic poles, the concept of poles should not be taken literally: it is merely a way of referring to the two different ends of a magnet. The magnet does not have distinct north or south particles on opposing sides. If a bar magnet is broken into two pieces, in an attempt to separate the north and south poles, the result will be two bar magnets, *each* of which has both a north and south pole. However, a version of the magnetic-pole approach is used by professional magneticians to design permanent magnets. In this approach, the divergence of the magnetization $\nabla \cdot M$ inside a magnet and the surface normal component $M \cdot n$ are treated as a distribution of magnetic monopoles.

This is a mathematical convenience and does not imply that there are actually monopoles in the magnet. If the magnetic-pole distribution is known, then the pole model gives the magnetic field H. Outside the magnet, the field B is proportional to H, while inside the magnetization must be added to H. An extension of this method that allows for internal magnetic charges is used in theories of ferromagnetism.

Another model is the Ampère model, where all magnetization is due to the effect of microscopic, or atomic, circular bound currents, also called Ampèrian currents, throughout the material. For a uniformly magnetized cylindrical bar magnet, the net effect of the microscopic bound currents is to make the magnet behave as if there is a macroscopic sheet of electric current flowing around the surface, with local flow direction normal to the cylinder axis. Microscopic currents in atoms inside the material are generally canceled by currents in neighbouring atoms, so only the surface makes a net contribution; shaving off the outer layer of a magnet will *not* destroy its magnetic field, but will leave a new surface of uncancelled currents from the circular currents throughout the material. The right-hand rule tells which direction the current flows.

Pole Naming Conventions

The north pole of a magnet is the pole that, when the magnet is freely suspended, points towards the Earth's North Magnetic Pole which is located in northern Canada. Since opposite poles (north and south) attract, the Earth's "*North Magnetic Pole*" is thus actually the *south* pole of the Earth's magnetic field. As a practical matter, in order to tell which pole of a magnet is north and which is south, it is not necessary to use the Earth's magnetic field at all. For example, one method would be to compare it to an electromagnet, whose poles can be identified by the right-hand rule. The magnetic field lines of a magnet are considered by convention to emerge from the magnet's north pole and reenter at the south pole.

Magnetism

Magnetism is a class of physical phenomena that includes forces exerted by magnets on other magnets. It has its origin in electric currents and the fundamental magnetic moments of elementary particles. These give rise to a magnetic field that acts on other currents and moments. All materials are influenced to some extent by a magnetic field. The strongest effect is on permanent magnets, which have persistent magnetic moments caused by ferromagnetism. Most materials do not have permanent moments. Some are attracted to a magnetic field (paramagnetism); others are repulsed by a magnetic field (diamagnetism); others have a much more complex relationship with an applied magnetic field (spin glass behaviour and antiferromagnetism). Substances that are negligibly affected by magnetic fields are known as *non-magnetic* substances. They include copper, aluminium, gases, and plastic. Pure oxygen exhibits magnetic properties when cooled to a liquid state.

The magnetic state (or phase) of a material depends on temperature (and other variables such as pressure and the applied magnetic field) so that a material may exhibit more than one form of magnetism depending on its temperature, etc.

History

Aristotle attributed the first of what could be called a scientific discussion on magnetism to Thales of Miletus, who lived from about 625 BC to about 545 BC. Around the same time, in ancient India, the Indian surgeon, Sushruta, was the first to make use of the magnet for surgical purposes.

In ancient China, the earliest literary reference to magnetism lies in a 4th-century BC book named after its author, *The Master of Demon Valley* (<>7ŒP[): "The lodestone makes iron come or it attracts it." The earliest mention of the attraction of a needle appears in a work composed between AD 20 and 100 (*Louen-heng*): "A lodestone attracts a needle." The Chinese scientist Shen Kuo (1031–1095) was the first person to write of the magnetic needle compass and that it improved the accuracy of navigation by employing the astronomical concept of true north *(Dream Pool Essays*, AD 1088), and by the 12th century the Chinese were known to use the lodestone compass for navigation. They sculpted a directional spoon from lodestone in such a way that the handle of the spoon always pointed south.

Alexander Neckam, by 1187, was the first in Europe to describe the compass and its use for navigation. In 1269, Peter Peregrinus de Maricourt wrote the *Epistola de magnete*, the first extant treatise describing the properties of magnets. In 1282, the properties of magnets and the dry compass were discussed by Al-Ashraf, a Yemeni physicist, astronomer, and geographer.

In 1600, William Gilbert published his *De Magnete, Magneticisque Corporibus, et de Magno Magnete Tellure* (*On the Magnet and Magnetic Bodies, and on the Great Magnet the Earth*). In this work he describes many of his experiments with his model earth called the terrella. From his experiments, he concluded that the Earth was itself magnetic and that this was the reason compasses pointed north (previously, some believed that it was the pole star (Polaris) or a large magnetic island on the north pole that attracted the compass).

An understanding of the relationship between electricity and magnetism began in 1819 with work by Hans Christian Oersted, a professor at the University of Copenhagen, who discovered more or less by accident that an electric current could influence a compass needle. This landmark experiment is known as Oersted's Experiment. Several other experiments followed, with André-Marie Ampère, who in 1820 discovered that the magnetic field circulating in a closed-path was related to the current flowing through the perimeter of the path; Carl Friedrich Gauss; Jean-Baptiste Biot and Félix Savart, both of which in 1820 came up with the Biot-Savart Law giving an equation for the magnetic field from a current-carrying wire; Michael Faraday, who in 1831 found that a time-varying magnetic flux through a loop of wire induced a voltage, and others finding further links between magnetism and electricity.

James Clerk Maxwell synthesized and expanded these insights into Maxwell's equations, unifying electricity, magnetism, and optics into the field of electromagnetism. In 1905, Einstein used these laws in motivating his theory of special relativity, requiring that the laws held true in all inertial reference frames.

Electromagnetism has continued to develop into the 21st century, being incorporated into the more fundamental theories of gauge theory, quantum electrodynamics, electroweak theory, and finally the standard model.

Sources of Magnetism

Magnetism, at its Root, Arises from two Sources: Electric current. The other sources of magnetism are the nuclear magnetic moments of the nuclei in the material which are typically thousands of times smaller than the electrons' magnetic moments, so they are negligible in the context of the magnetization of materials. Nuclear magnetic moments are very important in other contexts, particularly in nuclear magnetic resonance (NMR) and magnetic resonance imaging (MRI).

Ordinarily, the enormous number of electrons in a material are arranged such that their magnetic moments (both orbital and intrinsic) cancel out. This is due, to some extent, to electrons combining into pairs with opposite intrinsic magnetic moments as a result of the Pauli exclusion principle, or combining into filled subshells with zero net orbital motion. In both cases, the electron arrangement is so as to exactly cancel the magnetic moments from each electron. Moreover, even when the electron configuration *is* such that there are unpaired electrons and/or non-filled subshells, it is often the case that the various electrons in the solid will contribute magnetic moments that point in different, random directions, so that the material will not be magnetic.

However, sometimes — either spontaneously, or owing to an applied external magnetic field — each of the electron magnetic moments will be, on average, lined up. Then the material can produce a net total magnetic field, which can potentially be quite strong.

The magnetic behaviour of a material depends on its structure, particularly its electron configuration, for the reasons mentioned above, and also on the temperature. At high temperatures, random thermal motion makes it more difficult for the electrons to maintain alignment.

Paramagnetism

Paramagnetism is a form of magnetism whereby certain materials are attracted by an externally applied magnetic field. In contrast with this behaviour, diamagnetic materials are repelled by magnetic fields.

Paramagnetic materials include most chemical elements and some compounds; they have a relative magnetic permeability greater or equal to 1 (i.e., a positive magnetic susceptibility) and hence are attracted to magnetic fields. The magnetic moment induced by the applied field is linear in the field strength and rather weak. It typically requires a sensitive analytical balance to detect the effect and modern measurements on paramagnetic materials are often conducted with a SQUID magnetometer.

Paramagnetic materials have a small, positive susceptibility to magnetic fields. These materials are slightly attracted by a magnetic field and the material does not retain the magnetic properties when the external field is removed. Paramagnetic properties are due to the presence of some unpaired electrons, and from the realignment of the electron paths caused by the external magnetic field. Paramagnetic materials include magnesium, molybdenum, lithium, and tantalum.

Unlike ferromagnets, paramagnets do not retain any magnetization in the absence of an externally applied magnetic field because thermal motion randomizes the spin orientations. Some paramagnetic materials retain spin disorder at absolute zero, meaning they are paramagnetic in the ground state. Thus the total magnetization drops to zero when the applied field is removed. Even in the presence of the field there is only a small induced magnetization because only a small fraction of the spins will be oriented by the field. This fraction is proportional to the field strength and this explains the linear dependency. The attraction experienced by ferromagnetic materials is non-linear and much stronger, so that it is easily observed, for instance, by the attraction between a refrigerator magnet and the iron of the refrigerator itself.

Relation to Electron Spins

Constituent atoms or molecules of paramagnetic materials have permanent magnetic moments (dipoles), even in the absence of an applied field. The permanent moment generally is due to the spin of unpaired electrons in atomic or molecular electron orbitals. In pure paramagnetism, the dipoles do not interact with one another and are randomly oriented in the absence of an external field due to thermal agitation, resulting in zero net magnetic moment. When a magnetic field is applied, the dipoles will tend to align with the applied field, resulting in a net magnetic moment in the direction of the applied field. In the classical description, this alignment can be understood to occur due to a torque being provided on the magnetic moments by an applied field, which tries to align the dipoles parallel to the applied field. However, the true origins of the alignment can only be understood via the quantum-mechanical properties of spin and angular momentum.

If there is sufficient energy exchange between neighbouring dipoles they will interact, and may spontaneously align or anti-align and form magnetic domains, resulting in ferromagnetism (permanent magnets) or antiferromagnetism, respectively. Paramagnetic behaviour can also be observed in ferromagnetic materials that are above their Curie temperature, and in antiferromagnets above their Néel temperature. At these temperatures the available thermal energy simply overcomes the interaction energy between the spins.

In general paramagnetic effects are quite small: the magnetic susceptibility is of the order of 10^{-3} to 10^{-5} for most paramagnets, but may be as high as 10^{-1} for synthetic paramagnets such as ferrofluids.

Delocalization

Table: *Selected Pauli-paramagnetic metals*

Material	***Magnetic susceptibility ($\times 10^{-5}$)***
Tungsten	6.8
Cesium	5.1
Aluminium	2.2
Lithium	1.4
Magnesium	1.2
Sodium	0.72

In conductive materials the electrons are delocalized, that is, they travel through the solid more or less as an electron gas. Conductivity can be understood in a band structure picture as arising from the incomplete filling of energy bands.

In an ordinary nonmagnetic conductor the conduction band is identical for both spin-up and spin-down electrons. When a magnetic field is applied, the conduction band splits apart into a spin-up and a spin-down band due to the difference in magnetic potential energy for spin-up and spin-down electrons. Since the Fermi level must be identical for both bands, this means that there will be a small surplus of the type of spin in the band that moved downwards. This effect is a weak form of paramagnetism known as Pauli paramagnetism.

The effect always competes with a diamagnetic response of opposite sign due to all the core electrons of the atoms. Stronger forms of magnetism usually require localized rather than itinerant electrons. However in some cases a bandstructure can result in which there are two delocalized sub-bands with states of opposite spins that have different energies. If one subband is preferentially filled over the other, one can have itinerant ferromagnetic order. This situation usually only occurs in relatively narrow (d-)bands, which are poorly delocalized.

s and p Electrons

Generally, strong delocalization in a solid due to large overlap with neighbouring wave functions means that there will be a large Fermi velocity; this means that the number of electrons in a band is less sensitive to shifts in that band's energy, implying a weak magnetism. This is why s- and p-type metals are typically either Pauli-paramagnetic or as in the case of gold even diamagnetic. In the latter case the diamagnetic contribution from the closed shell inner electrons simply wins from the weak paramagnetic term of the almost free electrons.

d and f Electrons

Stronger magnetic effects are typically only observed when d- or f-electrons are involved. Particularly the latter are usually strongly localized. Moreover the size of the magnetic moment on a lanthanide atom can be quite large as it can carry up to 7 unpaired electrons in the case of gadolinium(III) (hence its use in MRI). This high magnetic moments associated with lanthanides is one reason why superstrong magnets are typically based on elements like neodymium or samarium.

Molecular Localization

Of course the above picture is a *generalization* as it pertains to materials with an extended lattice rather than a molecular structure. Molecular structure can also lead to localization of electrons. Although there are usually energetic reasons why a molecular structure results such that it does not exhibit partly filled orbitals (i.e. unpaired spins), some non-closed shell moieties do occur in nature. Molecular oxygen is a good example. Even in the frozen solid it contains di-radical molecules resulting in paramagnetic behaviour. The unpaired spins reside in orbitals derived from oxygen p wave functions, but the overlap is limited to the one neighbour in the O_2 molecules. The distances to other oxygen atoms in the lattice remain too large to lead to delocalization and the magnetic moments remain unpaired.

Curie's Law

For low levels of magnetization, the magnetization of paramagnets follows what is known as Curie's law, at least approximately. This law indicates that the susceptibility χ of paramagnetic materials is inversely proportional to their temperature, i.e. that materials become more magnetic at lower temperatures. The mathematical expression is:

$$M = \chi H = \frac{C}{T} H$$

where:

M is the resulting magnetization

χ is the magnetic susceptibility

H is the auxiliary magnetic field, measured in amperes/metre

T is absolute temperature, measured in kelvins

C is a material-specific Curie constant

Curie's law is valid under the commonly encountered conditions of low magnetization ($\mu_B H \lesssim k_B T$), but does not apply in the high-field/low-temperature regime where saturation of magnetization occurs ($\mu_B H \gtrsim k_B T$) and magnetic dipoles are all aligned with the applied field.

When the dipoles are aligned, increasing the external field will not increase the total magnetization since there can be no further alignment.

For a paramagnetic ion with noninteracting magnetic moments with angular momentum J, the Curie constant is related the individual ions' magnetic moments,

$$C = \frac{N_A}{3k_B} \mu_{\text{eff}}^2 \text{ where } \mu_{\text{eff}} = g_J \mu_B \sqrt{J(J+1)}.$$

The parameter μ_{eff} is interpreted as the effective magnetic moment per paramagnetic ion. If one uses a classical treatment with molecular magnetic moments represented as discrete magnetic dipoles, μ, a Curie Law expression of the same form will emerge with μ appearing in place of μ_{eff}.

When orbital angular momentum contributions to the magnetic moment are small, as occurs for most organic radicals or for octahedral transition metal complexes with d^3 or high-spin d^5 configurations, the effective magnetic moment takes the form ($g_e = 2.0023... \approx 2$),

$\mu_{eff} \simeq 2\sqrt{S(S+1)}\mu_B = \sqrt{n(n+2)}\mu_B$, where n is the number of unpaired electrons. In other transition metal complexes this yields a useful, if somewhat cruder, estimate.

Examples of Paramagnets

Materials that are called 'paramagnets' are most often those that exhibit, at least over an appreciable temperature range, magnetic susceptibilities that adhere to the Curie or Curie–Weiss laws. In principle any system that contains atoms, ions, or molecules with unpaired spins can be called a paramagnet, but the interactions between them need to be carefully considered.

Systems with Minimal Interactions

The narrowest definition would be: a system with unpaired spins that *do not interact* with each other. In this narrowest sense, the only pure paramagnet is a dilute gas of monatomic hydrogen atoms. Each atom has one non-interacting unpaired electron. Of course, the latter could be said about a gas of lithium atoms but these already possess two paired core electrons that produce a diamagnetic response of opposite sign.

Strictly speaking Li is a mixed system therefore, although admittedly the diamagnetic component is weak and often neglected. In the case of heavier elements the diamagnetic contribution becomes more important and in the case of metallic gold it dominates the properties. Of course, the element hydrogen is virtually never called 'paramagnetic' because the monatomic gas is stable only at extremely high temperature; H atoms combine to form molecular H_2 and in so doing, the magnetic moments are lost (*quenched*), because the spins pair.

Hydrogen is therefore *diamagnetic* and the same holds true for many other elements. Although the electronic configuration of the individual atoms (and ions) of most elements contain unpaired spins, they are not necessarily paramagnetic, because at ambient temperature quenching is very much the rule rather than the exception.

The quenching tendency is weakest for f-electrons because *f* (especially 4*f*) orbitals are radially contracted and they overlap only weakly with orbitals on adjacent atoms. Consequently, the lanthanide elements with incompletely filled 4f-orbitals are paramagnetic or magnetically ordered.

Table: μ_{eff} *values for typical* d^3 *and* d^5 *transition metal complexes.*

Material	μ_{eff}/m_B
$[Cr(NH_3)_6]Br_3$	3.77
$K_3[Cr(CN)_6]$	3.87
$K_3[MoCl_6]$	3.79
$K_4[V(CN)_6]$	3.78
$[Mn(NH_3)_6]Cl_2$	5.92
$(NH_4)_2[Mn(SO_4)_2]\cdot 6H_2O$	5.92
$NH_4[Fe(SO_4)_2]\cdot 12H_2O$	5.89

Thus, condensed phase paramagnets are only possible if the interactions of the spins that lead either to quenching or to ordering are kept at bay by structural isolation of the magnetic centres. There are two classes of materials for which this holds:

- Molecular materials with a (isolated) paramagnetic centre.
  - Good examples are coordination complexes of d- or f-metals or proteins with such centres, e.g. myoglobin. In such materials the organic part of the molecule acts as an envelope shielding the spins from their neighbours.
  - Small molecules can be stable in radical form, oxygen O_2 is a good example. Such systems are quite rare because they tend to be rather reactive.
- Dilute systems.
  - Dissolving a paramagnetic species in a diamagnetic lattice at small concentrations, e.g. Nd^{3+} in $CaCl_2$ will separate the neodymium ions at large enough distances that they do not interact. Such systems are of prime importance for what can be considered the most sensitive method to study paramagnetic systems: EPR.

Systems with Interactions

As stated above many materials that contain d- or f-elements do retain unquenched spins. Salts of such elements often show paramagnetic behaviour but at low enough temperatures the magnetic moments may order. It is not uncommon to call such materials 'paramagnets', when referring to their paramagnetic behaviour above their Curie or Néel-points, particularly if such temperatures are very low or have never been properly measured. Even for iron it is not uncommon to say that *iron becomes a paramagnet* above its relatively high Curie-point.

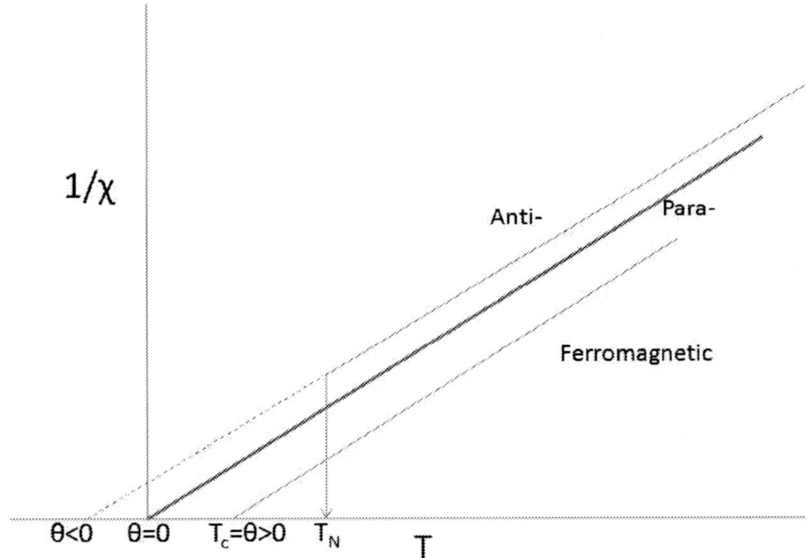

Figure: *Idealized Curie–Weiss behaviour; N.B. T_C=θ, but T_N is not θ. Paramagnetic regimes are denoted by solid lines. Close to T_N or T_C the behaviour usually deviates from ideal.*

In that case the Curie-point is seen as a phase transition between a ferromagnet and a 'paramagnet'. The word paramagnet now merely refers to the linear response of the system to an applied field, the temperature dependence of which requires an amended version of Curie's law, known as the Curie–Weiss law:

$$\mathrm{M} = \frac{C}{T - \theta}\mathrm{H}$$

This amended law includes a term θ that describes the exchange interaction that is present albeit overcome by thermal motion. The sign of θ depends on whether ferro- or antiferromagnetic interactions dominate and it is seldom exactly zero, except in the dilute, isolated cases mentioned above.

Obviously, the paramagnetic Curie–Weiss description above T_N or T_C is a rather different interpretation of the word 'paramagnet' as it does *not* imply the *absence* of interactions, but rather that the magnetic structure is random in the absence of an external field at these sufficiently high temperatures. Even if θ is close to zero this does not mean that there are no interactions, just that the aligning ferro- and the anti-aligning antiferromagnetic ones cancel. An additional complication is that the interactions are often different in different directions of the crystalline lattice (anisotropy), leading to complicated magnetic structures once ordered. Randomness of the structure also applies to the many metals that show a net paramagnetic response over a broad temperature range. They do not follow a Curie type law as function of temperature however, often they are more or less temperature independent. This type of behaviour is of an itinerant nature and better called Pauli-paramagnetism, but it is not unusual to see e.g. the metal aluminium called a 'paramagnet', even though interactions are strong enough to give this element very good electrical conductivity.

Superparamagnets

Some materials show induced magnetic behaviour that follows a Curie type law but with exceptionally large values for the Curie constants. These materials are known as superparamagnets. They are characterized by a strong ferromagnetic or ferrimagnetic type of coupling into domains of a limited size that behave independently from one another.

The bulk properties of such a system resembles that of a paramagnet, but on a microscopic level they are ordered. The materials do show an ordering temperature above which the behaviour reverts to ordinary paramagnetism (with interaction). Ferrofluids are a good example, but the phenomenon can also occur inside solids, e.g., when dilute paramagnetic centres are introduced in a strong itinerant medium of ferromagnetic coupling such as when Fe is substituted in $TlCu_2Se_2$ or the alloy AuFe. Such systems contain ferromagnetically coupled clusters that freeze out at lower temperatures. They are also called mictomagnets.

Diamagnetism

Diamagnetism is the property of an object or material that causes it to create a magnetic field in opposition to an externally applied magnetic field. It is a quantum mechanical effect that occurs in all materials; where it is the only contribution to the magnetism the material is called a *diamagnet.*

Unlike a ferromagnet, a diamagnet is not a permanent magnet. Its magnetic permeability is less than μ_0 (the permeability of free space). In most materials diamagnetism is a weak effect, but a superconductor repels the magnetic field entirely, apart from a thin layer at the surface.

Diamagnets were first discovered when Sebald Justinus Brugmans observed in 1778 that bismuth and antimony were repelled by magnetic fields. The term *diamagnetism* was coined by Michael Faraday in September 1845, when he realized that every material responded (in either a diamagnetic or paramagnetic way) to an applied magnetic field.

Materials

Table: *Notable diamagnetic materials*

Material	χ_v (× 10^{-5})
Superconductor	-10^5
Pyrolytic carbon	–40.0
Bismuth	–16.6
Mercury	–2.9
Silver	–2.6
Carbon (diamond)	–2.1
Lead	–1.8
Carbon (graphite)	–1.6
Copper	–1.0
Water	–0.91

Diamagnetism, to a greater or lesser degree, is a property of all materials and always makes a weak contribution to the material's response to a magnetic field. However, for materials that show some other form of magnetism (such as ferromagnetism or paramagnetism), the diamagnetic contribution becomes negligible. Substances that mostly display diamagnetic behaviour are termed diamagnetic materials, or diamagnets.

Materials called diamagnetic are those that non-physicists generally think of as *non-magnetic*, and include water, wood, most organic compounds such as petroleum and some plastics, and many metals including copper, particularly the heavy ones with many core electrons, such as mercury, gold and bismuth. The magnetic susceptibility of various molecular fragments are called Pascal's constants.

Diamagnetic materials have a relative magnetic permeability that is less than or equal to 1, and therefore a magnetic susceptibility less than or equal to 0, since susceptibility is defined as $\chi_v = \mu_v - 1$. This means that diamagnetic materials are repelled by magnetic fields. However, since diamagnetism is such a weak property its effects are not observable in everyday life. For example, the magnetic susceptibility of diamagnets such as water is $\chi_v = -9.05\times10^{-6}$. The most strongly diamagnetic material is bismuth, $\chi_v = -1.66\times10^{-4}$, although pyrolytic carbon may have a susceptibility of $\chi_v =$ “4.00×10^{-4} in one plane. Nevertheless, these values are orders of magnitudes smaller than the magnetism exhibited by paramagnets and ferromagnets. Note that because χ_v is derived from the ratio of the internal magnetic field to the applied field, it is a dimensionless value.

All conductors exhibit an effective diamagnetism when they experience a changing magnetic field. The Lorentz force on electrons causes them to circulate around forming eddy currents. The eddy currents then produce an induced magnetic field opposite the applied field, resisting the conductor's motion.

Superconductors

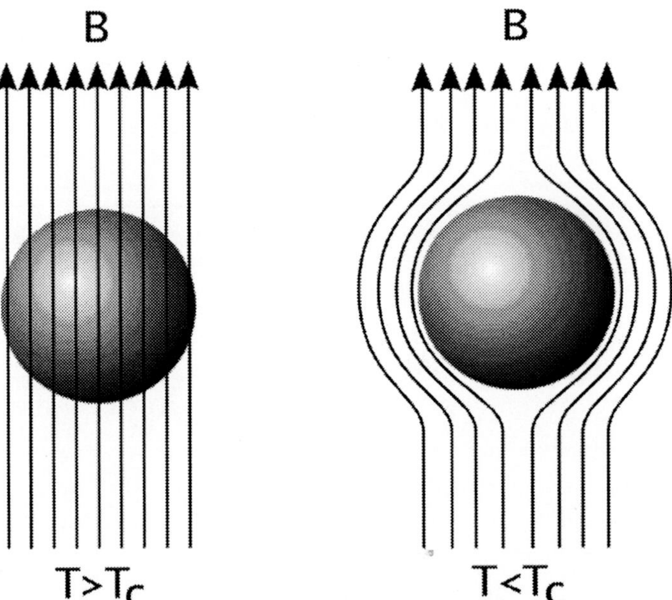

Figure: *Transition from ordinary conductivity (left) to superconductivity (right). At the transition, the superconductor expels the magnetic field and then acts as a perfect diamagnet.*

Superconductors may be considered perfect diamagnets ($\chi_v = -1$), since they expel all fields (except in a thin surface layer) due to the Meissner effect. However this effect is not due to eddy currents, as in ordinary diamagnetic materials.

Demonstrations

Curving water Surfaces: If a powerful magnet (such as a supermagnet) is covered with a layer of water (that is thin compared to the diameter of the magnet) then the field of the magnet significantly repels the water. This causes a slight dimple in the water's surface that may be seen by its reflection.

Levitation

Figure: *A live frog levitates inside a 32 mm diameter vertical bore of a Bitter solenoid in a magnetic field of about 16 teslas at the Nijmegen High Field Magnet Laboratory.*

Diamagnets may be levitated in stable equilibrium in a magnetic field, with no power consumption. Earnshaw's theorem seems to preclude the possibility of static magnetic levitation. However, Earnshaw's theorem only applies to objects with positive susceptibilities, such as ferromagnets (which have a permanent positive moment) and paramagnets (which induce a positive moment). These are attracted to field maxima, which do not exist in free space. Diamagnets (which induce a negative moment) are attracted to field minima, and there can be a field minimum in free space. A thin slice of pyrolytic graphite, which is an unusually strong diamagnetic material, can be stably floated in a magnetic field, such as that from rare earth permanent magnets. This can be done with all components at room temperature, making a visually effective demonstration of diamagnetism. The Radboud University Nijmegen, the Netherlands, has conducted experiments where water and other substances were successfully levitated. Most spectacularly, a live frog was levitated.

In September 2009, NASA's Jet Propulsion Laboratory in Pasadena, California announced they had successfully levitated mice using a superconducting magnet, an important step forward since mice are closer biologically to humans than frogs. They hope to perform experiments regarding the effects of microgravity on bone and muscle mass. Recent experiments studying the growth of protein crystals has led to a technique using powerful magnets to allow growth in ways that counteract Earth's gravity.

A simple homemade device for demonstration can be constructed out of bismuth plates and a few permanent magnets that levitate a permanent magnet.

Theory

The electrons in a material generally circulate in orbitals, with effectively zero resistance and act like current loops. Thus it might be imagined that diamagnetism effects in general would be very, very common, since any applied magnetic field would generate currents in these loops that would oppose the change, in a similar way to superconductors, which are essentially perfect diamagnets. However since the electrons are rigidly held in orbitals by the charge of the protons, and are further constrained by quantum mechanics, although many materials exhibit diamagnetism, they typically respond very little to the applied field.

The Bohr–van Leeuwen theorem proves that there cannot be any diamagnetism or paramagnetism in a purely classical system. Yet the classical theory for Langevin diamagnetism gives the same prediction as the quantum theory. The classical theory is given below.

Langevin Diamagnetism

The Langevin theory of diamagnetism applies to materials containing atoms with closed shells. A field with intensity B, applied to an electron with charge e and mass m, gives rise to Larmor precession with frequency $\omega = eB / 2m$. The number of revolutions per unit time is $\omega / 2\pi$, so the current for an atom with Z electrons is (in SI units)

$$I = -\frac{Ze^2B}{4\pi m}.$$

The magnetic moment of a current loop is equal to the current times the area of the loop. Suppose the field is aligned with the *z* axis. The average loop area can be given as $\pi\langle\rho^2\rangle$, where $\langle\rho^2\rangle$ is the mean square distance of the electrons perpendicular to the *z* axis. The magnetic moment is therefore

$$\mu = -\frac{Ze^2B}{4m}\langle\rho^2\rangle.$$

If the distribution of charge is spherically symmetric, we can suppose that the distribution of *x,y,z* coordinates are independent and identically distributed. Then $\langle x^2\rangle = \langle y^2\rangle = \langle z^2\rangle = \frac{1}{3}\langle r^2\rangle$, where $\langle r^2\rangle$ is the mean square distance of the electrons from the nucleus. Therefore $\langle\rho^2\rangle = \langle x^2\rangle + \langle y^2\rangle = \frac{2}{3}\langle r^2\rangle$. If *N* is the number of atoms per unit volume, the diamagnetic susceptibility in SI units is

$$\chi = \frac{\mu_0 N\mu}{B} = -\frac{\mu_0 NZe^2}{6m}\langle r^2\rangle.$$

In Metals

The Langevin theory does not apply to metals because they have non-localized electrons. The theory for the diamagnetism of a free electron gas is called Landau diamagnetism, and instead considers the weak counter-acting field that forms when their trajectories are curved due to the Lorentz force. Landau diamagnetism, however, should be contrasted with Pauli paramagnetism, an effect associated with the polarization of delocalized electrons' spins.

Ferromagnetism

Ferromagnetism is the basic mechanism by which certain materials (such as iron) form permanent magnets, or are attracted to magnets. In physics, several different types of magnetism are distinguished. Ferromagnetism (including ferrimagnetism) is the strongest type; it is the only type that creates forces strong enough to be felt, and is responsible for the common phenomena of magnetism encountered in everyday life. Other substances respond weakly to magnetic fields with two other types of magnetism, paramagnetism and diamagnetism, but the forces are so weak that they can only be detected by sensitive instruments in a laboratory. An everyday example of ferromagnetism is a refrigerator magnet used to hold notes on a refrigerator door. The attraction between a magnet and ferromagnetic material is "the quality of magnetism first apparent to the ancient world, and to us today".

Permanent magnets (materials that can be magnetized by an external magnetic field and remain magnetized after the external field is removed) are either ferromagnetic or ferrimagnetic, as are other materials that are noticeably attracted to them. Only a few substances are ferromagnetic. The common ones are iron, nickel, cobalt and most of their alloys, some compounds of rare earth metals, and a few naturally-occurring minerals such as lodestone.

Ferromagnetism is very important in industry and modern technology, and is the basis for many electrical and electromechanical devices such as electromagnets, electric motors, generators, transformers, and magnetic storage such as tape recorders, and hard disks.

History and Distinction from Ferrimagnetism

Historically, the term *ferromagnet* was used for any material that could exhibit spontaneous magnetization: a net magnetic moment in the absence of an external magnetic field. This general definition is still in common use. More recently, however, different classes of spontaneous magnetization have been identified when there is more than one magnetic ion per primitive cell of the material, leading to a stricter definition of "ferromagnetism" that is often used to distinguish it from ferrimagnetism. In particular, a material is "ferromagnetic" in this narrower sense only if *all* of its magnetic ions add a positive contribution to the net magnetization. If some of the magnetic ions *subtract* from the net magnetization (if they are partially *anti*-aligned), then the material is "ferrimagnetic". If the moments of the aligned and anti-aligned ions balance completely so as to have zero net magnetization, despite the magnetic ordering, then it is an antiferromagnet. These alignment effects only occur at temperatures below a certain critical temperature, called the Curie temperature (for ferromagnets and ferrimagnets) or the Néel temperature (for antiferromagnets).

Among the first investigations of ferromagnetism are the pioneering works of Aleksandr Stoletov on measurement of the magnetic permeability of ferromagnetics, known as the Stoletov curve.

Ferromagnetic Materials

Table: *Curie temperatures for some crystalline ferromagnetic (* = ferrimagnetic) materials*

Material	***Curietemp. (K)***
Co	1388
Fe	1043
$Fe_2O_3^*$	948
$FeOFe_2O_3^*$	858
$NiOFe_2O_3^*$	858
$CuOFe_2O_3^*$	728
$MgOFe_2O_3^*$	713
MnBi	630
Ni	627
MnSb	587
$MnOFe_2O_3^*$	573
$Y_3Fe_5O_{12}^*$	560
CrO_2	386
MnAs	318
Gd	292
Dy	88
EuO	69

The table on the right lists a selection of ferromagnetic and ferrimagnetic compounds, along with the temperature above which they cease to exhibit spontaneous magnetization.

Ferromagnetism is a property not just of the chemical make-up of a material, but of its crystalline structure and microscopic organization. There are ferromagnetic metal alloys whose constituents are not themselves ferromagnetic, called Heusler alloys, named after Fritz Heusler. Conversely there are non-magnetic alloys, such as types of stainless steel, composed almost exclusively of ferromagnetic metals.

One can also make amorphous (non-crystalline) ferromagnetic metallic alloys by very rapid quenching (cooling) of a liquid alloy. These have the advantage that their properties are nearly isotropic (not aligned along a crystal axis); this results in low coercivity, low hysteresis loss, high permeability, and high electrical resistivity. One such typical material is a transition metal-metalloid alloy, made from about 80% transition metal (usually Fe, Co, or Ni) and a metalloid component (B, C, Si, P, or Al) that lowers the melting point. A relatively new class of exceptionally strong ferromagnetic materials are the rare-earth magnets. They contain lanthanide elements that are known for their ability to carry large magnetic moments in well-localized f-orbitals.

Actinide Ferromagnets

A number of actinide compounds are ferromagnets at room temperature or become ferromagnets below the Curie temperature (T_C). PuP is one actinide pnictide that is a paramagnet and has cubic symmetry at room temperature, but upon cooling undergoes a lattice distortion to tetragonal when cooled to below its T_c = 125 K. PuP has an easy axis of <100>, so that

$$\frac{c}{a} - 1 = -(31 \pm 1) \times 10^{-4}$$

at 5 K. The lattice distortion is presumably a consequence of strain induced by the magnetoelastic interactions as the magnetic moments aligned parallel within magnetic domains.

In $NpFe_2$ the easy axis is <111>. Above T_C ~500 K $NpFe_2$ is also paramagnetic and cubic. Cooling below the Curie temperature produces a rhombohedral distortion wherein the rhombohedral angle changes from 60° (cubic phase) to 60.53°. An alternate description of this distortion is to consider the length c along the unique trigonal axis (after the distortion has begun) and a as the distance in the plane perpendicular to c. In the cubic phase this reduces to $\frac{c}{a} = 1.00$. Below the Curie temperature

$$\frac{c}{a} - 1 = -(120 \pm 5) \times 10^{-4}$$

which is the largest strain in any actinide compound. $NpNi_2$ undergoes a similar lattice distortion below T_C = 32 K, with a strain of (43 ± 5) × 10^{-4}. $NpCo_2$ is a ferrimagnet below 15 K.

Lithium Gas

In 2009, a team of MIT physicists demonstrated that a lithium gas cooled to less than one kelvin can exhibit ferromagnetism. The team cooled fermionic lithium-6 to less than 150 billionths of one kelvin above absolute zero using infrared laser cooling. This demonstration is the first time that ferromagnetism has been demonstrated in a gas.

Explanation

The Bohr–van Leeuwen theorem shows that magnetism cannot occur in purely classical solids. Without quantum mechanics, there would be no diamagnetism, paramagnetism or ferromagnetism. The property of ferromagnetism is due to the direct influence of two effects from quantum mechanics: spin and the Pauli exclusion principle.

Origin of Magnetism

One of the fundamental properties of an electron (besides that it carries charge) is that it has a magnetic dipole moment, i.e., it behaves itself as a tiny magnet. This dipole moment comes from the more fundamental property of the electron that it has quantum mechanical spin. The quantum mechanical nature of this spin causes the electron to only be able to be in two states, with the magnetic field either pointing "up" or "down" (for any choice of up and down). The spin of the electrons in atoms is the main source of ferromagnetism, although there is also a contribution from the orbital angular momentum of the electron about the nucleus. When these tiny magnetic dipoles are aligned in the same direction, their individual magnetic fields add together to create a measurable macroscopic field.

However, in materials with a filled electron shell, the total dipole moment of the electrons is zero because the spins are in up/down pairs. Only atoms with partially filled shells (i.e., unpaired spins) can have a net magnetic moment, so ferromagnetism only occurs in materials with partially filled shells. Because of Hund's rules, the first few electrons in a shell tend to have the same spin, thereby increasing the total dipole moment.

These unpaired dipoles (often called simply "spins" even though they also generally include angular momentum) tend to align in parallel to an external magnetic field, an effect called paramagnetism. Ferromagnetism involves an additional phenomenon, however: The dipoles tend to align spontaneously, giving rise to a spontaneous magnetization, even when there is no applied field.

Exchange Interaction

According to classical electromagnetism, two nearby magnetic dipoles will tend to align in *opposite* directions, so their magnetic fields will oppose one another and cancel out. However, this effect is very weak, because the magnetic fields generated by individual spins are small and the resulting alignment is easily destroyed by thermal fluctuations. In a few materials, a much stronger interaction between spins arises because the change in the direction of the spin leads to a change in electrostatic repulsion between neighbouring electrons, due to a particular quantum mechanical effect called the exchange interaction. At short distances, the exchange interaction is much stronger than the dipole-dipole magnetic interaction. As a result, in a few materials, the ferromagnetic ones, nearby spins tend to align in the same direction.

The exchange interaction is related to the Pauli exclusion principle, which says that two electrons with the same spin cannot also have the same "position". Therefore, under certain conditions, when the orbitals of the unpaired outer valence electrons from adjacent atoms overlap, the distributions of their electric charge in space are further apart when the electrons have parallel spins than when they have opposite spins. This reduces the electrostatic energy of the electrons when their spins are parallel compared to their energy when the spins are anti-parallel, so the parallel-spin state is more stable. In simple terms, the electrons, which repel one another, can move "further apart" by aligning their spins, so the spins of these electrons tend to line up. This difference in energy is called the exchange energy.

The materials in which the exchange interaction is much stronger than the competing dipole-dipole interaction are frequently called *magnetic materials*. For instance, in iron (Fe) the exchange force is about 1000 times stronger than the dipole interaction. Therefore below the Curie temperature virtually all of the dipoles in a ferromagnetic material will be aligned. The exchange interaction is also responsible for the other types of spontaneous ordering of atomic magnetic moments occurring in magnetic solids, antiferromagnetism and ferrimagnetism. There are different exchange interaction mechanisms which create the magnetism in different ferromagnetic, ferrimagnetic, and antiferromagnetic substances. These mechanisms include direct exchange, RKKY exchange, double exchange, and superexchange.

Magnetic Anisotropy

Although the exchange interaction keeps spins aligned, it does not align them in a particular direction. Without magnetic anisotropy, the spins in a magnet randomly change direction in response to thermal fluctuations and the magnet is superparamagnetic. There are several kinds of magnetic anisotropy, the most common of which is magnetocrystalline anisotropy. This is a dependence of the energy on the direction of magnetization relative to the crystallographic lattice. Another common source of anisotropy, inverse magnetostriction, is induced by internal strains. Single-domain magnets also can have a *shape anisotropy* due to the magnetostatic effects of the particle shape. As the temperature of a magnet increases, the anisotropy tends to decrease, and there is often a blocking temperature at which a transition to superparamagnetism occurs.

Magnetic Domains

The above would seem to suggest that every piece of ferromagnetic material should have a strong magnetic field, since all the spins are aligned, yet iron and other ferromagnets are often found in an "unmagnetized" state. The reason for this is that a bulk piece of ferromagnetic material is divided into tiny *magnetic domains* (also known as *Weiss domains*). Within each domain, the spins are aligned, but (if the bulk material is in its lowest energy configuration, i.e. *unmagnetized*), the spins of separate domains point in different directions and their magnetic fields cancel out, so the object has no net large scale magnetic field.

Ferromagnetic materials spontaneously divide into magnetic domains because the *exchange interaction* is a short-range force, so over long distances of many atoms the tendency of the magnetic dipoles to reduce their energy by orienting in opposite directions wins out. If all the dipoles in a piece of ferromagnetic material are aligned parallel, it creates a large magnetic field extending into the space around it. This contains a lot of magnetostatic energy. The material can reduce this energy by splitting into many domains pointing in different directions, so the magnetic field is confined to small local fields in the material, reducing the volume of the field. The domains are separated by thin domain walls a number of molecules thick, in which the direction of magnetization of the dipoles rotates smoothly from one domain's direction to the other.

Thus, a piece of iron in its lowest energy state ("unmagnetized") generally has little or no net magnetic field. However, if it is placed in a strong enough external magnetic field, the domain walls will move, reorienting the domains so more of the dipoles are aligned with the external field. The domains will remain aligned when the external field is removed, creating a magnetic field of their own extending into the space around the material, thus creating a "permanent" magnet.

The domains do not go back to their original minimum energy configuration when the field is removed because the domain walls tend to become 'pinned' or 'snagged' on defects in the crystal lattice, preserving their parallel orientation. This is shown by the Barkhausen effect: as the magnetizing field is changed, the magnetization changes in thousands of tiny discontinuous jumps as the domain walls suddenly "snap" past defects.

This magnetization as a function of the external field is described by a hysteresis curve. Although this state of aligned domains found in a piece of magnetized ferromagnetic material is not a minimal-energy configuration, it is metastable, and can persist for long periods, as shown by samples of magnetite from the sea floor which have maintained their magnetization for millions of years.

Alloys used for the strongest permanent magnets are "hard" alloys made with many defects in their crystal structure where the domain walls "catch" and stabilize. The net magnetization can be destroyed by heating and then cooling (annealing) the material without an external field, however. The thermal motion allows the domain boundaries to move, releasing them from any defects, to return to their low-energy unaligned state.

Curie Temperature

As the temperature increases, thermal motion, or entropy, competes with the ferromagnetic tendency for dipoles to align. When the temperature rises beyond a certain point, called the Curie temperature, there is a second-order phase transition and the system can no longer maintain a spontaneous magnetization, although it still responds paramagnetically to an external field. Below that temperature, there is a spontaneous symmetry breaking and magnetic moments become aligned with their neighbours. The Curie temperature itself is a critical point, where the magnetic susceptibility is theoretically infinite and, although there is no net magnetization, domain-like spin correlations fluctuate at all length scales.

The study of ferromagnetic phase transitions, especially via the simplified Ising spin model, had an important impact on the development of statistical physics. There, it was first clearly shown that mean field theory approaches failed to predict the correct behaviour at the critical point (which was found to fall under a *universality class* that includes many other systems, such as liquid-gas transitions), and had to be replaced by renormalization group theory.

Magnetic Domain

A magnetic domain is a region within a magnetic material which has uniform magnetization. This means that the individual magnetic moments of the atoms are aligned with one another and they point in the same direction. When cooled below a temperature called the Curie temperature, the magnetization of a piece of ferromagnetic material spontaneously divides into many small regions called magnetic domains.

The magnetization within each domain points in a uniform direction, but the magnetization of different domains may point in different directions. Magnetic domain structure is responsible for the magnetic behaviour of ferromagnetic materials like iron, nickel, cobalt and their alloys, ferrites etc. such as the formation of permanent magnets. The regions separating magnetic domains are called domain walls, where the magnetization rotates coherently from the direction in one domain to that in the next domain. The study of magnetic domains is called micromagnetics.

Development of Domain Theory

Magnetic domain theory was developed by French physicist Pierre-Ernest Weiss who in 1906 suggested existence of magnetic domains in ferromagnets. He suggested that large number of atomic magnetic moments (typically 10^{12}-10^{18}) were aligned parallel. The direction of alignment varies from domain to domain in a more or less random manner although certain crystallographic axis may be preferred by the magnetic moments, called easy axes.

Weiss still had to explain the reason for the spontaneous alignment of atomic moments within a ferromagnetic material, and he came up with the so-called Weiss mean field : he assumed that a given magnetic moment in a material experienced a very high effective magnetic field due to the magnetization of its neighbours. In the original Weiss theory the mean field was proportional to the bulk magnetization M, so that

$$H_e = \alpha M$$

where α is the mean field constant. However this is not applicable to ferromagnets due to the variation of magnetization from domain to domain. In this case, the interaction field is

$$H_e = \alpha M_s$$

Where M_s is the saturation magnetization at 0K.

Later, the quantum theory made it possible to understand the microscopic origin of the Weiss field. The exchange interaction between localized spins favoured a parallel (in ferromagnets) or an anti-parallel (in anti-ferromagnets) state of neighbouring magnetic moments

Domain Structure

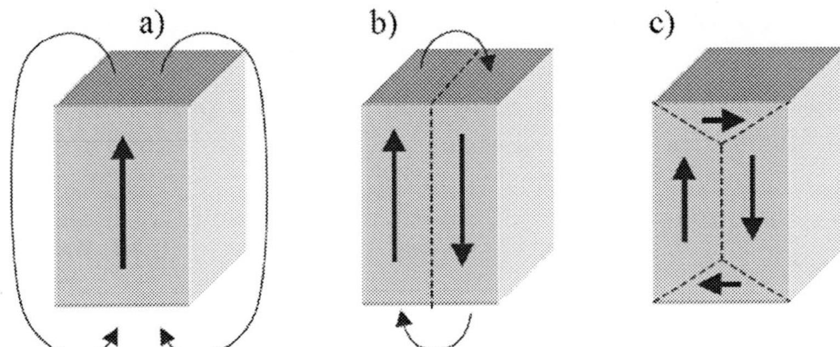

Figure: *How dividing a ferromagnetic material into magnetic domains reduces the magnetostatic energy*

Why Domains Form

The reason a piece of magnetic material such as iron spontaneously divides into separate domains, rather than exist in a state with magnetization in the same direction throughout the material, is to minimize its internal energy.

A large region of ferromagnetic material with a constant magnetization throughout will create a large magnetic field extending into the space outside itself *(diagram a, right)*. This requires a lot of *magnetostatic energy* stored in the field. To reduce this energy, the sample can split into two domains, with the magnetization in opposite directions in each domain *(diagram b right)*. The magnetic field lines pass in loops in opposite directions through each domain, reducing the field outside the material. To reduce the field energy further, each of these domains can split also, resulting in smaller parallel domains with magnetization in alternating directions, with smaller amounts of field outside the material.

The domain structure of actual magnetic materials does not usually form by the process of large domains splitting into smaller ones as described here. When a sample is cooled below the Curie temperature, for example, the equilibrium domain configuration simply appears. But the description of domains splitting is often used to reveal the energy tradeoffs in domain formation.

Size of Domains

A domain which is too big is unstable, and will divide into smaller domains. But a small enough domain will be stable and will not split, and this determines the size of the domains created in a material. This size depends on the balance of several energies within the material. Each time a region of magnetization splits into two domains, it creates a "*domain wall*" between the domains, where magnetic dipoles (molecules) with magnetization pointing in different directions are adjacent. The exchange interaction which creates the magnetization is a force which tends to align nearby dipoles so they point in the same direction. Forcing adjacent dipoles to point in different directions requires energy. Therefore creating a domain wall requires extra energy, called the "*exchange energy*", which is proportional to the area of the wall.

Thus the net amount that the energy is reduced when a domain splits is equal to the difference between the magnetic field energy saved, and the additional energy of the domain wall created. The field energy saved is proportional to the cube of the domain size, while the domain wall energy is proportional to the square of the domain size. So as the domains get smaller, the net energy saved by splitting decreases. The domains keep dividing into smaller domains until the energy cost of creating an additional domain wall is just equal to the field energy saved. Then the domains of this size are stable. In most materials the domains are microscopic in size, around 10^{-4} - 10^{-6} m.

Magnetic Anisotropy

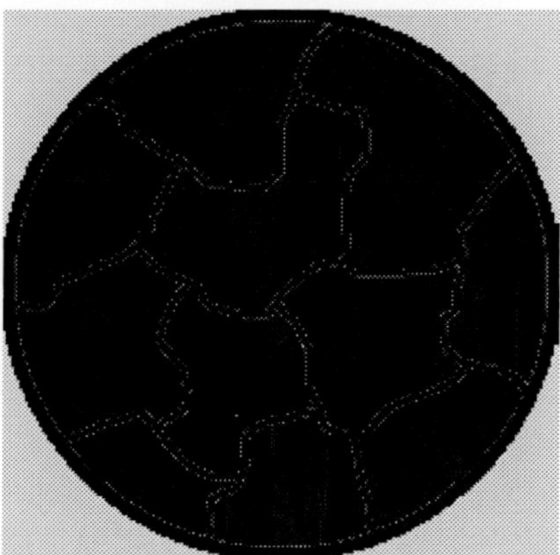

Figure: *Micrograph of surface of ferromagnetic material, showing the crystal grains, each divided into several domains parallel to its "easy" axis of magnetization, with the magnetization in alternating directions* (red and green areas).

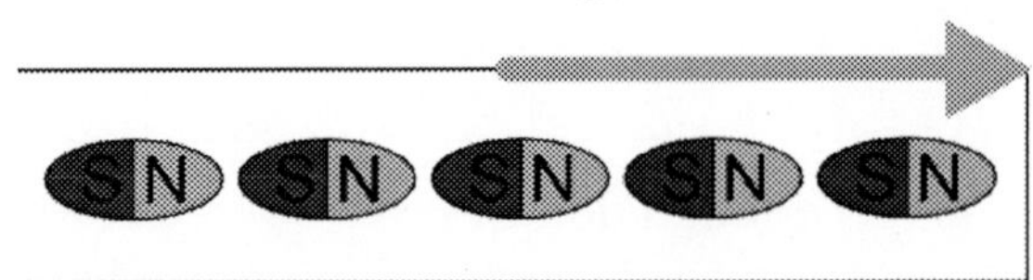

Figure: *Animation showing how magnetostriction works. A changing external magnetic field causes the magnetic dipoles to rotate, changing the dimensions of the crystal lattice.*

An additional way for the material to further reduce its magnetostatic energy is to form domains with magnetization at right angles to the other domains *(diagram c, right)*, instead of just in opposing parallel directions. These domains, called *flux closure domains*, allow the field lines to turn 180° within the material, forming closed loops entirely within the material, reducing the magnetostatic energy to zero. However, forming these domains incurs two additional energy costs. First, the crystal lattice of most magnetic materials has magnetic anisotropy, which means it has an "easy" direction of magnetization, parallel to one of the crystal axes. Changing the magnetization of the material to any other direction takes additional energy, called the "*magnetocrystalline anisotropy energy*".

Magnetostriction

The other energy cost to creating domains with magnetization at an angle to the "easy" direction is caused by the phenomenon called magnetostriction. When the magnetization of a piece of magnetic material is changed to a different direction, it causes a slight change in its shape. The change in magnetic field causes the magnetic dipole molecules to change shape slightly, making the crystal lattice longer in one dimension and shorter in other dimensions. However, since the magnetic domain is "squished in" with its boundaries held rigid by the surrounding material, it cannot actually change shape. So instead, changing the direction of the magnetization induces tiny mechanical stresses in the material, requiring more energy to create the domain. This is called "*magnetoelastic anisotropy energy*".

To form these closure domains with "sideways" magnetization requires additional energy due to the aforementioned two factors. So flux closure domains will only form where the magnetostatic energy saved is greater than the sum of the "exchange energy" to create the domain wall, the magnetocrystalline anisotropy energy, and the magnetoelastic anisotropy energy. Therefore most of the volume of the material is occupied by domains with magnetization either "up" or "down" along the "easy" direction, and the flux closure domains only form in small areas at the edges of the other domains where they are needed to provide a path for magnetic field lines to change direction *(diagram c, above).*

Grain Structure

The above describes magnetic domain structure in a perfect crystal lattice, such as would be found in a single crystal of iron. However most magnetic materials are polycrystalline, composed of microscopic crystalline grains. These grains are *not* the same as domains. Each grain is a little crystal, with the crystal lattices of separate grains oriented in random directions. In most materials, each grain is big enough to contain several domains. Each crystal has an "easy" axis of magnetization, and is divided into domains with the axis of magnetization parallel to this axis, in alternate directions.

"Magnetized" States

It can be seen that, although on a microscopic scale almost all the magnetic dipoles in a piece of ferromagnetic material are lined up parallel to their neighbours in domains, creating strong *local* magnetic fields, energy minimization results in a domain structure that minimizes the *large-scale* magnetic field. The domains point in different directions, confining the field lines to microscopic loops between neighbouring domains, so the combined fields cancel at a distance. Therefore a bulk piece of ferromagnetic material in its lowest energy state has little or no external magnetic field. The material is said to be "unmagnetized".

However, the domains can also exist in other configurations in which their magnetization mostly points in the same direction, creating an external magnetic field. Although these are not minimum energy configurations, due to a phenomenon where the domain walls become "pinned" to defects in the crystal lattice they can be *local* minimums of the energy, and therefore can be very stable. Applying an external magnetic field to the material can make the domain walls move, causing the domains aligned with the field to grow, and the opposing domains to shrink. When the external field is removed, the domain walls remain pinned in their new orientation and the aligned domains produce a magnetic field. This is what happens when a piece of ferromagnetic material is "magnetized" and becomes a permanent magnet.

Landau-Lifshitz Energy Equation

The contributions of the different internal energy factors described above is expressed by the free energy equation proposed by Lev Landau and Evgeny Lifshitz in 1935, which forms the basis of the modern theory of magnetic domains. The domain structure of a material is the one which minimizes the Gibbs free energy of the material. For a crystal of magnetic material, this is the Landau-Lifshitz free energy, E, which is the sum of these energy terms:

$$E = E_{ex} + E_D + E_\lambda + E_k + E_H$$

where

- E_{ex} *is exchange energy:* This is the energy due to the exchange interaction between magnetic dipole molecules in ferromagnetic, ferrimagnetic and antiferromagnetic materials. It is lowest when the dipoles are all pointed in the same direction, so it is responsible for magnetization of magnetic materials. When two domains with different directions of magnetization are next to each other, at the domain wall between them magnetic dipoles pointed in different directions lie next to each other, increasing this energy. This additional exchange energy is proportional to the total area of the domain walls.
- E_D *is magnetostatic energy:* This is a self-energy, due to the interaction of the magnetic field created by the magnetization in some part of the sample on other parts of the same sample. It is dependent on the volume occupied by the magnetic field extending outside the domain. This energy is reduced by minimizing the length of the loops of magnetic field lines outside the domain. For example, this tends to encourage the magnetization to be parallel to the surfaces of the sample, so the field lines won't pass outside the sample. Reducing this energy is the main reason for the creation of magnetic domains.
- E_λ *is magnetoelastic anisotropy energy:* This energy is due to the effect of magnetostriction, a slight change in the dimensions of the crystal when magnetized. This causes elastic strains in the lattice, and the direction of magnetization that minimizes these strain energies will be favoured. This energy tends to be minimized when the axis of magnetization of the domains in a crystal are all parallel.
- E_k *is magnetocrystalline anisotropy energy:* Due to its magnetic anisotropy, the crystal lattice is "easy" to magnetize in one direction, and "hard" to magnetize in others. This energy is minimized when the magnetization is along the "easy" crystal axis, so the magnetization of most of the domains in a crystal grain tend to be in either direction along the "easy" axis. Since the crystal lattice in separate grains of the material is usually oriented in different random directions, this causes the dominant domain magnetization in different grains to be pointed in different directions.

- E_H *is Zeeman energy:* This is energy which is added to or subtracted from the magnetostatic energy, due to the interaction between the magnetic material and an externally applied magnetic field. It is proportional to the negative of the cosine of the angle between the field and magnetization vectors. Domains with their magnetic field oriented parallel to the applied field reduce this energy, while domains with their magnetic field oriented opposite to the applied field increase this energy. So applying a magnetic field to a ferromagnetic material generally causes the domain walls to move so as to increase the size of domains lying mostly parallel to the field, at the cost of decreasing the size of domains opposing the field. This is what happens when ferromagnetic materials are "magnetized". With a strong enough external field, the domains opposing the field will be swallowed up and disappear; this is called saturation.

Figure: *Rotation of orientation and increase in size of magnetic domains to an externally applied field (compare Zeeman energy).*

Some sources define a wall energy E_W equal to the sum of the exchange energy and the magnetocrystalline anisotropy energy, which replaces E_{ex} and E_k in the above equation.

A stable domain structure is a magnetization function *M(X)*, considered as a continuous vector field, which minimizes the total energy *E* throughout the material. To find the minimums a variational method is used, resulting in a set of nonlinear differential equations, called *Brown's equations* after William Fuller Brown Jr. Although in principle these equations can be solved for the stable domain configurations *M(X)*, in practice only the simplest examples can be solved. Analytic solutions do not exist, and numerical solutions calculated by the finite element method are computationally intractable because of the large difference in scale between the domain size and the wall size.

Therefore micromagnetics has evolved approximate methods which assume that the magnetization of dipoles in the bulk of the domain, away from the wall, all point in the same direction, and numerical solutions are only used near the domain wall, where the magnetization is changing rapidly.

Observing Domains

There are a number of microscopy methods which can make the magnetization at a surface of a magnetic material visible, revealing the magnetic domains. Each method has a different application because not all domains are the same. In magnetic materials, domains can be circular, square, irregular, elongated, and striped, all of which have varied sizes and dimensions. Large domains, within the range of 25-100 micrometers can be easily seen by Kerr microscopy, which uses the magneto-optic Kerr effect, which is the rotation of the polarization of light reflected from a magnetized surface. Smaller domains, down to the scale of a few nanometers, can be viewed by the use of magnetic force microscopy.

Ferrimagnetism

In physics, a ferrimagnetic material is one in which the magnetic moments of the atoms on different sublattices are opposed, as in antiferromagnetism; however, in ferrimagnetic materials, the opposing moments are unequal and a spontaneous magnetization remains. This happens when the sublattices consist of different materials or ions (such as Fe^{2+} and Fe^{3+}). Ferrimagnetism is exhibited by ferrites and magnetic garnets. The oldest known magnetic substance, magnetite (iron(II,III) oxide; Fe_3O_4), is a ferrimagnet; it was originally classified as a ferromagnet before Néel's discovery of ferrimagnetism and antiferromagnetism in 1948.

Some ferrimagnetic materials are YIG (yttrium iron garnet) and ferrites composed of iron oxides and other elements such as aluminum, cobalt, nickel, manganese and zinc.

Effects of Temperature

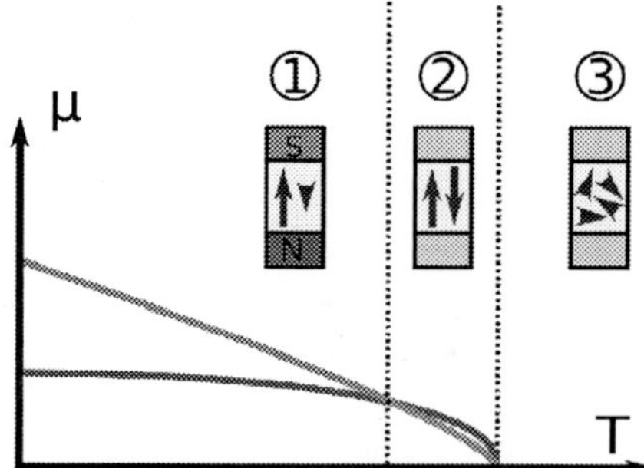

Below the magnetization compensation point, ferrimagnetic material is magnetic. At the compensation point, the magnetic components cancel each other and the total magnetic moment is zero. 3. Above the Curie point, material loses magnetism.

Ferrimagnetic materials are like ferromagnets in that they hold a spontaneous magnetization below the Curie temperature, and show no magnetic order (are paramagnetic) above this temperature. However, there is sometimes a temperature *below* the Curie temperature at which the two sublattices have equal moments, resulting in a net magnetic moment of zero; this is called the *magnetization compensation point*. This compensation point is observed easily in garnets and rare earth - transition metal alloys (RE-TM). Furthermore, ferrimagnets may also exhibit an *angular momentum compensation point* at which the angular momentum of the magnetic sublattices is compensated. This compensation point is a crucial point for achieving high speed magnetization reversal in magnetic memory devices.

Properties

Ferrimagnetic materials have high resistivity and have anisotropic properties. The anisotropy is actually induced by an external applied field. When this applied field aligns with the magnetic dipoles it causes a net magnetic dipole moment and causes the magnetic dipoles to precess at a frequency controlled by the applied field, called *Larmor* or *precession frequency*. As a particular example, a microwave signal circularly polarized in the same direction as this precession strongly interacts with the magnetic dipole moments; when it is polarized in the opposite direction the interaction is very low. When the interaction is strong, the microwave signal can pass through the material. This directional property is used in the construction of microwave devices like isolators, circulators and gyrators. Ferrimagnetic materials are also used to produce optical isolators and circulators.

Molecular Ferrimagnets

Ferrimagnetism can also occur in molecular magnets. A classic example is a dodecanuclear Manganese molecule with an effective spin of S = 10 derived from antiferromagnetic interaction on Mn(IV) metal centres with Mn(III) and Mn(II) metal centres.

Antiferromagnetism

In materials that exhibit antiferromagnetism, the magnetic moments of atoms or molecules, usually related to the spins of electrons, align in a regular pattern with neighbouring spins (on different sublattices) pointing in opposite directions. This is, like ferromagnetism and ferrimagnetism, a manifestation of ordered magnetism. Generally, antiferromagnetic order may exist at sufficiently low temperatures, vanishing at and above a certain temperature, the Néel temperature (named after Louis Néel, who had first identified this type of magnetic ordering).

Measurement

When no external field is applied, the antiferromagnetic structure corresponds to a vanishing total magnetization. In an external magnetic field, a kind of ferrimagnetic behaviour may be displayed in the antiferromagnetic phase, with the absolute value of one of the sublattice magnetizations differing from that of the other sublattice, resulting in a nonzero net magnetization. Although the net magnetization should be zero at a temperature of absolute zero, the effect of spin canting often causes a small net magnetization to develop.

The magnetic susceptibility of an antiferromagnetic material typically shows a maximum at the Néel temperature. In contrast, at the transition between the ferromagnetic to the paramagnetic phases the susceptibility will diverge. In the antiferromagnetic case, a divergence is observed in the *staggered susceptibility*.

Various microscopic (exchange) interactions between the magnetic moments or spins may lead to antiferromagnetic structures. In the simplest case, one may consider an Ising model on an bipartite lattice, e.g. the simple cubic lattice, with couplings between spins at nearest neighbour sites. Depending on the sign of that interaction, ferromagnetic or antiferromagnetic order will result. Geometrical frustration or competing ferro- and antiferromagnetic interactions may lead to different and, perhaps, more complicated magnetic structures.

Antiferromagnetic Materials

Antiferromagnetic materials occur commonly among transition metal compounds, especially oxides. Examples include hematite, metals such as chromium, alloys such as iron manganese (FeMn), and oxides such as nickel oxide (NiO). There are also numerous examples among high nuclearity metal clusters. Organic molecules can also exhibit antiferromagnetic coupling under rare circumstances.

Antiferromagnets can couple to ferromagnets, for instance, through a mechanism known as exchange bias, in which the ferromagnetic film is either grown upon the antiferromagnet or annealed in an aligning magnetic field, causing the surface atoms of the ferromagnet to align with the surface atoms of the antiferromagnet. This provides the ability to "pin" the orientation of a ferromagnetic film, which provides one of the main uses in so-called spin valves, which are the basis of magnetic sensors including modern hard drive read heads. The temperature at or above which an antiferromagnetic layer loses its ability to "pin" the magnetization direction of an adjacent ferromagnetic layer is called the blocking temperature of that layer and is usually lower than the Néel temperature.

Geometric Frustration

Unlike ferromagnetism, anti-ferromagnetic interactions can lead to multiple optimal states (ground states—states of minimal energy). In one dimension, the anti-ferromagnetic ground state is an alternating series of spins: up, down, up, down, etc. Yet in two dimensions, multiple ground states can occur.

Consider an equilateral triangle with three spins, one on each vertex. If each spin can take on only two values (up or down), there are $2^3 = 8$ possible states of the system, six of which are ground states. The two situations which are not ground states are when all three spins are up or are all down. In any of the other six states, there will be two favourable interactions and one unfavourable one. This illustrates frustration: the inability of the system to find a single ground state. This type of magnetic behaviour has been found in minerals that have a crystal stacking structure such as a Kagome lattice or hexagonal lattice.

Other Properties

Synthetic antiferromagnets (often abbreviated by SAF) are artificial antiferromagnets consisting of two or more thin ferromagnetic layers separated by a nonmagnetic layer. Due to dipole coupling of the ferromagnetic layers which results in antiparallel alignment of the magnetization of the ferromagnets.

Antiferromagnetism plays a crucial role in giant magnetoresistance, as had been discovered in 1988 by the Nobel prize winners Albert Fert and Peter Grünberg (awarded in 2007) using synthetic antiferromagnets.

There are also examples of disordered materials (such as iron phosphate glasses) that become antiferromagnetic below their Néel temperature.

These disordered networks 'frustrate' the antiparallelism of adjacent spins; i.e. it is not possible to construct a network where each spin is surrounded by opposite neighbour spins. It can only be determined that the average correlation of neighbour spins is antiferromagnetic. This type of magnetism is sometimes called *speromagnetism*.

Chapter 8

Superparamagnetism

Superparamagnetism is a form of magnetism, which appears in small ferromagnetic or ferrimagnetic nanoparticles. In sufficiently small nanoparticles, magnetization can randomly flip direction under the influence of temperature.

The typical time between two flips is called the Néel relaxation time. In the absence of external magnetic field, when the time used to measure the magnetization of the nanoparticles is much longer than the *Néel relaxation time*, their magnetization appears to be in average zero: they are said to be in the superparamagnetic state.

In this state, an external magnetic field is able to magnetize the nanoparticles, similarly to a paramagnet. However, their magnetic susceptibility is much larger than the one of paramagnets.

The Néel Relaxation in the Absence of Magnetic Field

Normally, any ferromagnetic or ferrimagnetic material undergoes a transition to a paramagnetic state above its Curie temperature. Superparamagnetism is different from this standard transition since it occurs below the Curie temperature of the material.

Superparamagnetism occurs in nanoparticles which are single-domain, i.e. composed of a single magnetic domain. This is possible when their diameter is below 3–50 nm, depending on the materials.

In this condition, it is considered that the magnetization of the nanoparticles is a single giant magnetic moment, sum of all the individual magnetic moments carried by the atoms of the nanoparticle. Those in the field of superparamagnetism call this "macro-spin approximation".

Because of the nanoparticle's magnetic anisotropy, the magnetic moment has usually only two stable orientations antiparallel to each other, separated by an energy barrier.

The stable orientations define the nanoparticle's so called "easy axis". At finite temperature, there is a finite probability for the magnetization to flip and reverse its direction.

The mean time between two flips is called the Néel relaxation time τ_N and is given by the following Néel-Arrhenius equation:

$$\tau_N = \tau_0 \exp\left(\frac{KV}{k_B T}\right),$$

where:

- τ_N is thus the average length of time that it takes for the nanoparticle's magnetization to randomly flip as a result of thermal fluctuations.
- τ_0 is a length of time, characteristic of the material, called the *attempt time* or *attempt period* (its reciprocal is called the *attempt frequency*); its typical value is 10^{-9}–10^{-10} second.
- K is the nanoparticle's magnetic anisotropy energy density and V its volume. KV is therefore the energy barrier associated with the magnetization moving from its initial easy axis direction, through a "hard plane", to the other easy axis direction.
- k_B is the Boltzmann constant.
- T is the temperature.

This length of time can be anywhere from a few nanoseconds to years or much longer.

In particular, it can be seen that the Néel relaxation time is an exponential function of the grain volume, which explains why the flipping probability becomes rapidly negligible for bulk materials or large nanoparticles.

Blocking Temperature

Let us imagine that the magnetization of a single superparamagnetic nanoparticle is measured and let us define τ_m as the measurement time.

If $\tau_m \gg \tau_N$, the nanoparticle magnetization will flip several times during the measurement, then the measured magnetization will average to zero.

If $\tau_m \ll \tau_N$, the magnetization will not flip during the measurement, so the measured magnetization will be what the instantaneous magnetization was at the beginning of the measurement.

In the former case, the nanoparticle will appear to be in the superparamagnetic state whereas in the latter case it will appear to be "blocked" in its initial state.

The state of the nanoparticle (superparamagnetic or blocked) depends on the measurement time. A transition between superparamagnetism and blocked state occurs when $\tau_m = \tau_N$.

In several experiments, the measurement time is kept constant but the temperature is varied, so the transition between superparamagnetism and blocked state is seen as a function of the temperature.

The temperature for which $\tau_m = \tau_N$ is called the blocking temperature:

$$T_B = \frac{KV}{k_B \ln\left(\frac{\tau_m}{\tau_0}\right)}$$

For typical laboratory measurements, the value of the logarithm in the previous equation is in the order of 20–25.

Effect of a Magnetic Field

When an external magnetic field is applied to an assembly of superparamagnetic nanoparticles, their magnetic moments tend to align along the applied field, leading to a net magnetization.

The magnetization curve of the assembly, i.e. the magnetization as a function of the applied field, is a reversible S-shaped increasing function. This function is quite complicated but for some simple cases:

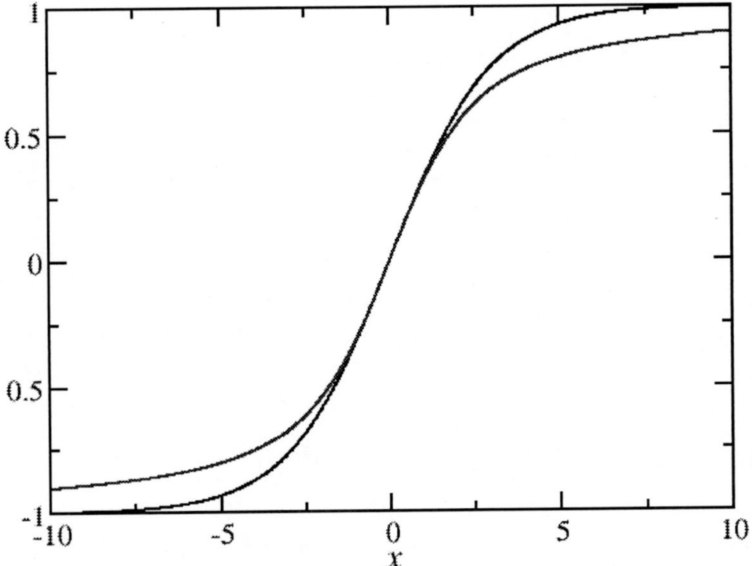

Figure: *Langevin function (red line), compared with* $\tanh(x/3)$ *(blue line).*

1. If all the particles are identical (same energy barrier and same magnetic moment), their easy axes are all oriented parallel to the applied field and the temperature is low enough ($T_B < T \lesssim KV/(10\,k_B)$), then the magnetization of the assembly is

$$M(H) \approx n\mu \tanh\left(\frac{\mu_0 H\mu}{k_B T}\right).$$

2. If all the particles are identical and the temperature is high enough ($T \gtrsim KV/k_B$), then, irrespective of the orientations of the easy axes:

$$M(H) \approx n\mu L\left(\frac{\mu_0 H\mu}{k_B T}\right)$$

In the above equations:

- n in the density of nanoparticles in the sample
- μ_0 is the magnetic permeability of vacuum
- μ is the magnetic moment of a nanoparticle
- $L(x) = 1/\tanh(x) - 1/x$ is the Langevin function

The initial slope of the $M(H)$ function is the magnetic susceptibility of the sample χ:

$$\chi = \frac{n\mu_0\mu^2}{k_B T} \text{ in the first case}$$

$$\chi = \frac{n\mu_0\mu^2}{3k_B T} \text{ in the second case.}$$

The later susceptibility is also valid for all temperatures $T > T_B$ if the easy axes of the nanoparticles are randomly oriented.

It can be seen from these equations that large nanoparticles have a larger μ and so a larger susceptibility. This explains why superparamagnetic nanoparticles have a much larger susceptibility than standard paramagnets: they behave exactly as a paramagnet with a huge magnetic moment.

Time Dependence of the Magnetization

There is no time-dependence of the magnetization when the nanoparticles are either completely blocked ($T \ll T_B$) or completely superparamagnetic ($T \gg T_B$). There is, however, a narrow window around T_B where the measurement time and the relaxation time have comparable magnitude. In this case, a frequency-dependence of the susceptibility can be observed. For a randomly oriented sample, the complex susceptibility is:

$$\chi(\omega) = \frac{\chi_{sp} + i\omega\tau\chi_b}{1 + i\omega\tau}$$

where

- $\frac{\omega}{2\pi}$ is the frequency of the applied field
- $\chi_{sp} = \frac{n\mu_0\mu^2}{3k_B T}$ is the susceptibility in the superparamagnetic state
- $\chi_b = \frac{n\mu_0\mu^2}{3KV}$ is the susceptibility in the blocked state
- $\tau = \frac{\tau_N}{2}$ is the relaxation time of the assembly

From this frequency-dependent susceptibility, the time-dependence of the magnetization for low-fields can be derived:

$$\tau\frac{\mathrm{d}M}{\mathrm{d}t} + M = \tau\chi_b\frac{\mathrm{d}H}{\mathrm{d}t} + \chi_{sp}H$$

Measurements

A superparamagnetic system can be measured with AC susceptibility measurements, where an applied magnetic field varies in time, and the magnetic response of the system is measured. A superparamagnetic system will show a characteristic frequency dependence: When the frequency is much higher than $1/\tau_N$, there will be a different magnetic response than when the frequency is much lower than $1/\tau_N$, since in the latter case, but not the former, the ferromagnetic clusters will have time to respond to the field by flipping their magnetization. The precise dependence can be calculated from the Néel-Arrhenius equation, assuming that the neighbouring clusters behave independently of one another (if clusters interact, their behaviour becomes more complicated).

Effect on Hard Drives

Superparamagnetism sets a limit on the storage density of hard disk drives due to the minimum size of particles that can be used. This limit is known as the superparamagnetic limit.

- Older hard disk technology uses longitudinal recording. It has an estimated limit of 100 to 200 Gbit/in^2
- Current hard disk technology uses perpendicular recording. As of August 2010 drives with densities of 667Gb/in^2 are available commercially. Perpendicular recording is predicted to allow information densities of up to around 1 Tbit/in^2 (1024 Gbit/in^2).
- Future hard disk technologies currently in development include: heat-assisted magnetic recording (HAMR), which use materials that are stable at much smaller sizes. They require heating before the magnetic orientation of a bit can be changed; and bit-patterned recording (BPR).

Applications of Superparamagnetism

General Applications:

- Ferrofluid: tunable viscosity

Biomedical Applications

- Imaging: Contrast agents in Magnetic Resonance Imaging (MRI)
- Magnetic separation: cell-, DNA-, protein- separation, RNA fishing
- Treatments: targeted drug delivery, magnetic hyperthermia, magnetofection

Magnetism, Electricity, and Special Relativity

As a consequence of Einstein's theory of special relativity, electricity and magnetism are fundamentally interlinked. Both magnetism lacking electricity, and electricity without magnetism, are inconsistent with special relativity, due to such effects as length contraction, time dilation, and the fact that the magnetic force is velocity-dependent. However, when both electricity and magnetism are taken into account, the resulting theory (electromagnetism) is fully consistent with special relativity. In particular, a phenomenon that appears purely electric to one observer may be purely magnetic to another, or more generally the relative contributions of electricity and magnetism are dependent on the frame of reference. Thus, special relativity "mixes" electricity and magnetism into a single, inseparable phenomenon called electromagnetism, analogous to how relativity "mixes" space and time into spacetime.

Magnetic Fields in a Material

In a vacuum,

$$B = \mu_0 H,$$

where μ_0 is the vacuum permeability.

In a material,

$$B = \mu_0 (H + M).$$

The quantity $\mu_0 M$ is called *magnetic polarization.*

If the field H is small, the response of the magnetization M in a diamagnet or paramagnet is approximately linear:

$$M = \chi H,$$

the constant of proportionality being called the magnetic susceptibility. If so,

$$\mu_0 (H + M) = \mu_0 (1 + \chi) H = \mu_r \mu_0 H = \mu H.$$

In a hard magnet such as a ferromagnet, M is not proportional to the field and is generally nonzero even when H is zero.

Force Due to Magnetic Field - The Magnetic Force

The phenomenon of magnetism is "mediated" by the magnetic field. An electric current or magnetic dipole creates a magnetic field, and that field, in turn, imparts magnetic forces on other particles that are in the fields.

Maxwell's equations, which simplify to the Biot-Savart law in the case of steady currents, describe the origin and behaviour of the fields that govern these forces. Therefore magnetism is seen whenever electrically charged particles are in motion—for example, from movement of electrons in an electric current, or in certain cases from the orbital motion of electrons around an atom's nucleus. They also arise from "intrinsic" magnetic dipoles arising from quantum-mechanical spin.

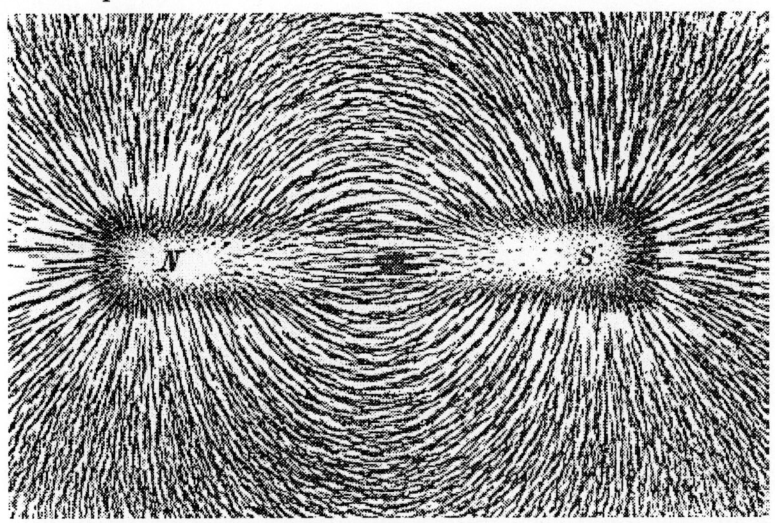

Figure: *Magnetic lines of force of a bar magnet shown by iron filings on paper*

The same situations that create magnetic fields — charge moving in a current or in an atom, and intrinsic magnetic dipoles — are also the situations in which a magnetic field has an effect, creating a force. Following is the formula for moving charge; for the forces on an intrinsic dipole.

When a charged particle moves through a magnetic field B, it feels a Lorentz force F given by the cross product:

$$\mathrm{F} = q(\mathrm{v} \times \mathrm{B})$$

where

$q$is the electric charge of the particle, and v is the velocity vector of the particle

Because this is a cross product, the force is perpendicular to both the motion of the particle and the magnetic field. It follows that the magnetic force does no work on the particle; it may change the direction of the particle's movement, but it cannot cause it to speed up or slow down. The magnitude of the force is

$$F = qvB\sin\theta$$

where θ is the angle between v and B.

One tool for determining the direction of the velocity vector of a moving charge, the magnetic field, and the force exerted is labelling the index finger "V", the middle finger "B", and the thumb "F" with your right hand. When making a gun-like configuration, with the middle finger crossing under the index finger, the fingers represent the velocity vector, magnetic field vector, and force vector, respectively.

Magnetic Dipoles

A very common source of magnetic field shown in nature is a dipole, with a "South pole" and a "North pole", terms dating back to the use of magnets as compasses, interacting with the Earth's magnetic field to indicate North and South on the globe. Since opposite ends of magnets are attracted, the north pole of a magnet is attracted to the south pole of another magnet. The Earth's North Magnetic Pole (currently in the Arctic Ocean, north of Canada) is physically a south pole, as it attracts the north pole of a compass.

A magnetic field contains energy, and physical systems move toward configurations with lower energy. When diamagnetic material is placed in a magnetic field, a *magnetic dipole* tends to align itself in opposed polarity to that field, thereby lowering the net field strength. When ferromagnetic material is placed within a magnetic field, the magnetic dipoles align to the applied field, thus expanding the domain walls of the magnetic domains.

Magnetic Monopoles

Since a bar magnet gets its ferromagnetism from electrons distributed evenly throughout the bar, when a bar magnet is cut in half, each of the resulting pieces is a smaller bar magnet. Even though a magnet is said to have a north pole and a south pole, these two poles cannot be separated from each other.

A monopole — if such a thing exists — would be a new and fundamentally different kind of magnetic object. It would act as an isolated north pole, not attached to a south pole, or vice versa. Monopoles would carry "magnetic charge" analogous to electric charge. Despite systematic searches since 1931, as of 2010, they have never been observed, and could very well not exist.

Nevertheless, some theoretical physics models predict the existence of these magnetic monopoles. Paul Dirac observed in 1931 that, because electricity and magnetism show a certain symmetry, just as quantum theory predicts that individual positive or negative electric charges can be observed without the opposing charge, isolated South or North magnetic poles should be observable. Using quantum theory Dirac showed that if magnetic monopoles exist, then one could explain the quantization of electric charge—that is, why the observed elementary particles carry charges that are multiples of the charge of the electron.

Certain grand unified theories predict the existence of monopoles which, unlike elementary particles, are solitons (localized energy packets). The initial results of using these models to estimate the number of monopoles created in the big bang contradicted cosmological observations — the monopoles would have been so plentiful and massive that they would have long since halted the expansion of the universe. However, the idea of inflation (for which this problem served as a partial motivation) was successful in solving this problem, creating models in which monopoles existed but were rare enough to be consistent with current observations.

Quantum-mechanical Origin of Magnetism

In principle all kinds of magnetism originate (similar to Superconductivity) from specific quantum-mechanical phenomena (e.g. Mathematical formulation of quantum mechanics, in particular the chapters on spin and on the Pauli principle).

A successful model was developed already in 1927, by Walter Heitler and Fritz London, who derived quantum-mechanically, how hydrogen molecules are formed from hydrogen atoms, i.e. from the atomic hydrogen orbitals u_A and u_B centred at the nuclei A and B. That this leads to magnetism, is not at all obvious, but will be explained in the following.

According the Heitler-London theory, so-called two-body molecular σ-orbitals are formed, namely the resulting orbital is:

$$\psi(\mathrm{r}_1,\mathrm{r}_2)=\frac{1}{\sqrt{2}}\left(u_A(\mathrm{r}_1)u_B(\mathrm{r}_2)+u_B(\mathrm{r}_1)u_A(\mathrm{r}_2)\right)$$

Here the last product means that a first electron, r_1, is in an atomic hydrogen-orbital centred at the second nucleus, whereas the second electron runs around the first nucleus. This “exchange” phenomenon is an expression for the quantum-mechanical property that particles with identical properties cannot be distinguished. It is specific not only for the formation of chemical bonds, but as we will see, also for magnetism, i.e. in this connection the term exchange interaction arises, a term which is essential for the origin of magnetism, and which is stronger, roughly by factors 100 and even by 1000, than the energies arising from the electrodynamic dipole-dipole interaction.

As for the *spin function* $\chi(s_1, s_2)$, which is responsible for the magnetism, we have the already mentioned Pauli's principle, namely that a symmetric orbital (i.e. with the + sign as above) must be multiplied with an antisymmetric spin function (i.e. with a " sign), and *vice versa*. Thus:

$$\chi(s_1, s_2) = \frac{1}{\sqrt{2}}\left(\alpha(s_1)\beta(s_2) - \beta(s_1)\alpha(s_2)\right),$$

I.e., not only u_A and u_B must be substituted by α and β, respectively (the first entity means "spin up", the second one "spin down"), but also the sign + by the " sign, and finally r_i by the discrete values s_i (= ±½); thereby we $\alpha(+1/2) = \beta(-1/2) = 1$ have and $\alpha(-1/2) = \beta(+1/2) = 0$. The "singlet state", i.e. the " sign, means: the spins are *antiparallel*, i.e. for the solid we have antiferromagnetism, and for two-atomic molecules one has diamagnetism. The tendency to form a (homoeopolar) chemical bond (this means: the formation of a *symmetric* molecular orbital, i.e. with the + sign) results through the Pauli principle automatically in an *antisymmetric* spin state (i.e. with the – sign). In contrast, the Coulomb repulsion of the electrons, i.e. the tendency that they try to avoid each other by this repulsion, would lead to an *antisymmetric* orbital function (i.e. with the – sign) of these two particles, and complementary to a *symmetric* spin function (i.e. with the + sign, one of the so-called "triplet functions"). Thus, now the spins would be *parallel* (ferromagnetism in a solid, paramagnetism in two-atomic gases).

The last-mentioned tendency dominates in the metals iron, cobalt and nickel, and in some rare earths, which are *ferromagnetic*. Most of the other metals, where the first-mentioned tendency dominates, are *nonmagnetic* (e.g. sodium, aluminium, and magnesium) or *antiferromagnetic* (e.g. manganese). Diatomic gases are also almost exclusively diamagnetic, and not paramagnetic. However, the oxygen molecule, because of the involvement of π-orbitals, is an exception important for the life-sciences.

The Heitler-London considerations can be generalized to the Heisenberg model of magnetism (Heisenberg 1928).

The explanation of the phenomena is thus essentially based on all subtleties of quantum mechanics, whereas the electrodynamics covers mainly the phenomenology.

Other Units

- gauss — The gauss is the centimeter-gram-second (CGS) unit of magnetic field (denoted B).
- oersted — The oersted is the CGS unit of magnetizing field (denoted H).
- maxwell — The maxwell is the CGS unit for magnetic flux.
- gamma — is a unit of *magnetic flux density* that was commonly used before the tesla came into use (1.0 gamma = 1.0 nanotesla)
- μ_0 — common symbol for the permeability of free space ($4\pi\times10^{-7}$ newton/(ampere-turn)^2).

Living Things

Some organisms can detect magnetic fields, a phenomenon known as magnetoception. Magnetobiology studies magnetic fields as a medical treatment; fields naturally produced by an organism are known as biomagnetism.

Magnetic Fields or Quantities, Types of Magnetism

A magnetic field is a mathematical description of the magnetic influence of electric currents and magnetic materials. The magnetic field at any given point is specified by both a *direction* and a *magnitude* (or strength); as such it is a vector field. The magnetic field is most commonly defined in terms of the Lorentz force it exerts on moving electric charges. *Magnetic field* can refer to two separate but closely related fields which are denoted by the symbols B and H.

Magnetic fields are produced by moving electric charges and the intrinsic magnetic moments of elementary particles associated with a fundamental quantum property, their spin. In special relativity, electric and magnetic fields are two interrelated aspects of a single object, called the electromagnetic tensor; the split of this tensor into electric and magnetic fields depends on the relative velocity of the observer and charge. In quantum physics, the electromagnetic field is quantized and electromagnetic interactions result from the exchange of photons.

In everyday life, magnetic fields are most often encountered as an invisible force created by permanent magnets which pulls on iron objects and attracts or repels other magnets. Magnetic fields are very widely used throughout modern technology, particularly in electrical engineering and electromechanics. The Earth produces its own magnetic field, which is important in navigation. Rotating magnetic fields are utilized in both electric motors and generators. Magnetic forces give information about the charge carriers in a material through the Hall effect. The interaction of magnetic fields in electric devices such as transformers is studied in the discipline of magnetic circuits.

History

Although magnets and magnetism were known much earlier, the study of magnetic fields began in 1269 when French scholar Petrus Peregrinus de Maricourt mapped out the magnetic field on the surface of a spherical magnet using iron needles. Noting that the resulting field lines crossed at two points he named those points 'poles' in analogy to Earth's poles. Almost three centuries later, William Gilbert of Colchester replicated Petrus Peregrinus' work and was the first to state explicitly that Earth is a magnet. Published in 1600, Gilbert's work, *De Magnete*, helped to establish magnetism as a science.

In 1750, John Michell stated that magnetic poles attract and repel in accordance with an inverse square law. Charles-Augustin de Coulomb experimentally verified this in 1785 and stated explicitly that the North and South poles cannot be separated. Building on this force between poles, Siméon-Denis Poisson (1781–1840) created the first successful model of the magnetic field, which he presented in 1824. In this model, a magnetic H-field is produced by 'magnetic poles' and magnetism is due to small pairs of north/south magnetic poles.

Three discoveries challenged this foundation of magnetism, though. First, in 1819, Hans Christian Oersted discovered that an electric current generates a magnetic field encircling it. Then in 1820, André-Marie Ampère showed that parallel wires having currents in the same direction attract one another. Finally, Jean-Baptiste Biot and Félix Savart discovered the Biot–Savart law in 1820, which correctly predicts the magnetic field around any current-carrying wire.

Extending these experiments, Ampère published his own successful model of magnetism in 1825. In it, he showed the equivalence of electrical currents to magnets and proposed that magnetism is due to perpetually flowing loops of current instead of the dipoles of magnetic charge in Poisson's model. This has the additional benefit of explaining why magnetic charge can not be isolated. Further, Ampère derived both Ampère's force law describing the force between two currents and Ampère's law which, like the Biot–Savart law, correctly described the magnetic field generated by a steady current. Also in this work,

Ampère introduced the term electrodynamics to describe the relationship between electricity and magnetism. In 1831, Michael Faraday discovered electromagnetic induction when he found that a changing magnetic field generates an encircling electric field. He described this phenomenon in what is known as Faraday's law of induction. Later, Franz Ernst Neumann proved that, for a moving conductor in a magnetic field, induction is a consequence of Ampère's force law . In the process he introduced the magnetic vector potential, which was later shown to be equivalent to the underlying mechanism proposed by Faraday.

In 1850, Lord Kelvin, then known as William Thomson, distinguished between two magnetic fields now denoted H and B. The former applied to Poisson's model and the latter to Ampère's model and induction. Further, he derived how H and B relate to each other.

Between 1861 and 1865, James Clerk Maxwell developed and published Maxwell's equations, which explained and united all of classical electricity and magnetism. The first set of these equations was published in a paper entitled *On Physical Lines of Force* in 1861. These equations were valid although incomplete. He completed Maxwell's set of equations in his later 1865 paper *A Dynamical Theory of the Electromagnetic Field* and demonstrated the fact that light is an electromagnetic wave. Heinrich Hertz experimentally confirmed this fact in 1887. Although implicit in Ampère's force law the force due to a magnetic field on a moving electric charge was not correctly and explicitly stated until 1892 by Hendrik Lorentz who theoretically derived it from Maxwell's equations. With this last piece of the puzzle, the classical theory of electrodynamics was essentially complete.

The twentieth century extended electrodynamics to include relativity and quantum mechanics. Albert Einstein, in his paper of 1905 that established relativity, showed that both the electric and magnetic fields are part of the same phenomena viewed from different reference frames. Finally, the emergent field of quantum mechanics was merged with electrodynamics to form quantum electrodynamics (QED).

Definitions, Units, and Measurement

Definition:

Alternative names for B

- Magnetic flux density
- Magnetic induction
- Magnetic field

Alternative names for H

- Magnetic field intensity
- Magnetic field strength
- Magnetic field
- Magnetizing field

The magnetic field can be defined in several equivalent ways based on the effects it has on its environment.

Often the magnetic field is defined by the force it exerts on a moving charged particle. It is known from experiments in electrostatics that a particle of charge q in an electric field E experiences a force $F = qE$. However, in other situations, such as when a charged particle moves in the vicinity of a current-carrying wire, the force also depends on the velocity of that particle. Fortunately, the velocity dependent portion can be separated out such that the force on the particle satisfies the *Lorentz force law*,

$$F = q(E + v \times B).$$

Here v is the particle's velocity and × denotes the cross product. The vector B is termed the magnetic field, and it is *defined* as the vector field necessary to make the Lorentz force law correctly describe the motion of a charged particle. This definition allows the determination of B in the following way

[T]he command, "Measure the direction and magnitude of the vector B at such and such a place," calls for the following operations: Take a particle of known charge q. Measure the force on q at rest, to determine E. Then measure the force on the particle when its velocity is v; repeat with v in some other direction. Now find a B that makes [the Lorentz force law] fit all these results—that is the magnetic field at the place in question.

Alternatively, the magnetic field can be defined in terms of the torque it produces on a magnetic dipole.

In addition to B, there is a quantity H which is also sometimes called the "magnetic field". In a vacuum, B and H are proportional to each other, with the multiplicative constant depending on the physical units. Inside a material they are different. The term *magnetic field* is historically reserved for H while using other terms for B. Informally, though, and formally for some recent textbooks mostly in physics, the term 'magnetic field' is used to describe B as well as or in place of H. There are many alternative names for both.

Units

In SI units, B is measured in teslas (symbol: T) and correspondingly Φ_B (magnetic flux) is measured in webers (symbol: Wb) so that a flux density of 1 Wb/m^2 is 1 tesla. The SI unit of tesla is equivalent to (newton·second)/(coulomb·metre). In Gaussian-cgs units, B is measured in gauss (symbol: G). (The conversion is 1 T = 10,000 G.) The H-field is measured in amperes per metre (A/m) in SI units, and in oersteds (Oe) in cgs units.

Measurement

The smallest precision level for a magnetic field measurement is on the order of attoteslas (10^{-18} teslas); the largest magnetic field produced in a laboratory is 2.8 kT (VNIIEF in Sarov, Russia, 1998). The magnetic field of some astronomical objects such as magnetars are much higher; magnetars range from 0.1 to 100 GT (10^8 to 10^{11} T).

Devices used to measure the local magnetic field are called magnetometers. Important classes of magnetometers include using a rotating coil, Hall effect magnetometers, NMR magnetometers, SQUID magnetometers, and fluxgate magnetometers. The magnetic fields of distant astronomical objects are measured through their effects on local charged particles. For instance, electrons spiraling around a field line produce synchrotron radiation that is detectable in radio waves.

Magnetic Field Lines

Mapping the magnetic field of an object is simple in principle. First, measure the strength and direction of the magnetic field at a large number of locations (or at every point in space). Then, mark each location with an arrow (called a vector) pointing in the direction of the local magnetic field with its magnitude proportional to the strength of the magnetic field.

An alternative method to map the magnetic field is to 'connect' the arrows to form magnetic *field lines*. The direction of the magnetic field at any point is parallel to the direction of nearby field lines, and the local density of field lines can be made proportional to its strength.

Magnetic field lines are like the contour lines (constant altitude) on a topographic map in that they represent something continuous, and a different mapping scale would show more or fewer lines. An advantage of using magnetic field lines as a representation is that many laws of magnetism (and electromagnetism) can be stated completely and concisely using simple concepts such as the 'number' of field lines through a surface. These concepts can be quickly 'translated' to their mathematical form. For example, the number of field lines through a given surface is the surface integral of the magnetic field.

Various phenomena have the effect of "displaying" magnetic field lines as though the field lines are physical phenomena. For example, iron filings placed in a magnetic field line up to form lines that correspond to 'field lines'. Magnetic fields' "lines" are also visually displayed in polar auroras, in which plasma particle dipole interactions create visible streaks of light that line up with the local direction of Earth's magnetic field. Field lines can be used as a qualitative tool to visualize magnetic forces. In ferromagnetic substances like iron and in plasmas, magnetic forces can be understood by imagining that the field lines exert a tension, (like a rubber band) along their length, and a pressure perpendicular to their length on neighbouring field lines. 'Unlike' poles of magnets attract because they are linked by many field lines; 'like' poles repel because their field lines do not meet, but run parallel, pushing on each other. The rigorous form of this concept is the electromagnetic stress–energy tensor.

Magnetic Field and Permanent Magnets

Permanent magnets are objects that produce their own persistent magnetic fields. They are made of ferromagnetic materials, such as iron and nickel, that have been magnetized, and they have both a north and a south pole.

Magnetic Field of Permanent Magnets

The magnetic field of permanent magnets can be quite complicated, especially near the magnet. The magnetic field of a small straight magnet is proportional to the magnet's *strength* (called its magnetic dipole moment m). The equations are non-trivial and also depend on the distance from the magnet and the orientation of the magnet. For simple magnets, m points in the direction of a line drawn from the south to the north pole of the magnet. Flipping a bar magnet is equivalent to rotating its m by 180 degrees.

The magnetic field of larger magnets can be obtained by modelling them as a collection of a large number of small magnets called dipoles each having their own m. The magnetic field produced by the magnet then is the net magnetic field of these dipoles. And, any net force on the magnet is a result of adding up the forces on the individual dipoles.

There are two competing models for the nature of these dipoles. These two models produce two different magnetic fields, H and B. Outside a material, though, the two are identical (to a multiplicative constant) so that in many cases the distinction can be ignored. This is particularly true for magnetic fields, such as those due to electric currents, that are not generated by magnetic materials.

Magnetic Pole Model and the H-field

It is sometimes useful to model the force and torques between two magnets as due to magnetic poles repelling or attracting each other in the same manner as the Coulomb force between electric charges. In this model, a magnetic H-field is produced by *magnetic charges* that are 'smeared' around each pole. The H-field, therefore, is analogous to the electric field E, which starts at a positive electric charge and ends at a negative electric charge. Near the north pole, therefore, all H-field lines point away from the north pole (whether inside the magnet or out) while near the south pole (whether inside the magnet or out) all H-field lines point toward the south pole. A north pole, then, feels a force in the direction of the H-field while the force on the south pole is opposite to the H-field.

In the magnetic pole model, the elementary magnetic dipole m is formed by two opposite magnetic poles of pole strength q_m separated by a very small distance vector d, such that m = q_m d.

Magnetic poles cannot exist apart from each other; all magnets have north/south pairs that cannot be separated without creating two magnets each having a north/south pair. The magnetic pole model does not account for magnetism that is produced by electric currents, nor the force that a magnetic field applies to moving electric charges.

Amperian Loop Model and the B-field

After Oersted discovered that electric currents produce a magnetic field and Ampere discovered that electric currents attracted and repelled each other similar to magnets, it was natural to hypothesize that all magnetic fields are due to electric current loops. In this model developed by Ampere, the elementary magnetic dipole that makes up all magnets is a sufficiently small Amperian loop of current I. The dipole moment of this loop is $m = IA$ where A is the area of the loop.

These magnetic dipoles produce a magnetic B-field. One important property of the B-field produced this way is that magnetic B-field lines neither start nor end (mathematically, B is a solenoidal vector field);

a field line either extends to infinity or wraps around to form a closed curve. To date no exception to this rule has been found. Magnetic field lines exit a magnet near its north pole and enter near its south pole, but inside the magnet B-field lines continue through the magnet from the south pole back to the north. If a B-field line enters a magnet somewhere it has to leave somewhere else; it is not allowed to have an end point. Magnetic poles, therefore, always come in N and S pairs.

More formally, since all the magnetic field lines that enter any given region must also leave that region, subtracting the 'number' of field lines that enter the region from the number that exit gives identically zero. Mathematically this is equivalent to:

$$\oint_S \mathrm{B} \cdot \mathrm{dA} = 0,$$

where the integral is a surface integral over the closed surface S (a closed surface is one that completely surrounds a region with no holes to let any field lines escape). Since dA points outward, the dot product in the integral is positive for B-field pointing out and negative for B-field pointing in.

There is also a corresponding differential form of this equation covered in Maxwell's equations below.

Force Between Magnets

The force between two small magnets is quite complicated and depends on the strength and orientation of both magnets and the distance and direction of the magnets relative to each other. The force is particularly sensitive to rotations of the magnets due to magnetic torque. The force on each magnet depends on its magnetic moment and the magnetic field of the other.

To understand the force between magnets, it is useful to examine the *magnetic pole model* given above. In this model, the *H-field* of one magnet pushes and pulls on *both* poles of a second magnet. If this H-field is the same at both poles of the second magnet then there is no net force on that magnet since the force is opposite for opposite poles. If, however, the magnetic field of the first magnet is *nonuniform* (such as the H near one of its poles), each pole of the second magnet sees a different field and is subject to a different force. This difference in the two forces moves the magnet in the direction of increasing magnetic field and may also cause a net torque. This is a specific example of a general rule that magnets are attracted (or repulsed depending on the orientation of the magnet) into regions of higher magnetic field. Any non-uniform magnetic field, whether caused by permanent magnets or electric currents, exerts a force on a small magnet in this way.

The details of the Amperian loop model are different and more complicated but yield the same result: that magnetic dipoles are attracted/repelled into regions of higher magnetic field. Mathematically, the force on a small magnet having a magnetic moment m due to a magnetic field B is:

$$F = \nabla(m \cdot B),$$

where the gradient ∇ is the change of the quantity m · B per unit distance and the direction is that of maximum increase of m · B. To understand this equation, note that the dot product m · B = $mB\cos(\theta)$, where m and B represent the magnitude of the m and B vectors and θ is the angle between them. If m is in the same direction as B then the dot product is positive and the gradient points 'uphill' pulling the magnet into regions of higher B-field (more strictly larger m · B). This equation is strictly only valid for magnets of zero size, but is often a good approximation for not too large magnets. The magnetic force on larger magnets is determined by dividing them into smaller regions having their own m then summing up the forces on each of these regions.

Magnetic Torque on Permanent Magnets

If two like poles of two separate magnets are brought near each other, and one of the magnets is allowed to turn, it promptly rotates to align itself with the first. In this example, the magnetic field of the stationary magnet creates a *magnetic torque* on the magnet that is free to rotate. This magnetic torque τ tends to align a magnet's poles with the magnetic field lines. A compass, therefore, turns to align itself with Earth's magnetic field.

Magnetic torque is used to drive electric motors. In one simple motor design, a magnet is fixed to a freely rotating shaft and subjected to a magnetic field from an array of electromagnets. By continuously switching the electric current through each of the electromagnets, thereby flipping the polarity of their magnetic fields, like poles are kept next to the rotor; the resultant torque is transferred to the shaft.

As is the case for the force between magnets, the magnetic pole model leads more readily to the correct equation. Here, two equal and opposite magnetic charges experiencing the same H also experience equal and opposite forces. Since these equal and opposite forces are in different locations, this produces a torque proportional to the distance (perpendicular to the force) between them. With the definition of m as the pole strength times the distance between the poles, this leads to $\tau = \mu_0 m H \sin\theta$, where μ_0 is a constant called the vacuum permeability, measuring $4\pi\times10^{-7}$ V·s/(A·m) and θ is the angle between H and m.

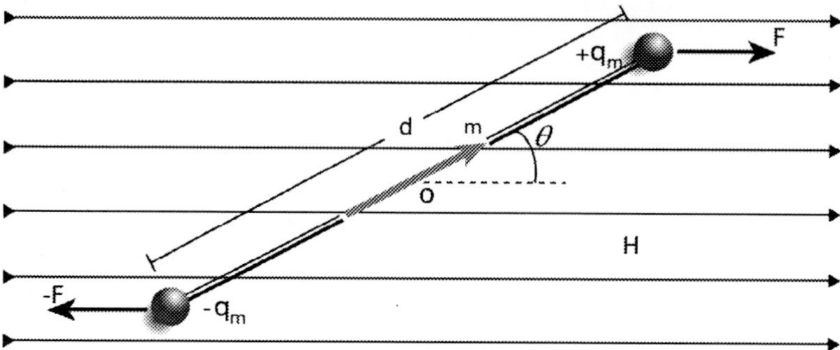

Figure: *torque on a dipole: An H field (to right) causes equal but opposite forces on a N pole (+q) and a S pole ("q) creating a torque.*

The Amperian loop model also predicts the same magnetic torque. Here, it is the B field interacting with the Amperian current loop through a Lorentz force described below. Again, the results are the same although the models are completely different.

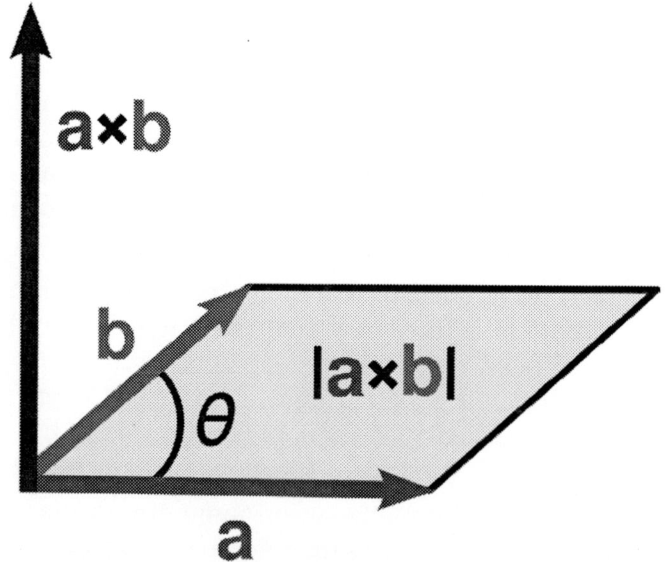

Figure: *Cross product:* |*a* × *b*| = a b *sinè.*

Mathematically, the torque τ on a small magnet is proportional both to the applied magnetic field and to the magnetic moment m of the magnet:

$$\tau = m \times B = \mu_0 m \times H,$$

where × represents the vector cross product. Note that this equation includes all of the qualitative information included above. There is no torque on a magnet if m is in the same direction as the magnetic field. (The cross product is zero for two vectors that are in the same direction.) Further, all other orientations feel a torque that twists them toward the direction of magnetic field.

Magnetic Field and Electric Currents

Currents of electric charges both generate a magnetic field and feel a force due to magnetic B-fields.

Magnetic Field Due to Moving Charges and Electric Currents

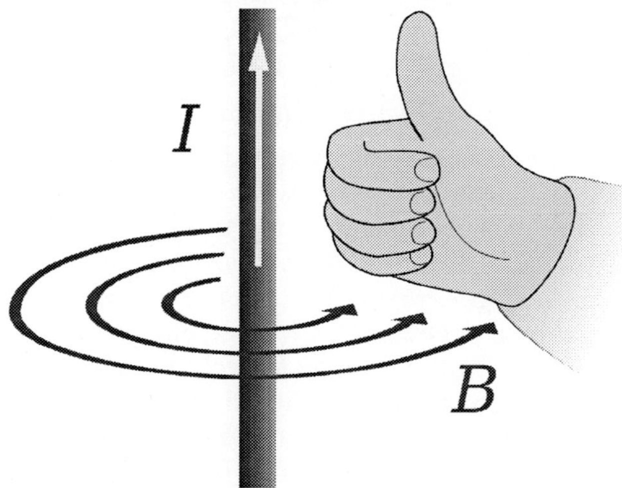

Figure: *Right hand grip rule: a current flowing in the direction of the white arrow produces a magnetic field shown by the red arrows.*

All moving charged particles produce magnetic fields. Moving point charges, such as electrons, produce complicated but well known magnetic fields that depend on the charge, velocity, and acceleration of the particles. Magnetic field lines form in concentric circles around a cylindrical current-carrying conductor, such as a length of wire. The direction of such a magnetic field can be determined by using the "right hand grip rule". The strength of the magnetic field decreases with distance from the wire. (For an infinite length wire the strength decreases inversely proportional to the distance.)

Figure: *Solenoid*

Bending a current-carrying wire into a loop concentrates the magnetic field inside the loop while weakening it outside. Bending a wire into multiple closely spaced loops to form a coil or "solenoid" enhances this effect. A device so formed around an iron core may act as an *electromagnet*, generating a strong, well-controlled magnetic field. An infinitely long cylindrical electromagnet has a uniform magnetic field inside, and no magnetic field outside. A finite length electromagnet produces a magnetic field that looks similar to that produced by a uniform permanent magnet, with its strength and polarity determined by the current flowing through the coil.

The magnetic field generated by a steady current I (a constant flow of electric charges, in which charge neither accumulates nor is depleted at any point) is described by the *Biot–Savart law*:

$$\mathrm{B} = \frac{\mu_0 I}{4\pi} \int_{\mathrm{wire}} \frac{d\ell \times \hat{\mathrm{r}}}{r^2},$$

where the integral sums over the wire length where vector *d–l* is the vector line element with direction in the same sense as the current I, μ_0 is the magnetic constant, r is the distance between the location of *d–l* and the location at which the magnetic field is being calculated, and $\hat{r}$ is a unit vector in the direction of r.

A slightly more general way of relating the current I to the B-field is through Ampère's law:

$$\oint B \cdot d\ell = \mu_0 I_{enc},$$

where the line integral is over any arbitrary loop and I_{enc} is the current enclosed by that loop. Ampère's law is always valid for steady currents and can be used to calculate the B-field for certain highly symmetric situations such as an infinite wire or an infinite solenoid.

In a modified form that accounts for time varying electric fields, Ampère's law is one of four Maxwell's equations that describe electricity and magnetism.

Force on Moving Charges and Current

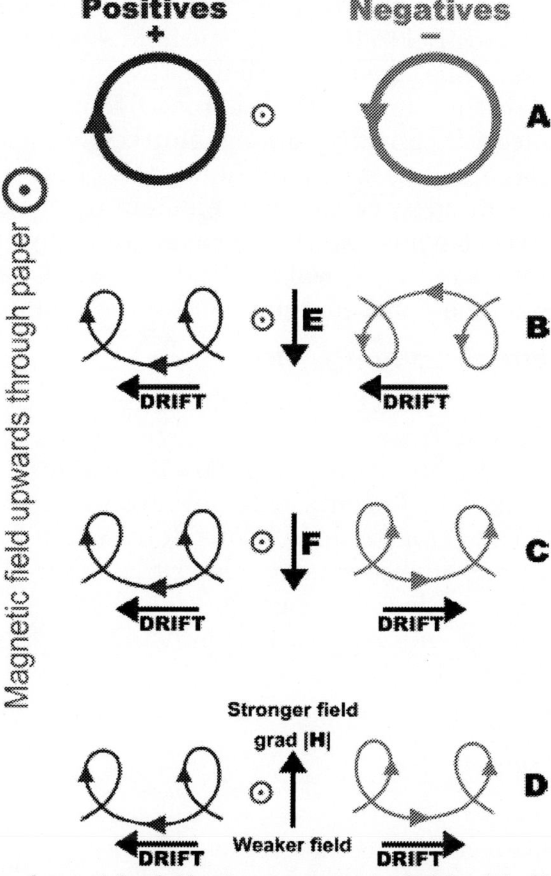

Figure: *Charged particle drifts in a magnetic field with (A) no net force, (B) an electric field, E, (C) a charge independent force, F (e.g. gravity), and (D) an inhomogeneous magnetic field, grad H.*

Force on a Charged Particle

A charged particle moving in a B-field experiences a *sideways* force that is proportional to the strength of the magnetic field, the component of the velocity that is perpendicular to the magnetic field and the charge of the particle. This force is known as the Lorentz force, and is given by

$$F = q\mathrm{v} \times B,$$

where F is the force, q is the electric charge of the particle, v is the instantaneous velocity of the particle, and B is the magnetic field (in teslas).

The Lorentz force is always perpendicular to both the velocity of the particle and the magnetic field that created it. When a charged particle moves in a static magnetic field, it traces a helical path in which the helix axis is parallel to the magnetic field, and in which the speed of the particle remains constant. Because the magnetic force is always perpendicular to the motion, the magnetic field can do no work on an isolated charge. It can only do work indirectly, via the electric field generated by a changing magnetic field. It is often claimed that the magnetic force can do work to a non-elementary magnetic dipole, or to charged particles whose motion is constrained by other forces, but this is incorrect because the work in those cases is performed by the electric forces of the charges deflected by the magnetic field.

Force on Current-carrying Wire

The force on a current carrying wire is similar to that of a moving charge as expected since a charge carrying wire is a collection of moving charges. A current-carrying wire feels a force in the presence of a magnetic field. The Lorentz force on a macroscopic current is often referred to as the *Laplace force.* Consider a conductor of length ℓ, cross section A, and charge q due to electric current i. If this conductor is placed in a magnetic field of magnitude B that makes an angle θ with the velocity of charges in the conductor, the force exerted on a single charge q is

$$F = qvB \sin\theta,$$

so, for N charges where

$$N = n\ell A,$$

the force exerted on the conductor is

$$f = FN = qvBn\ell A \sin\theta = Bi\ell \sin\theta,$$

where $i = nqvA$.

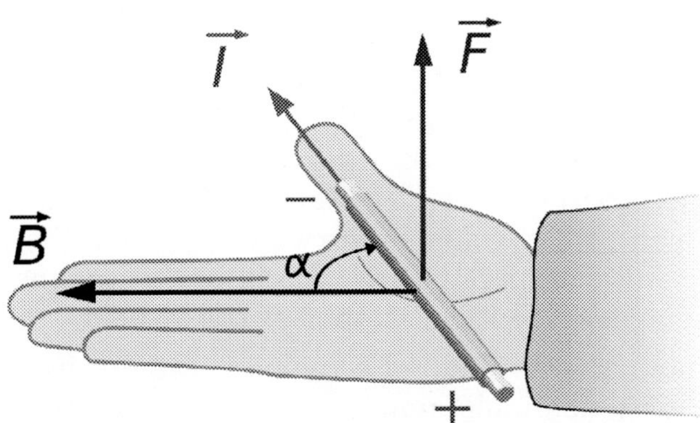

Figure: *The right-hand rule: Pointing the thumb of the right hand in the direction of the conventional current and the fingers in the direction of B the force on the current points out of the palm. The force is reversed for a negative charge.*

Direction of Force

The direction of force on a charge or a current can be determined by a mnemonic known as the *right-hand rule*. Using the right hand and pointing the thumb in the direction of the moving positive charge or positive current and the fingers in the direction of the magnetic field the resulting force on the charge points outwards from the palm. The force on a negatively charged particle is in the opposite direction. If both the speed and the charge are reversed then the direction of the force remains the same. For that reason a magnetic field measurement (by itself) cannot distinguish whether there is a positive charge moving to the right or a negative charge moving to the left. (Both of these cases produce the same current.) On the other hand, a magnetic field combined with an electric field *can* distinguish between these. An alternative mnemonic to the right hand rule Flemings's left hand rule.

Relation Between H and B

The formulas derived for the magnetic field above are correct when dealing with the entire current. A magnetic material placed inside a magnetic field, though, generates its own bound current, which can be a challenge to calculate. (This bound current is due to the sum of atomic sized current loops and the spin of the subatomic particles such as electrons that make up the material.) The H-field as defined above helps factor out this bound current; but to see how, it helps to introduce the concept of *magnetization* first.

Magnetization

The magnetization vector field M represents how strongly a region of material is magnetized. It is defined as the net magnetic dipole moment per unit volume of that region. The magnetization of a uniform magnet, therefore, is a constant in the material equal to its magnetic moment, m, divided by its volume. Since the SI unit of magnetic moment is ampere metre^2, the SI unit of magnetization M is ampere per metre, identical to that of the H-field.

The magnetization M field of a region points in the direction of the average magnetic dipole moment in that region. Magnetization field lines, therefore, begin near the magnetic south pole and ends near the magnetic north pole. (Magnetization does not exist outside of the magnet.)

In the Amperian loop model, the magnetization is due to combining many tiny Amperian loops to form a resultant current called *bound current*. This bound current, then, is the source of the magnetic B field due to the magnet. Given the definition of the magnetic dipole, the magnetization field follows a similar law to that of Ampere's law:

$$\oint \mathrm{M} \cdot d\ell = I_{\mathrm{b}},$$

where the integral is a line integral over any closed loop and I_{b} is the 'bound current' enclosed by that closed loop.

In the magnetic pole model, magnetization begins at and ends at magnetic poles. If a given region, therefore, has a net positive 'magnetic pole strength' (corresponding to a north pole) then it has more magnetization field lines entering it than leaving it. Mathematically this is equivalent to:

$$\oint_S \mu_0 \mathrm{M} \cdot \mathrm{dA} = -q_M$$

where the integral is a closed surface integral over the closed surface S and q_M is the 'magnetic charge' (in units of magnetic flux) enclosed by S. (A closed surface completely surrounds a region with no holes to let any field lines escape.) The negative sign occurs because the magnetization field moves from south to north.

H-field and Magnetic Materials

The H-field is defined as:

$$\mathrm{H} \equiv \frac{\mathrm{B}}{\mu_0} - \mathrm{M}, \text{ (definition of H in SI units)}$$

With this definition, Ampere's law becomes:

$$\oint H\cdot d\ell = \oint\left(\frac{B}{\mu_0} - M\right)\cdot d\ell = I_{\text{tot}} - I_{\text{b}} = I_{\text{f}},$$

where I_f represents the 'free current' enclosed by the loop so that the line integral of H does not depend at all on the bound currents.

$$(H_1^{\|} - H_2^{\|}) = K_f \times \hat{n},$$

where K_f is the surface free current density and the unit normal $\hat{n}$ points in the direction from medium 2 to medium 1.

Similarly, a surface integral of H over any closed surface is independent of the free currents and picks out the 'magnetic charges' within that closed surface:

$$\oint_S \mu_0 H\cdot dA = \oint_S (B - \mu_0 M)\cdot dA = (0 - (-q_M)) = q_M,$$

which does not depend on the free currents.

The H-field, therefore, can be separated into two independent parts:

$$H = H_0 + H_d,$$

where H_0 is the applied magnetic field due only to the free currents and H_d is the demagnetizing field due only to the bound currents.

The magnetic H-field, therefore, re-factors the bound current in terms of 'magnetic charges'. The H field lines loop only around 'free current' and, unlike the magnetic B field, begins and ends near magnetic poles as well.

Magnetism

Most materials respond to an applied B-field by producing their own magnetization M and therefore their own B-field. Typically, the response is very weak and exists only when the magnetic field is applied.

The term *magnetism* describes how materials respond on the microscopic level to an applied magnetic field and is used to categorize the magnetic phase of a material. Materials are divided into groups based upon their magnetic behaviour:

- Diamagnetic materials produce a magnetization that opposes the magnetic field.
- Paramagnetic materials produce a magnetization in the same direction as the applied magnetic field.

- Ferromagnetic materials and the closely related ferrimagnetic materials and antiferromagnetic materials can have a magnetization independent of an applied B-field with a complex relationship between the two fields.
- Superconductors (and ferromagnetic superconductors) are materials that are characterized by perfect conductivity below a critical temperature and magnetic field. They also are highly magnetic and can be perfect diamagnets below a lower critical magnetic field. Superconductors often have a broad range of temperatures and magnetic fields (the so named mixed state) under which they exhibit a complex hysteretic dependence of M on B.

In the case of paramagnetism and diamagnetism, the magnetization M is often proportional to the applied magnetic field such that:

$$B = \mu H,$$

where μ is a material dependent parameter called the permeability. In some cases the permeability may be a second rank tensor so that H may not point in the same direction as B. These relations between B and H are examples of constitutive equations. However, superconductors and ferromagnets have a more complex B to H relation.

Energy Stored in Magnetic Fields

Energy is needed to generate a magnetic field both to work against the electric field that a changing magnetic field creates and to change the magnetization of any material within the magnetic field. For non-dispersive materials this same energy is released when the magnetic field is destroyed so that this energy can be modelled as being stored in the magnetic field.

For linear, non-dispersive, materials (such that $B = \mu H$ where μ is frequency-independent), the energy density is:

$$u = \frac{B \cdot B}{2\mu} = \frac{\mu H \cdot H}{2}.$$

If there are no magnetic materials around then μ can be replaced by μ_0. The above equation cannot be used for nonlinear materials, though; a more general expression given below must be used.

In general, the incremental amount of work per unit volume δW needed to cause a small change of magnetic field δB is:

$$\delta W = H \cdot \delta B.$$

Once the relationship between H and B is known this equation is used to determine the work needed to reach a given magnetic state. For hysteretic materials such as ferromagnets and superconductors, the work needed also depends on how the magnetic field is created. For linear non-dispersive materials, though, the general equation leads directly to the simpler energy density equation given above.

Faraday's Law: Electric Force Due to a Changing B-field

A changing magnetic field, such as a magnet moving through a conducting coil, generates an electric field (and therefore tends to drive a current in the coil). This is known as *Faraday's law* and forms the basis of many electrical generators and electric motors.

Mathematically, Faraday's law is:

$$\mathcal{E} = -\frac{d\Phi_{\mathrm{m}}}{dt},$$

where $\mathcal{E}$ is the electromotive force (or *EMF*, the voltage generated around a closed loop) and Φ_{m} is the *magnetic flux*—the product of the area times the magnetic field normal to that area. (This definition of magnetic flux is why B is often referred to as magnetic flux density.)

The negative sign is necessary and represents the fact that any current generated by a changing magnetic field in a coil produces a magnetic field that *opposes* the change in the magnetic field that induced it. This phenomenon is known as Lenz's Law.

This integral formulation of Faraday's law can be converted into a differential form, which applies under slightly different conditions. This form is covered as one of Maxwell's equations below.

Maxwell's Correction to Ampère's Law: The magnetic field due to a Changing Electric Field

Similar to the way that a changing magnetic field generates an electric field, a changing electric field generates a magnetic field. This fact is known as *Maxwell's correction to Ampère's law.* Maxwell's correction to Ampère's Law bootstrap together with Faraday's law of induction to form electromagnetic waves, such as light. Thus, a changing electric field generates a changing magnetic field, which generates a changing electric field again.

Maxwell's correction to Ampère law is applied as an additive term to Ampere's law given above. This additive term is proportional to the time rate of change of the electric flux and is similar to Faraday's law above but with a different and positive constant out front. (The electric flux through an area is proportional to the area times the perpendicular part of the electric field.)

This full Ampère law including the correction term is known as the Maxwell–Ampère equation. It is not commonly given in integral form because the effect is so small that it can typically be ignored in most cases where the integral form is used. The Maxwell term *is* critically important in the creation and propagation of electromagnetic waves. These, though, are usually described using the differential form of this equation given below.

Maxwell's Equations

Like all vector fields, a magnetic field has two important mathematical properties that relates it to its *sources*. (For B the *sources* are currents and changing electric fields.) These two properties, along with the two corresponding properties of the electric field, make up *Maxwell's Equations*. Maxwell's Equations together with the Lorentz force law form a complete description of classical electrodynamics including both electricity and magnetism.

The first property is the divergence of a vector field A, $\nabla \cdot A$, which represents how A 'flows' outward from a given point. As discussed above, a B-field line never starts or ends at a point but instead forms a complete loop. This is mathematically equivalent to saying that the divergence of B is zero. (Such vector fields are called solenoidal vector fields.) This property is called Gauss's law for magnetism and is equivalent to the statement that there are no isolated magnetic poles or magnetic monopoles. The electric field on the other hand begins and ends at electric charges so that its divergence is non-zero and proportional to the charge density.

The second mathematical property is called the curl, such that $\nabla \times A$ represents how A curls or 'circulates' around a given point. The result of the curl is called a 'circulation source'. The equations for the curl of B and of E are called the Ampère–Maxwell equation and Faraday's law respectively. They represent the differential forms of the integral equations given above.

The complete set of Maxwell's equations then are:

$$\nabla \cdot B = 0,$$

$$\nabla \cdot E = \frac{\rho}{\epsilon_0},$$

$$\nabla \times \mathrm{B} = \mu_0 \mathrm{J} + \mu_0 \varepsilon_0 \frac{\partial \mathrm{E}}{\partial t},$$

$$\nabla \times \mathrm{E} = -\frac{\partial \mathrm{B}}{\partial t},$$

where J = complete microscopic current density and ρ is the charge density.

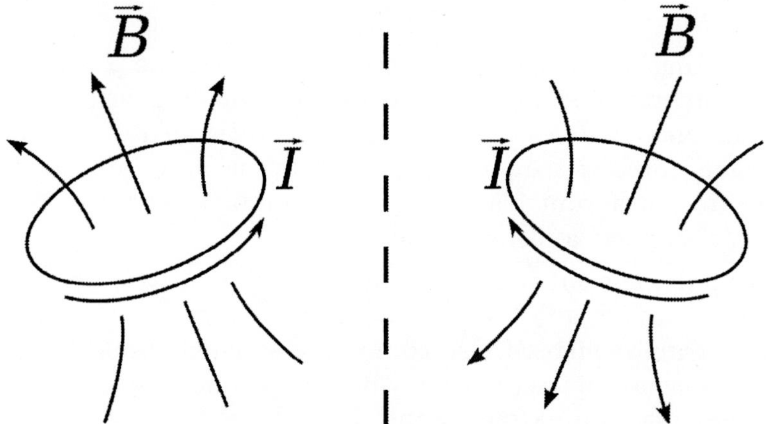

Figure: *Magnetic field, like all pseudovectors, changes sign when reflected in a mirror: When a current carrying loop (black) is reflected in a mirror (dotted line), its magnetic field (blue) is reflected* and *reversed.*

Technically, B is a pseudovector (also called an *axial vector*) due to being defined by a vector cross product.

As discussed above, materials respond to an applied electric E field and an applied magnetic B field by producing their own internal 'bound' charge and current distributions that contribute to E and B but are difficult to calculate.

To circumvent this problem, H and D fields are used to re-factor Maxwell's equations in terms of the *free current density* $\mathrm{J_f}$ and *free charge density* ρ_f:

$$\nabla \cdot \mathrm{B} = 0,$$

$$\nabla \cdot \mathrm{D} = \rho_\mathrm{f},$$

$$\nabla \times \mathrm{H} = \mathrm{J_f} + \frac{\partial \mathrm{D}}{\partial t},$$

$$\nabla \times \mathrm{E} = -\frac{\partial \mathrm{B}}{\partial t}.$$

These equations are not any more general than the original equations (if the 'bound' charges and currents in the material are known). They also need to be supplemented by the relationship between B and H as well as that between E and D. On the other hand, for simple relationships between these quantities this form of Maxwell's equations can circumvent the need to calculate the bound charges and currents.

Electric and Magnetic Fields: Different Aspects of the Same Phenomenon

According to the special theory of relativity, the partition of the electromagnetic force into separate electric and magnetic components is not fundamental, but varies with the observational frame of reference: An electric force perceived by one observer may be perceived by another (in a different frame of reference) as a magnetic force, or a mixture of electric and magnetic forces.

Formally, special relativity combines the electric and magnetic fields into a rank-2 tensor, called the electromagnetic tensor. Changing reference frames *mixes* these components. This is analogous to the way that special relativity *mixes* space and time into spacetime, and mass, momentum and energy into four-momentum.

Magnetic Vector Potential

In advanced topics such as quantum mechanics and relativity it is often easier to work with a potential formulation of electrodynamics rather than in terms of the electric and magnetic fields. In this representation, the *vector potential* A, and the scalar potential φ, are defined such that:

$$B = \nabla \times A,$$

$$E = -\nabla \varphi - \frac{\partial A}{\partial t}.$$

The vector potential A may be interpreted as a *generalized potential momentum per unit charge* just as φ is interpreted as a *generalized potential energy per unit charge.*

Maxwell's equations when expressed in terms of the potentials can be cast into a form that agrees with special relativity with little effort. In relativity A together with φ forms the four-potential, analogous to the four-momentum that combines the momentum and energy of a particle. Using the four potential instead of the electromagnetic tensor has the advantage of being much simpler—and it can be easily modified to work with quantum mechanics.

Quantum Electrodynamics

In modern physics, the electromagnetic field is understood to be not a *classical* field, but rather a quantum field; it is represented not as a vector of three numbers at each point, but as a vector of three quantum operators at each point. The most accurate modern description of the electromagnetic interaction (and much else) is *quantum electrodynamics* (QED), which is incorporated into a more complete theory known as the *Standard Model of particle physics.*

In QED, the magnitude of the electromagnetic interactions between charged particles (and their antiparticles) is computed using perturbation theory; these rather complex formulas have a remarkable pictorial representation as Feynman diagrams in which virtual photons are exchanged. Predictions of QED agree with experiments to an extremely high degree of accuracy: currently about 10^{-12} (and limited by experimental errors). This makes QED one of the most accurate physical theories constructed thus far. All equations in this article are in the classical approximation, which is less accurate than the quantum description mentioned here. However, under most everyday circumstances, the difference between the two theories is negligible.

Important uses and Examples of Magnetic Field

Earth's magnetic field:

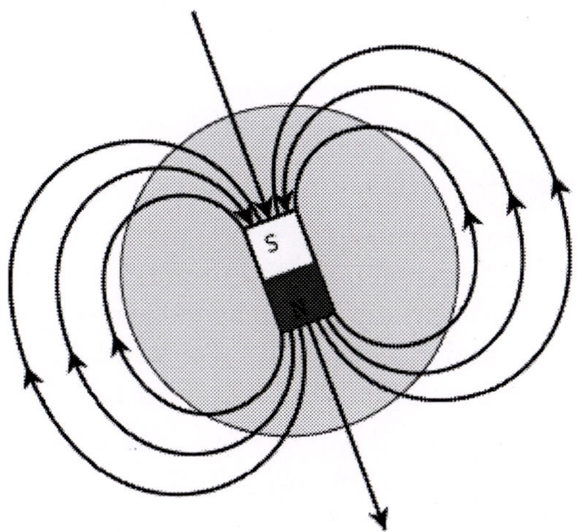

Figure: *A sketch of Earth's magnetic field representing the source of the field as a magnet. The geographic north pole of Earth is near the top of the diagram, the south pole near the bottom. The south pole of that magnet is deep in Earth's interior below Earth's North Magnetic Pole.*

The Earth's magnetic field is thought to be produced by convection currents in the outer liquid of Earth's core. The Dynamo theory proposes that these movements produce electric currents that, in turn, produce the magnetic field.

The presence of this field causes a compass, placed anywhere within it, to rotate so that the "north pole" of the magnet in the compass points roughly north, toward Earth's north magnetic pole. This is the traditional definition of the "north pole" of a magnet, although other equivalent definitions are also possible.

One confusion that arises from this definition is that, if Earth itself is considered as a magnet, the *south* pole of that magnet would be the one nearer the north magnetic pole, and vice-versa. The north magnetic pole is so-named not because of the polarity of the field there but because of its geographical location. The north and south poles of a permanent magnet are so-called because they are "north-seeking" and "south-seeking", respectively.

The figure is a sketch of Earth's magnetic field represented by field lines. For most locations, the magnetic field has a significant up/down component in addition to the North/South component. (There is also an East/West component; Earth's magnetic poles do not coincide exactly with Earth's geological pole.) The magnetic field can be visualised as a bar magnet buried deep in Earth's interior.

Earth's magnetic field is not constant—the strength of the field and the location of its poles vary. Moreover, the poles periodically reverse their orientation in a process called geomagnetic reversal. The most recent reversal occurred 780,000 years ago.

Rotating Magnetic Fields

The *rotating magnetic field* is a key principle in the operation of alternating-current motors. A permanent magnet in such a field rotates so as to maintain its alignment with the external field. This effect was conceptualized by Nikola Tesla, and later utilized in his, and others', early AC (alternating-current) electric motors.

A rotating magnetic field can be constructed using two orthogonal coils with 90 degrees phase difference in their AC currents. However, in practice such a system would be supplied through a three-wire arrangement with unequal currents. This inequality would cause serious problems in standardization of the conductor size and so, to overcome it, three-phase systems are used where the three currents are equal in magnitude and have 120 degrees phase difference. Three similar coils having mutual geometrical angles of 120 degrees create the rotating magnetic field in this case. The ability of the three-phase system to create a rotating field, utilized in electric motors, is one of the main reasons why three-phase systems dominate the world's electrical power supply systems. Because magnets degrade with time, synchronous motors use DC voltage fed rotor windings, which allows the excitation of the machine to be controlled and induction motors use short-circuited rotors (instead of a magnet) following the rotating magnetic field of a multicoiled stator. The short-circuited turns of the rotor develop eddy currents in the rotating field of the stator, and these currents in turn move the rotor by the Lorentz force. In 1882, Nikola Tesla identified the concept of the rotating magnetic field. In 1885, Galileo Ferraris independently researched the concept. In 1888, Tesla gained U.S. Patent 381,968 for his work. Also in 1888, Ferraris published his research in a paper to the *Royal Academy of Sciences* in Turin.

Hall Effect

The charge carriers of a current carrying conductor placed in a transverse magnetic field experience a sideways Lorentz force; this results in a charge separation in a direction perpendicular to the current and to the magnetic field. The resultant voltage in that direction is proportional to the applied magnetic field. This is known as the *Hall effect*. The *Hall effect* is often used to measure the magnitude of a magnetic field. It is used as well to find the sign of the dominant charge carriers in materials such as semiconductors (negative electrons or positive holes).

Magnetic Circuits

An important use of H is in *magnetic circuits* where B = μH inside a linear material. Here, μ is the magnetic permeability of the material. This result is similar in form to Ohm's law {{{1}, where J is the current density, σ is the conductance and E is the electric field. Extending this analogy, the counterpart to the macroscopic Ohm's law (*I* = *V*D *R*) is:

$$\Phi = \frac{F}{R_m},$$

where $\Phi = \int B \cdot dA$ is the magnetic flux in the circuit, $F = \int H \cdot d\ell$ is the magnetomotive force applied to the circuit, and R_m is the reluctance of the circuit. Here the reluctance R_m is a quantity similar in nature to resistance for the flux.

Using this analogy it is straightforward to calculate the magnetic flux of complicated magnetic field geometries, by using all the available techniques of circuit theory.

Magnetic Field Shape Descriptions

Figure: *Schematic quadrupole magnet (*"four-pole"*) magnetic field. There are four steel pole tips, two opposing magnetic north poles and two opposing magnetic south poles.*

- An *azimuthal* magnetic field is one that runs east-west.
- A *meridional* magnetic field is one that runs north-south. In the solar dynamo model of the Sun, differential rotation of the solar plasma causes the meridional magnetic field to stretch into an azimuthal magnetic field, a process called the *omega-effect*. The reverse process is called the *alpha-effect*.
- A *dipole* magnetic field is one seen around a bar magnet or around a charged elementary particle with nonzero spin.
- A *quadrupole* magnetic field is one seen, for example, between the poles of four bar magnets. The field strength grows linearly with the radial distance from its longitudinal axis.
- A *solenoidal* magnetic field is similar to a dipole magnetic field, except that a solid bar magnet is replaced by a hollow electromagnetic coil magnet.
- A *toroidal* magnetic field occurs in a doughnut-shaped coil, the electric current spiraling around the tube-like surface, and is found, for example, in a tokamak.
- A *poloidal* magnetic field is generated by a current flowing in a ring, and is found, for example, in a tokamak.

- A *radial* magnetic field is one in which the field lines are directed from the centre outwards, similar to the spokes in a bicycle wheel. An example can be found in a loudspeaker transducers (driver).
- A *helical* magnetic field is corkscrew-shaped, and sometimes seen in space plasmas such as the Orion Molecular Cloud.

Magnetic Dipoles

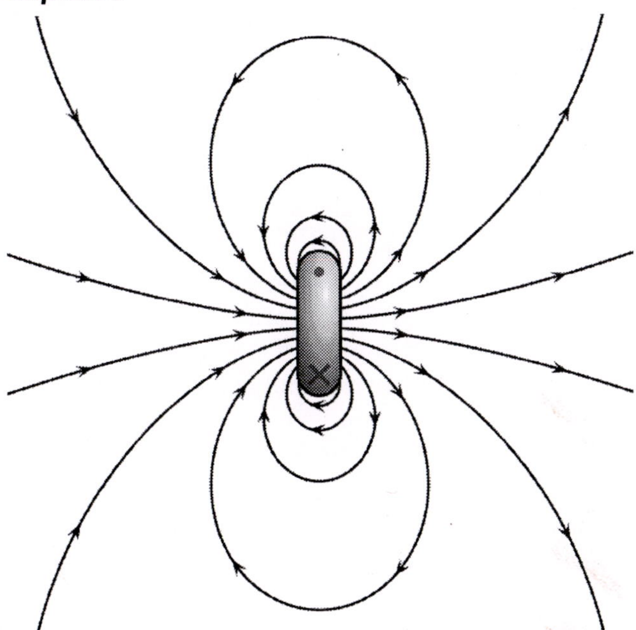

Figure: *Magnetic field lines around a "magnetostatic dipole" pointing to the right.*

The magnetic field of a magnetic dipole is depicted in the figure. From outside, the ideal magnetic dipole is identical to that of an ideal electric dipole of the same strength. Unlike the electric dipole, a magnetic dipole is properly modelled as a current loop having a current I and an area a. Such a current loop has a magnetic moment of:

$$m = Ia,$$

where the direction of m is perpendicular to the area of the loop and depends on the direction of the current using the right-hand rule. An ideal magnetic dipole is modelled as a real magnetic dipole whose area a has been reduced to zero and its current I increased to infinity such that the product $m = Ia$ is finite. This model clarifies the connection between angular momentum and magnetic moment, which is the basis of the Einstein-de Haas effect *rotation by magnetization* and its inverse, the Barnett effect or *magnetization by rotation*. Rotating the loop faster (in the same direction) increases the current and therefore the magnetic moment, for example.

It is sometimes useful to model the magnetic dipole similar to the electric dipole with two equal but opposite magnetic charges (one south the other north) separated by distance *d*. This model produces an H-field not a B-field. Such a model is deficient, though, both in that there are no magnetic charges and in that it obscures the link between electricity and magnetism. Further, as discussed above it fails to explain the inherent connection between angular momentum and magnetism.

Magnetic Monopole (Hypothetical)

A *magnetic monopole* is a hypothetical particle (or class of particles) that has, as its name suggests, only one magnetic pole (either a north pole or a south pole). In other words, it would possess a "magnetic charge" analogous to an electric charge. Magnetic field lines would start or end on magnetic monopoles, so if they exist, they would give exceptions to the rule that magnetic field lines neither start nor end.

Modern interest in this concept stems from particle theories, notably Grand Unified Theories and superstring theories, that predict either the existence, or the possibility, of magnetic monopoles. These theories and others have inspired extensive efforts to search for monopoles. Despite these efforts, no magnetic monopole has been observed to date.

In recent research, materials known as spin ices can simulate monopoles, but do not contain actual monopoles.

See Also

General:

- *Magnetohydrodynamics* – the study of the dynamics of electrically conducting fluids.
- *Magnetic nanoparticles* – extremely small magnetic particles that are tens of atoms wide
- *Magnetic reconnection* – an effect that causes solar flares and auroras.
- *Magnetic potential* – the vector and scalar potential representation of magnetism.

- *SI electromagnetism units* – common units used in electromagnetism.
- *Orders of magnitude (magnetic field)* – list of magnetic field sources and measurement devices from smallest magnetic fields to largest detected.
- Upward continuation

Mathematics

- *Magnetic helicity* – extent to which a magnetic field wraps around itself.

Applications

- *Dynamo theory* – a proposed mechanism for the creation of the Earth's magnetic field.
- *Helmholtz coil* – a device for producing a region of nearly uniform magnetic field.
- *Magnetic field viewing film* – Film used to view the magnetic field of an area.
- *Maxwell coil* – a device for producing a large volume of an almost constant magnetic field.
- *Stellar magnetic field* – a discussion of the magnetic field of stars.
- *Teltron Tube* – device used to display an electron beam and demonstrates effect of electric and magnetic fields on moving charges.

Magnetic Materials: Soft Magnets

Soft magnetic materials are those materials that are easily magnetised and demagnetised. They typically have intrinsic coercivity less than 1000 Am-1. They are used primarily to enhance and/or channel the flux produced by an electric current. The main parameter, often used as a figure of merit for soft magnetic materials, is the relative permeability (μr, where μr = B/ μoH, which is a measure of how readily the material responds to the applied magnetic field. The other main parameters of interest are the coercivity, the saturation magnetisation and the electrical conductivity.

The types of applications for soft magnetic materials fall into two main categories: AC and DC. In DC applications the material is magnetised in order to perform an operation and then demagnetised at the conclusion of the operation, e.g. an electromagnet on a crane at a scrap yard will be switched on to attract the scrap steel and then switched off to drop the steel.

In AC applications the material will be continuously cycled from being magnetised in one direction to the other, throughout the period of operation, e.g. a power supply transformer. A high permeability will be desirable for each type of application but the significance of the other properties varies.

For DC applications the main consideration for material selection is most likely to be the permeability. This would be the case, for example, in shielding applications where the flux must be channelled through the material. Where the material is used to generate a magnetic field or to create a force then the saturation magnetisation may also be significant.

For AC applications the important consideration is how much energy is lost in the system as the material is cycled around its hysteresis loop. The energy loss can originate from three different sources: 1. hysteresis loss, which is related to the area contained within the hysteresis loop; 2. eddy current loss, which is related to the generation of electric currents in the magnetic material and the associated resistive losses and 3. anomalous loss, which is related to the movement of domain walls within the material.

Hysteresis losses can be reduced by the reduction of the intrinsic coercivity, with a consequent reduction in the area contained within the hysteresis loop. Eddy current losses can be reduced by decreasing the electrical conductivity of the material and by laminating the material, which has an influence on overall conductivity and is important because of skin effects at higher frequency.

Finally, the anomalous losses can be reduced by having a completely homogeneous material, within which there will be no hindrance to the motion of domain walls.

Nano Materials

Nanomaterials is a field that takes a materials science-based approach on nanotechnology. It studies materials with morphological features on the nanoscale, and especially those that have special properties stemming from their nanoscale dimensions. Nanoscale is usually defined as smaller than a one tenth of a micrometer in at least one dimension, though sometimes includes up to a micrometer.

On 18 October 2011, the European Commission adopted the following definition of a nanomaterial:

A natural, incidental or manufactured material containing particles, in an unbound state or as an aggregate or as an agglomerate and where, for 50% or more of the particles in the number size distribution, one or more external dimensions is in the size range 1 nm – 100 nm. In specific cases and where warranted by concerns for the environment, health, safety or competitiveness the number size distribution threshold of 50% may be replaced by a threshold between 1 and 50%.

An important aspect of nanotechnology is the vastly increased ratio of surface area to volume present in many nanoscale materials, which makes possible new quantum mechanical effects. One example is the "quantum size effect" where the electronic properties of solids are altered with great reductions in particle size. Nanoparticles, for example, take advantage of their dramatically increased surface area to volume ratio. Their optical properties, e.g. fluorescence, become a function of the particle diameter. This effect does not come into play by going from macro to micro dimensions. However, it becomes pronounced when the nanometer size range is reached.

A certain number of physical properties also alter with the change from macroscopic systems. Novel mechanical properties of nanomaterials is a subject of nanomechanics research. When brought into a bulk material, nanoparticles can strongly influence the mechanical properties of the material, like stiffness or elasticity. For example, traditional polymers can be reinforced by nanoparticles resulting in novel materials which can be used as lightweight replacements for metals. Such nanotechnologically enhanced materials may enable a weight reduction accompanied by an increase in stability and improved functionality. Catalytic activities also reveal new behaviour in the interaction with biomaterials.

Classification

Materials referred to as "nanomaterials" generally fall into two categories: fullerenes, and inorganic nanoparticles.

Fullerenes

The fullerenes are a class of allotropes of carbon which conceptually are graphene sheets rolled into tubes or spheres. These include the carbon nanotubes (or silicon nanotubes) which are of interest both because of their mechanical strength and also because of their electrical properties. For the past decade, the chemical and physical properties of fullerenes have been a hot topic in the field of research and development, and are likely to continue to be for a long time. In April 2003, fullerenes were under study for potential medicinal use: binding specific antibiotics to the structure of resistant bacteria and even target certain types of cancer cells such as melanoma. The October 2005 issue of Chemistry and Biology contains an article describing the use of fullerenes as light-activated antimicrobial agents. In the field of nanotechnology, heat resistance and superconductivity are among the properties attracting intense research.

A common method used to produce fullerenes is to send a large current between two nearby graphite electrodes in an inert atmosphere. The resulting carbon plasma arc between the electrodes cools into sooty residue from which many fullerenes can be isolated.

There are many calculations that have been done using ab-initio Quantum Methods applied to fullerenes. By DFT and TDDFT methods one can obtain IR, Raman and UV spectra. Results of such calculations can be compared with experimental results.

Nanoparticles

Nanoparticles or nanocrystals made of metals, semiconductors, or oxides are of particular interest for their mechanical, electrical, magnetic, optical, chemical and other properties. Nanoparticles have been used as quantum dots and as chemical catalysts such as nanomaterial-based catalysts.

Nanoparticles are of great scientific interest as they are effectively a bridge between bulk materials and atomic or molecular structures. A bulk material should have constant physical properties regardless of its size, but at the nano-scale this is often not the case. Size-dependent properties are observed such as quantum confinement in semiconductor particles, surface plasmon resonance in some metal particles and superparamagnetism in magnetic materials.

Nanoparticles exhibit a number of special properties relative to bulk material. For example, the bending of bulk copper (wire, ribbon, etc.) occurs with movement of copper atoms/clusters at about the 50 nm scale. Copper nanoparticles smaller than 50 nm are considered super hard materials that do not exhibit the same malleability and ductility as bulk copper. The change in properties is not always desirable. Ferroelectric materials smaller than 10 nm can switch their magnetisation direction using room temperature thermal energy, thus making them useless for memory storage. Suspensions of nanoparticles are possible because the interaction of the particle surface with the solvent is strong enough to overcome differences in density, which usually result in a material either sinking or floating in a liquid. Nanoparticles often have unexpected visual properties because they are small enough to confine their electrons and produce quantum effects. For example gold nanoparticles appear deep red to black in solution.

The often very high surface area to volume ratio of nanoparticles provides a tremendous driving force for diffusion, especially at elevated temperatures. Sintering is possible at lower temperatures and over shorter durations than for larger particles. This theoretically does not affect the density of the final product, though flow difficulties and the tendency of nanoparticles to agglomerate do complicate matters. The surface effects of nanoparticles also reduces the incipient melting temperature.

Sol-gel

The sol-gel process is a wet-chemical technique commonly used to synthesise a wide variety of nanomaterials.

Characterization

The first observations and size measurements of nano-particles were made during the first decade of the 20th century. They are mostly associated with the name of Zsigmondy who made detailed studies of gold sols and other nanomaterials with sizes down to 10 nm and less. He published a book in 1914. He used an ultramicroscope that employs a *dark field* method for seeing particles with sizes much less than light wavelength.

There are traditional techniques developed during 20th century in Interface and Colloid Science for characterizing nanomaterials. These are widely used for *first generation* passive nanomaterials specified in the next section.

These methods include several different techniques for characterizing particle size distribution. This characterization is imperative because many materials that are expected to be nano-sized are actually aggregated in solutions. Some of methods are based on light scattering. Others apply ultrasound, such as ultrasound attenuation spectroscopy for testing concentrated nano-dispersions and microemulsions.

There is also a group of traditional techniques for characterizing surface charge or zeta potential of nano-particles in solutions. This information is required for proper system stabilzation, preventing its aggregation or flocculation. These methods include microelectrophoresis, electrophoretic light scattering and electroacoustics. The last one, for instance colloid vibration current method is suitable for characterizing concentrated systems.

Uniformity

The chemical processing and synthesis of high performance technological components for the private, industrial and military sectors requires the use of high purity ceramics, polymers, glass-ceramics and material composites. In condensed bodies formed from fine powders, the irregular sizes and shapes of nanoparticles in a typical powder often lead to non-uniform packing morphologies that result in packing density variations in the powder compact.

Uncontrolled agglomeration of powders due to attractive van der Waals forces can also give rise to in microstructural inhomogeneities. Differential stresses that develop as a result of non-uniform drying shrinkage are directly related to the rate at which the solvent can be removed, and thus highly dependent upon the distribution of porosity. Such stresses have been associated with a plastic-to-brittle transition in consolidated bodies, and can yield to crack propagation in the unfired body if not relieved.

In addition, any fluctuations in packing density in the compact as it is prepared for the kiln are often amplified during the sintering process, yielding inhomogeneous densification. Some pores and other structural defects associated with density variations have been shown to play a detrimental role in the sintering process by growing and thus limiting end-point densities. Differential stresses arising from inhomogeneous densification have also been shown to result in the propagation of internal cracks, thus becoming the strength-controlling flaws.

It would therefore appear desirable to process a material in such a way that it is physically uniform with regard to the distribution of components and porosity, rather than using particle size distributions which will maximize the green density. The containment of a uniformly dispersed assembly of strongly interacting particles in suspension requires total control over particle-particle interactions. It should be noted here that a number of dispersants such as ammonium citrate (aqueous) and imidazoline or oleyl alcohol (nonaqueous) are promising solutions as possible additives for enhanced dispersion and deagglomeration. Monodisperse nanoparticles and colloids provide this potential.

Monodisperse powders of colloidal silica, for example, may therefore be stabilized sufficiently to ensure a high degree of order in the colloidal crystal or polycrystalline colloidal solid which results from aggregation. The degree of order appears to be limited by the time and space allowed for longer-range correlations to be established. Such defective polycrystalline colloidal structures would appear to be the basic elements of sub-micrometer colloidal materials science, and, therefore, provide the first step in developing a more rigorous understanding of the mechanisms involved in microstructural evolution in high performance materials and components.

Safety

Nanomaterials behave differently than other similarly-sized particles. It is therefore necessary to develop specialized approaches to testing and monitoring their effects on human health and on the environment. The OECD Chemicals Committee has established the Working Party on Manufactured Nanomaterials to address this issue and to study the practices of OECD member countries in regards to nanomaterial safety.

While nanomaterials and nanotechnologies are expected to yield numerous health and health care advances, such as more targeted methods of delivering drugs, new cancer therapies, and methods of early detection of diseases, they also may have unwanted effects. Increased rate of absorption is the main concern associated with manufactured nanoparticles.

When materials are made into nanoparticles, their surface area to volume ratio increases. The greater specific surface area (surface area per unit weight) may lead to increased rate of absorption through the skin, lungs, or digestive tract and may cause unwanted effects to the lungs as well as other organs. However, the particles must be absorbed in sufficient quantities in order to pose health risks.

As the use of nanomaterials increases worldwide, concerns for worker and user safety are mounting. To address such concerns, the Swedish Karolinska Institute conducted a study in which various nanoparticles were introduced to human lung epithelial cells. The results, released in 2008, showed that iron oxide nanoparticles caused little DNA damage and were non-toxic. Zinc oxide nanoparticles were slightly worse. Titanium dioxide caused only DNA damage. Carbon nanotubes caused DNA damage at low levels. Copper oxide was found to be the worst offender, and was the only nanomaterial identified by the researchers as a clear health risk.

Silicon carbide

Silicon carbide (SiC), also known as carborundum, is a compound of silicon and carbon with chemical formula SiC. It occurs in nature as the extremely rare mineral moissanite. Silicon carbide powder has been mass-produced since 1893 for use as an abrasive. Grains of silicon carbide can be bonded together by sintering to form very hard ceramics that are widely used in applications requiring high endurance, such as car brakes, car clutches and ceramic plates in bulletproof vests. Electronic applications of silicon carbide as light emitting diodes (LEDs) and detectors in early radios were first demonstrated around 1907, and today SiC is widely used in high-temperature/high-voltage semiconductor electronics. Large single crystals of silicon carbide can be grown by the Lely method; they can be cut into gems known as synthetic moissanite. Silicon carbide with high surface area can be produced from SiO2 contained in plant material.

Discovery and Early Production

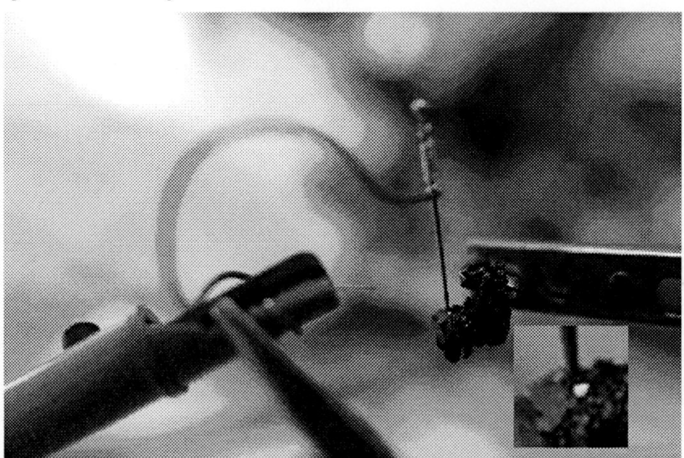

Figure: *A replication of H.J. Round's LED experiments*

Early Experiments

Non-systematic, less-recognized, and often unverified syntheses of silicon carbide include

- J. J. Berzelius's reduction of potassium fluorosilicate by potassium (1810)
- César-Mansuète Despretz's passing an electric current through a carbon rod embedded in sand (1849)
- Robert Sydney Marsden's (1856–1919) dissolution of silica in molten silver in a graphite crucible (1881)
- Albert Colson's heating of silicon under a stream of ethylene (1882)
- Paul Schuetzenberger's heating of a mixture of silicon and silica in a graphite crucible (1881).

Wide Scale Production

Wide-scale production is credited to Edward Goodrich Acheson in 1890. Acheson was attempting to prepare artificial diamonds when he heated a mixture of clay (aluminum silicate) and powdered coke (carbon) in an iron bowl. He called the blue crystals that formed *Carborundum*, believing it to be a new compound of carbon and aluminum, similar to corundum. In 1893, Henri Moissan discovered the very rare naturally-occurring SiC mineral while examining rock samples found in the Canyon Diablo meteorite in Arizona. The mineral was named moissanite in his honour. Moissan also synthesized SiC by several routes, including dissolution of carbon in molten silicon, melting a mixture of calcium carbide and silica, and by reducing silica with carbon in an electric furnace.

Acheson patented the method for making silicon carbide powder on February 28, 1893. Acheson also developed the electric batch furnace by which SiC is still made today and formed The Carborundum Company to manufacture bulk SiC, initially for use as an abrasive. In 1900 the company settled with the Electric Smelting and Aluminum Company when a judge's decision gave "priority broadly" to its founders "for reducing ores and other substances by the incandescent method". It is said that Acheson was trying to dissolve carbon in molten corundum (alumina) and discovered the presence of hard, blue-black crystals which he believed to be a compound of carbon and corundum: hence carborundum. It may be that he named the material "carborundum" by analogy to corundum, which is another very hard substance (9 on the Mohs scale).

The first use of SiC was as an abrasive. This was followed by electronic applications. In the beginning of the 20th century, silicon carbide was used as a detector in the first radios, and in 1907 Henry

Joseph Round produced the first LED by applying a voltage to a SiC crystal and observing yellow, green and orange emission at the cathode. Those experiments were later repeated by O. V. Losev in the Soviet Union in 1923.

Natural Occurrence

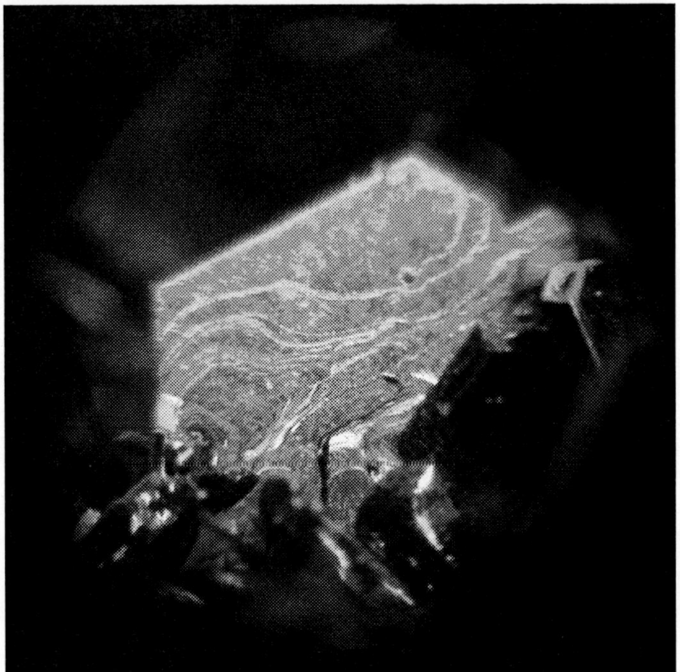

Figure: *Moissanite single crystal (≈1 mm in size)*

Naturally occurring moissanite is found in only minute quantities in certain types of meteorite and in corundum deposits and kimberlite. Virtually all the silicon carbide sold in the world, including moissanite jewels, is synthetic. Natural moissanite was first found in 1893 as a small component of the Canyon Diablo meteorite in Arizona by Dr. Ferdinand Henri Moissan, after whom the material was named in 1905. Moissan's discovery of naturally occurring SiC was initially disputed because his sample may have been contaminated by silicon carbide saw blades that were already on the market at that time.

While rare on Earth, silicon carbide is remarkably common in space. It is a common form of stardust found around carbon-rich stars, and examples of this stardust have been found in pristine condition in primitive (unaltered) meteorites. The silicon carbide found in space and in meteorites is almost exclusively the beta-polymorph. Analysis of SiC grains found in the Murchison meteorite, a carbonaceous chondrite meteorite, has revealed anomalous isotopic ratios of carbon and silicon, indicating an origin from outside the solar system; 99% of these SiC grains originate around carbon-rich asymptotic giant branch stars. SiC is commonly found around these stars as deduced from their infrared spectra.

Production

Because of the rarity of natural moissanite, most silicon carbide is synthetic. It is used as an abrasive, and more recently as a semiconductor and diamond simulant of gem quality.

The simplest manufacturing process is to combine silica sand and carbon in an Acheson graphite electric resistance furnace at a high temperature, between 1,600 °C (2,910 °F) and 2,500 °C (4,530 °F). Fine SiO_2 particles in plant material (e.g. rice husks) can be converted to SiC by heating in the excess carbon from the organic material. The silica fume, which is a byproduct of producing silicon metal and ferrosilicon alloys, also can be converted to SiC by heating with graphite at 1,500 °C (2,730 °F).

Figure: *Synthetic SiC crystals ~3 mm in diameter*

Figure: *Synthetic SiC Lely crystals*

The material formed in the Acheson furnace varies in purity, according to its distance from the graphite resistor heat source. Colourless, pale yellow and green crystals have the highest purity and are found closest to the resistor. The colour changes to blue and black at greater distance from the resistor, and these darker crystals are less pure. Nitrogen and aluminium are common impurities, and they affect the electrical conductivity of SiC.

Pure silicon carbide can be made by the so-called Lely process, in which SiC powder is sublimated into high-temperature species of silicon, carbon, silicon dicarbide (SiC_2), and disilicon carbide (Si_2C) in an argon gas ambient at 2500 °C and redeposited into flake-like single crystals, sized up to 2×2 cm^2, at a slightly colder substrate. This process yields high-quality single crystals, mostly of 6H-SiC phase (because of high growth temperature). A modified Lely process involving induction heating in graphite crucibles yields even larger single crystals of 4 inches (10 cm) in diameter, having a section 81 times larger compared to the conventional Lely process. Cubic SiC is usually grown by the more expensive process of chemical vapor deposition (CVD). Homoepitaxial and heteroepitaxial SiC layers can be grown employing both gas and liquid phase approaches. Pure silicon carbide can also be prepared by the thermal decomposition of a polymer, poly(methylsilyne), under an inert atmosphere at low temperatures. Relative to the CVD process, the pyrolysis method is advantageous because the polymer can be formed into various shapes prior to thermalization into the ceramic.

Electrical Conductivity

Silicon carbide is a semiconductor, which can be doped n-type by nitrogen or phosphorus and p-type by aluminium, boron, gallium or beryllium. Metallic conductivity has been achieved by heavy doping with boron, aluminium or nitrogen. Superconductivity has been detected in 3C-SiC:Al, 3C-SiC:B and 6H-SiC:B at the same temperature of 1.5 K. A crucial difference is however observed for the magnetic field behaviour between aluminium and boron doping: SiC:Al is type-II, same as Si:B. On the contrary, SiC:B is type-I. In attempt to explain this difference, it was noted that Si sites are m ore important than carbon sites for superconductivity in SiC. Whereas boron substitutes carbon in SiC, Al substitutes Si sites. Therefore, Al and B "see" different environment that might explain different properties of SiC:Al and SiC:B.

Uses

Abrasive and cutting tools: In the arts, silicon carbide is a popular abrasive in modern lapidary due to the durability and low cost of the material. In manufacturing, it is used for its hardness in abrasive machining processes such as grinding, honing, water-jet cutting and sandblasting. Particles of silicon carbide are laminated to paper to create sandpapers and the grip tape on skateboards.

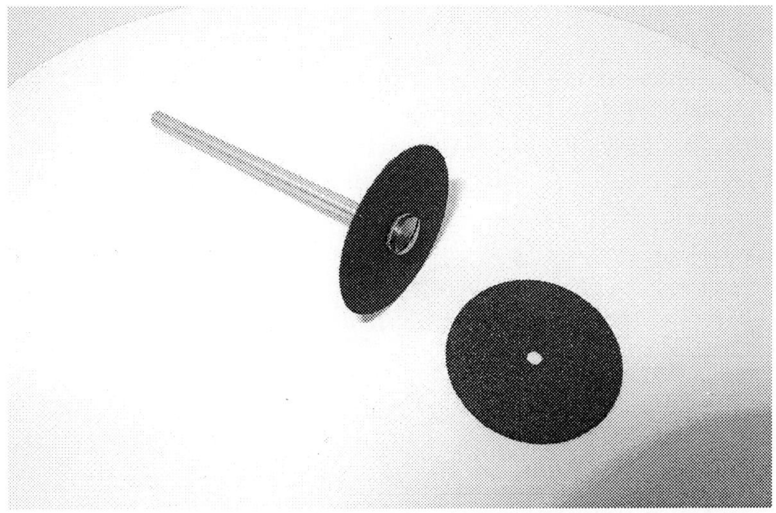

Figure: *Cutting disks made of SiC*

In 1982 an exceptionally strong composite of aluminium oxide and silicon carbide whiskers was discovered. Development of this laboratory-produced composite to a commercial product took only three years.

In 1985, the first commercial cutting tools made from this alumina and silicon carbide whisker-reinforced composite were introduced by the Advanced Composite Materials Corporation (ACMC) and Greenleaf Corporation.

Structural Material

In the 1980s and 1990s, silicon carbide was studied in several research programmes for high-temperature gas turbines in Europe, Japan and the United States. The components were intended to replace nickel superalloy turbine blades or nozzle vanes. However, none of these projects resulted in a production quantity, mainly because of its low impact resistance and its low fracture toughness.

Like other hard ceramics (namely alumina and boron carbide), silicon carbide is used in composite armor (e.g. Chobham armor), and in ceramic plates in bulletproof vests. Dragon Skin, which is produced by Pinnacle Armor, uses disks of silicon carbide.

Silicon carbide is used as a support and shelving material in high temperature kilns such as for firing ceramics, glass fusing, or glass casting. SiC kiln shelves are considerably lighter and more durable than traditional alumina shelves.

Automobile Parts

Silicon-infiltrated carbon-carbon composite is used for high performance "ceramic" brake discs, as it is able to withstand extreme temperatures. The silicon reacts with the graphite in the carbon-carbon composite to become carbon-fibre-reinforced silicon carbide (C/SiC). These discs are used on some road-going sports cars, including the Porsche Carrera GT, the Bugatti Veyron, the Chevrolet Corvette ZR1, Bentleys, Ferraris, Lamborghinis, and some specific high performance Audis. Silicon carbide is also used in a sintered form for diesel particulate filters.

Electric Systems

The earliest electrical application of SiC was in lightning arresters in electric power systems. These devices must exhibit high resistance until the voltage across them reaches a certain threshold V_T, at which point their resistance must drop to a lower level and maintain this level until the applied voltage drops below V_T.

It was recognized early on that SiC had such a voltage-dependent resistance, and so columns of SiC pellets were connected between high-voltage power lines and the earth. When a lightning strike to the line raises the line voltage sufficiently, the SiC column will conduct, allowing strike current to pass harmlessly to the earth instead of along the power line. Such SiC columns proved to conduct significantly at normal power-line operating voltages and thus had to be placed in series with a spark gap. This spark gap is ionized and rendered conductive when lightning raises the voltage of the power line conductor, thus effectively connecting the SiC column between the power conductor and the earth. Spark gaps used in lightning arresters are unreliable, either failing to strike an arc when needed or failing to turn off afterwards, in the latter case due to material failure or contamination by dust or salt. Usage of SiC columns was originally intended to eliminate the need for the spark gap in a lightning arrester. Gapped SiC lightning arresters were used as lightning-protection tool and sold under GE and Westinghouse brand names, among others. The gapped SiC arrester has been largely displaced by no-gap varistors that use columns of zinc oxide pellets.

Power Electronic Devices

Silicon carbide is a semiconductor in research and early mass-production providing advantages for fast, high-temperature and/or high-voltage devices. First devices available were Schottky diodes, followed by Junction-gate FETs and MOSFETs for high-power switching. Bipolar transistors and thyristors are currently developed. A major problem for SiC commercialization has been the elimination of defects: edge dislocations, screw dislocations (both hollow and closed core), triangular defects and basal plane dislocations. As a result, devices made of SiC crystals initially displayed poor reverse blocking performance though researchers have been tentatively finding solutions to improving the breakdown performance. Apart from crystal quality, problems with the interface of SiC with silicon dioxide have hampered the development of SiC-based power MOSFETs and insulated-gate bipolar transistors. Although the mechanism is still unclear, nitridation has dramatically reduced the defects causing the interface problems. In 2008, the first commercial JFETs rated at 1200 V were introduced to the market, followed in 2011 by the first commercial MOSFETs rated at 1200 V. Beside SiC switches and SiC Schottky diodes (also Schottky barrrier diode – SBD) in the popular TO-247 package, companies started even earlier to implement the bare chips into their power modules. SiC SBD diodes found wide market spread being used in PFC circuits and IGBT power modules.

LEDs

The history of SiC LEDs is quite remarkable: the first LED action was demonstrated in 1907 using SiC and the first commercial LEDs were again based on SiC. Yellow LEDs made from 3C-SiC were manufactured in the Soviet Union in the 1970s, and blue ones (6H-SiC) worldwide in the 1980s. The production was soon stopped because gallium nitride showed 10–100 times brighter emission. This difference in efficiency is due to the unfavourable indirect bandgap of SiC, whereas GaN has a direct bandgap which favours light emission. However, SiC is still one of the important LED components – it is a popular substrate for growing GaN devices, and it also serves as a heat spreader in high-power LEDs.

Astronomy

The low thermal expansion coefficient, high hardness, rigidity and thermal conductivity make silicon carbide a desirable mirror material for astronomical telescopes. The growth technology (chemical vapor deposition) has been scaled up to produce disks of polycrystalline silicon carbide up to 3.5 metres in diameter, and several telescopes (like the Herschel Space Telescope) are already equipped with SiC optics.

Thin Filament Pyrometry

Silicon carbide fibres are used to measure gas temperatures in an optical technique called thin filament pyrometry. It involves the placement of a thin filament in a hot gas stream. Radiative emissions from the filament can be correlated with filament temperature. Filaments are SiC fibres with a diameter of 15 micrometers; about one fifth that of a human hair. Because the fibres are so thin, they do little to disturb the flame and their temperature remains close to that of the local gas. Temperatures of about 800–2500 K can be measured.

Heating Elements

References to silicon carbide heating elements exist from the early 20th century when they were produced by Acheson's Carborundum Co. in the U.S. and EKL in Berlin. Silicon carbide offered increased operating temperatures compared with metallic heaters. Silicon carbide elements are used today in the melting of glass and non-ferrous metal, heat treatment of metals, float glass production, production of ceramics and electronics components, igniters in pilot lights for gas heaters, etc.

Nuclear Fuel Particles

Silicon carbide is an important material in TRISO-coated fuel particles, the type of nuclear fuel found in high temperature gas cooled reactors (such as the Pebble Bed Reactor). A layer of silicon carbide gives coated fuel particles structural support and is the main diffusion barrier to the release of fission products.

Nuclear Fuel Cladding

Silicon carbide composite material has been investigated for use as a replacement for Zircaloy cladding in light water reactors. The composite consists of SiC fibres wrapped around a SiC inner layer and surrounded by an SiC outer layer. Problems have been reported with the ability to join the pieces of the SiC composite.

Jewellery

As a gemstone used in jewellery, silicon carbide is called "synthetic moissanite" or just "moissanite" after the mineral name. Moissanite is similar to diamond in several important respects: it is transparent and hard (9–9.5 on the Mohs scale, compared to 10 for diamond), with a refractive index between 2.65 and 2.69 (compared to 2.42 for diamond).

Moissanite is somewhat harder than common cubic zirconia. Unlike diamond, moissanite can be strongly birefringent. This quality is desirable in some optical applications, but not in gemstones. For this reason, moissanite jewels are cut along the optic axis of the crystal to minimize birefringent effects. It is lighter (density 3.21 g/cm^3 vs. 3.53 g/cm^3), and much more resistant to heat than diamond. This results in a stone of higher luster, sharper facets and good resilience. Loose moissanite stones may be placed directly into wax ring moulds for lost-wax casting; unlike diamond, which burns at 800 °C, moissanite remains undamaged by temperatures up to 1800 °C. Moissanite has become popular as a diamond substitute, and may be misidentified as diamond, since its thermal conductivity is much closer to that of diamond than any other diamond substitute. Many thermal diamond-testing devices cannot distinguish moissanite from diamond, but the gem is distinct in its birefringence and a very slight green or yellow fluorescence under ultraviolet light. Some moissanite stones also have curved string-like inclusions, which diamonds never have.

Steel Production

Silicon carbide, dissolved in a basic oxygen furnace used for making steel, acts as a fuel. The additional energy liberated allows the furnace to process more scrap with the same charge of hot metal. It can also be used to raise tap temperatures and adjust the carbon and silicon content. Silicon carbide is cheaper than a combination of ferrosilicon and carbon, produces cleaner steel and less emissions due to low levels of trace elements, has a low gas content, and does not lower the temperature of steel.

Catalyst Support

The natural resistance to oxidation exhibited by silicon carbide, as well as the discovery of new ways to synthesize the cubic β-SiC form, with its larger surface area, has led to significant interest in its use as a heterogeneous catalyst support. This form has already been employed as a catalyst support for the oxidation of hydrocarbons, such as n-butane, to maleic anhydride.

Carborundum Printmaking

Silicon carbide is used in carborundum printmaking – a collagraph printmaking technique. Carborundum grit is applied in a paste to the surface of an aluminium plate. When the paste is dry, ink is applied and trapped in its granular surface, then wiped from the bare areas of the plate. The ink plate is then printed onto paper in a rolling-bed press used for intaglio printmaking. The result is a print of painted marks embossed into the paper.

Graphene Production

Silicon carbide is used to produce epitaxial graphene by graphitization at high temperatures. This is considered as one of the promising methods to synthesize graphene at large scale for practical applications.

Nanomagnet

A nanomagnet is a submicrometric system that presents spontaneous magnetic order (magnetization) at zero applied magnetic field (remanence).

The small size of nanomagnets prevents the formation of magnetic domains. The magnetization dynamics of sufficiently small nanomagnets at low temperatures, typically single-molecule magnets, presents quantum phenomena, such as macroscopic spin tunnelling. At larger temperatures, the magnetization undergoes random thermal fluctuations (superparamagnetism) which present a limit for the use of nanomagnets for permanent information storage.

Canonical examples of nanomagnets are grains of ferromagnetic metals (iron, cobalt, and nickel) and single-molecule magnets. The vast majority of nanomagnets feature transition metal (titanium, vanadium, chromium, manganese, iron, cobalt or nickel) or rare earth (Gd, Eu, Er) magnetic atoms. Attending to the number of magnetic atoms, the smallest nanomagnets reported so far are double decker phthalocyanes with only one rare earth atom. Canonical single-molecule magnets are the so called Mn12 and Fe8 systems, with 12 and 8 transition metal atoms each and both with spin 10 (S=10) ground states.

The phenomenon of zero field magnetization requires three conditions:

- (1) A ground state with finite spin
- (2) a magnetic anisotropy energy barrier
- (3) long spin relaxation time.

Conditions (1) and (2), but not (3), have been demonstrated in a number of nanostructures, such as nanoparticles, nanoislands, and quantum dots with a controlled number of magnetic atoms (between 1 and 10).

A nanomagnet can have enhanced electronic properties due to size effect, such as long spin relaxation time of conduction electron, which may be useful for nano-scale spintronic device.

Nanoelectronics

Nanoelectronics refer to the use of nanotechnology in electronic components. The term covers a diverse set of devices and materials, with the common characteristic that they are so small that inter-atomic interactions and quantum mechanical properties need to be studied extensively. Some of these candidates include: hybrid molecular/semiconductor electronics, one dimensional nanotubes/nanowires, or advanced molecular electronics. Recent silicon CMOS technology generations, such as the 22nm node, are already within this regime. Nanoelectronics are sometimes considered as disruptive technology because present candidates are significantly different from traditional transistors.

Fundamental Concepts

In 1965 Gordon Moore observed that silicon transistors were undergoing a continual process of scaling downward, an observation which was later codified as Moore's law. Since his observation transistor minimum feature sizes have decreased from 10 micrometers to the 28-22 nm range in 2011.

The field of nanoelectronics aims to enable the continued realization of this law by using new methods and materials to build electronic devices with feature sizes on the nanoscale.

The volume of an object decreases as the third power of its linear dimensions, but the surface area only decreases as its second power. This somewhat subtle and unavoidable principle has huge ramifications. For example the power of a drill (or any other machine) is proportional to the volume, while the friction of the drill's bearings and gears is proportional to their surface area.

For a normal-sized drill, the power of the device is enough to handily overcome any friction. However, scaling its length down by a factor of 1000, for example, decreases its power by 1000^3 (a factor of a billion) while reducing the friction by only 1000^2 (a factor of "only" a million). Proportionally it has 1000 times less power per unit friction than the original drill. If the original friction-to-power ratio was, say, 1%, that implies the smaller drill will have 10 times as much friction as power. The drill is useless.

For this reason, while super-miniature electronic integrated circuits are fully functional, the same technology cannot be used to make working mechanical devices beyond the scales where frictional forces start to exceed the available power. So even though you may see microphotographs of delicately etched silicon gears, such devices are currently little more than curiosities with limited real world applications, for example, in moving mirrors and shutters. Surface tension increases in much the same way, thus magnifying the tendency for very small objects to stick together.

This could possibly make any kind of "micro factory" impractical: even if robotic arms and hands could be scaled down, anything they pick up will tend to be impossible to put down. The above being said, molecular evolution has resulted in working cilia, flagella, muscle fibres and rotary motors in aqueous environments, all on the nanoscale. These machines exploit the increased frictional forces found at the micro or nanoscale. Unlike a paddle or a propeller which depends on normal frictional forces (the frictional forces perpendicular to the surface) to achieve propulsion, cilia develop motion from the exaggerated drag or laminar forces (frictional forces parallel to the surface) present at micro and nano dimensions. To build meaningful "machines" at the nanoscale, the relevant forces need to be considered. We are faced with the development and design of intrinsically pertinent machines rather than the simple reproductions of macroscopic ones. All scaling issues therefore need to be assessed thoroughly when evaluating nanotechnology for practical applications.

Bibliography

Ashby, Michael; Hugh Shercliff and David Cebon: *Materials: Engineering, Science, Processing and Design,* Butterworth-Heinemann, Oxford, UK, 2007.

Beer, F.P., Johnston, Jr., E.R.: *Mechanics of Materials*, McGraw-Hill, Inc., New York, 1981.

Brady, G.S.: *Materials Handbook,* McGraw-Hill, New York, 1997.

Callister, Jr., William D.: *Materials Science and Engineering – An Introduction,* John Wiley and Sons, London, 2000.

Carter, G.F., Paul, D.E.: *Materials Science and Engineering*, ASM International, materials Park, OH, 1991.

Davis, J.R.: *ASM Materials Engineering Dictionary*, ASM International, materials Park, OH, 1992.

Eberhart, Mark: *Why Things Break: Understanding the World by the Way It Comes Apart*, Harmony, NY, 2003.

Gere J.M., Timoshenko, S.P.: *Mechanics of Materials,* Brooks/Cole, Monterey, CA, 1984.

Gordon, James Edward: *The New Science of Strong Materials or Why You Don't Fall Through the Floor,* Princeton University Press, NY, 1984.

Kaplan, W.A.: *Modern Plastics Encyclopedia '99,* McGraw-Hill, New York, 1999.

Lewis, P.R., Reynolds, K. & Gagg, C.: *Forensic Materials Engineering: Case Studies*, CRC Press, Boca Raton, 2003.

Lovesey, S. W.: *Theory of Neutron Scattering from Condensed Matter; Volume 1: Neutron Scattering*, Clarendon Press, Oxford, 1984.

Mathews, F.L. & Rawlings, R.D.: *Composite Materials: Engineering and Science*, CRC Press, Boca Raton, 1999.

Oberg, E., Jones, F.D., Horton, H.L., Ryffel, H.H.: *Machinery's Handbook,* Industrial Press, New York, 1996.

Parker, S.P.: *Dictionary of Science and Technical Terms,* McGraw-Hill, New York, 1994.

Rahn, Armin: *The Basics of Soldering,* John Wiley and Sons, New York, 1993.

Squires, G.L.: *Introduction to the Theory of Thermal Neutron Scattering,* Mineola, Dover Publications Inc., New York, 1996.

Tres, P.A.: *Designing Plastic Parts for Assembly,* Hanser Publishers, Munich, 1998.

van Vlack, L.H.: *Elements of Materials Science and Engineering,* Addison-Wesley, Reading, MA, 1985.

Walker, P.: *Chambers Dictionary of Materials Science and Technology,* Chambers Publishing, UK, 1993.

Index

❑❑❑